Trends and Prospects in Pig Genomics and Genetics

Trends and Prospects in Pig Genomics and Genetics

Guest Editors

Katarzyna Piórkowska
Katarzyna Ropka-Molik

Basel • Beijing • Wuhan • Barcelona • Belgrade • Novi Sad • Cluj • Manchester

Guest Editors

Katarzyna Piórkowska
Animal Molecular Biology
National Research Institute
of Animal Production
Kraków
Poland

Katarzyna Ropka-Molik
Animal Molecular Biology
National Research Institute
of Animal Production
Kraków
Poland

Editorial Office
MDPI AG
Grosspeteranlage 5
4052 Basel, Switzerland

This is a reprint of the Special Issue, published open access by the journal *Genes* (ISSN 2073-4425), freely accessible at: www.mdpi.com/journal/genes/special_issues/Pig_Genetics_2.

For citation purposes, cite each article independently as indicated on the article page online and using the guide below:

Lastname, A.A.; Lastname, B.B. Article Title. *Journal Name* **Year**, *Volume Number*, Page Range.

ISBN 978-3-7258-3006-0 (Hbk)
ISBN 978-3-7258-3005-3 (PDF)
https://doi.org/10.3390/books978-3-7258-3005-3

Contents

Preface

Special Issue 'Trends and Prospects in Pigs Genomics and Genetics' has intended to collect the newest information in pig genetics and genomics to help solve breeding problems. This SI is directed to researchers focusing on pig breeding and genetics, including pig diseases. I want to thank all the authors of the current SI publications for their contribution and presentation of their valuable results.

Katarzyna Piórkowska and Katarzyna Ropka-Molik
Guest Editors

Editorial

Trends and Prospects in Pig Genomics and Genetics

Katarzyna Piórkowska * **and Katarzyna Ropka-Molik**

National Research Institute of Animal Production, Animal Molecular Biology, 31-047 Cracow, Poland; katarzyna.ropka@iz.edu.pl
* Correspondence: katarzyna.piorkowska@iz.edu.pl

Citation: Piórkowska, K.; Ropka-Molik, K. Trends and Prospects in Pig Genomics and Genetics. *Genes* **2024**, *15*, 1292. https://doi.org/10.3390/genes15101292

Received: 26 September 2024
Accepted: 28 September 2024
Published: 30 September 2024

Pork is one of the most commonly consumed meat in the world. The domestication of pigs in Europe dates back to 5000 BC, from which the selection over decades has led to a significant reduction in backfat levels and improvements in growth rates. In pigs, with the development of registered and genetic evaluation methods, breeding has moved from genetic improvement through open herd books to breeding-specific lines [1]. Breeding goals in Europe have changed for economic reasons, consumer tastes and customs, and also due to the introduction of new methods for measuring phenotypic traits. Today's goals focus on reproduction, resistance, and meat quality. Nevertheless, work is still underway on new breeding goals based on important new traits, as well as their study and implementation in measurement tools [2]. Genomic selection used in many countries can increased the genetic progress rate by up to 35% per year for all traits. However, not all European countries have introduced this requirement; in many, activities are still underway, e.g., preparing reference populations (Relationship-Based Genomic Selection (RBGS)) [3]. Proposed the next step in pig breeding is to use full information on the DNA sequence, which will enable an understanding of specific genotypes and their underlying biology, as well as identifying genes that may affect resistance to diseases that pose a challenge to the industry [4]. Another advance in the genetic improvement of pigs was the attempt at gene editing, where based on these methods, pigs resistant to PRRSv were produced [5]. Geneticists' current challenge is developing pigs resistant to ASF, a disease brought from Africa that causes huge economic losses, mainly among the populations of central and eastern Europe [6,7].

There is still a lot going on in pig genetics and genomics, not only in Europe. The latest research on Chinese pig populations, based on the genome at the chromosomal level of the Chenghua pig, has redefined pig introgression [8]. It has been determined that the migration routes of pigs from China, where pig domestication occurred much earlier than in Europe, approximately 12,000 years ago, led gradually through different geographical areas of China and only then reached Europe. In addition, the researchers identified two genes, FBN1 and NR6A1, associated with evolutionary adaptation to different geographical latitudes in Chinese pigs [8]. Another Chinese study based on 250 sequences from 32 Eurasian pig breeds constructed a pangenome in which the so-called PAVs (presence/absence variants), non-reference sequences, and over 3000 new genes were identified, as well as unidentified features of the pig mobilome, including several transposable elements (TE) candidates as adaptive insertions that were co-opted into genes responsible for hypoxia responses, skeletal development, regulation of heart contraction and development of neurons, probably contributing to the local adaptation of Tibetan wild boars [9].

This special issue, entitled Trends and Prospects in Pig Genomics and Genetics, examines various directions for improving this species in terms of utility, health, and welfare traits, as well as the use of porcine cells to model human disease states.

One of the main directions of the current SI was the assessment of molecular mechanisms related to the deposition of adipose tissue, which is associated with the taste of pork, but is also considered in the context of modelling processes generally related to obesity.

Blińska et al. [10] used an in vitro model of differentiation of mesenchymal stem cells into adipocytes and digital PCR to identify significant extra- and intracellular microRNA potential involved in adipogenesis in their study. They observed that only microRNAs related to the inflammatory process were highly expressed and secreted by differentiating adipocytes. On the other hand, Wang et al. [11] focused on LncPLAAT3-AS transcript, which promotes the transcription of PLAAT3, a regulator of adipogenesis. Using dual luciferase assay, the authors revealed that LncPLAAT3-AS sponging miR-503-5p leads to the inhibition of PLAAT3 expression. Moreover, additional observation showed that LncPLAAT3-AS and PLAAT3 were significantly more expressed in fatty pig breeds' adipose tissue than in lean pigs. In turn, using RT-PCR and RACE methods, Zhu et al. [12] characterized in fat tissue cis-SAGe product BCL2L2–PABPN1 (PB), generated based on the fusion of both proteins. It was found that PB promotes proliferation and inhibits the differentiation of primary porcine preadipocytes. Moreover, in porcine preadipocytes with overexpression of PB, identified numerous DEGs (differentially expressed genes) and DEmiRNAs related to fat-associated pathways such as MAPK and PI3K-AKt and miRNAs ssc-miR-339-3p is critical for adipogenesis regulation throughout PB. Lin et al. [13] recognized the role of SESN3 in regulating adipogenesis and found that Sestrin-3 inhibits porcine preadipocyte proliferation. The mechanism of activation was through SMAD3 involved in the development of non-alcoholic steatohepatitis (NASH), where SESN3 inhibited SMAD3, thus improving ssc-miR124 activity, and then suppressed C/EBPα and GR to regulate pre-adipocytes adipogenesis. Piórkowska et al. [14] using variant calling and $\chi2$ analyses based on liver RNA-seq data, identified genetic markers related to the FGL1 gene, which probably play a significant role in lipid metabolism because it is related to the regeneration of the liver organ, in addition, its abundance is expressed in brown adipose tissue and associated with proper plasma lipid profiles in mice blood. The authors suggested that the FGL1 rs340465447_A allele can be used as a target in pig selection focused on elevated fat levels.

Two reports of SI considered a problem of fertility—litter size. Srihi et al. [15] estimated the additive and dominance variances of the purebred (Retinto and Entrepelado pigs) and CASTÚA crossbreed populations for litter size (including total number born (TNB) and number born alive (NBA)) for and calculated the additive genetic correlations between the purebred and crossbred performances. The authors identified four genomic regions (containing 30 SNPs) that each explained >2% of the additive genetic variance in chromosomes 6, 8 and 12. In turn, Sell-Kubiak et al. [16] based on 2000 SNP associated with pig litter size traits and reported based on previous genome-wide association studies (GWASs), gathered and integrated associations between SNPs and these traits. Authors selected the most interesting candidate genes reported in multiple populations such as PRKD1 involved in angiogenesis and associated with stillborn and TNB, and two new not previously reported—FAM13C and AGMO related to TNB, and the most promising candidate genes for litter size—SOSTDC1, which was described before as associated with male fertile in rats.

Porcine cells are commonly used as models of human disease, which is under consideration in medical science. The present SI contains two studies [17,18] using porcine atrial cardiomyocytes during the primary in vitro culture as a model that describes molecular mechanisms occurring in these cell types related to heart failure. The heart was considered a non-regenerated organ, but a few reports suggest it has modest intrinsic regenerative potential. Therefore, Bryl et al. [18] characterized cell cultures from the right atrial appendage and right atrial wall during cell culture cultivation duration of up to 30 days, based on a microarrays approach, they observed DEG enrichment in GO of stem cell population maintenance" and "stem cell proliferation", which suggested for previously described regenerative heart potential. In turn, Nawrocki et al. [17], using in vitro cell culture, extracted from the myocardium, revealed that DEGs were classified as involved in ontological groups such as: "cellular component assembly", "cellular component organization", "cellular component biogenesis", and "cytoskeleton organization with significantly increased expression of

COL5A2, COL8A1, and COL12A1 encoding different collagen subunits, pivotal in cardiac extracellular matrix (ECM) and significant down-regulated related to cellular architecture such as ABLIM1, TMOD1, XIRP1, and PHACTR1. Using porcine cells as a model allows for a better understanding of underlying molecular mechanisms of cardiovascular pathologies, which seems crucial to developing effective therapeutic options.

In this special issue, we present the latest advances in pig genetics and genomics, including identifying mechanisms related to adipose tissue deposition and lipid metabolism and using pig cells to model processes related to circulatory system abnormalities, including identifying candidate genes. The special issue also includes manuscripts on identifying markers for reproductive traits in pigs, which is an important economic aspect, and several items presenting non-standard research methods. We warmly encourage you to familiarize yourself with our new SI entitled "Trends and Prospects in Pig Genomics and Genetics".

Author Contributions: K.P. writing—original draft preparation; K.R.-M. writing—review and editing. All authors have read and agreed to the published version of the manuscript.

Conflicts of Interest: None of the authors has a financial or other relationship with other people or organizations that may inappropriately influence this work.

References

1. Mateos, G.G.; Corrales, N.L.; Talegón, G.; Aguirre, L. Pig meat production in the European Union-27: Current status, challenges, and future trends. *Anim. Biosci.* **2024**, *37*, 755. [CrossRef] [PubMed]
2. Merks, J.W.M.; Mathur, P.K.; Knol, E.F. New phenotypes for new breeding goals in pigs. *Animal* **2012**, *6*, 535–543. [CrossRef] [PubMed]
3. Georges, M.; Charlier, C.; Hayes, B. Harnessing genomic information for livestock improvement. *Nat. Rev. Genet.* **2019**, *20*, 135–156. [CrossRef] [PubMed]
4. Quer, J.; Colomer-Castell, S.; Campos, C.; Andrés, C.; Piñana, M.; Cortese, M.F.; González-Sánchez, A.; Garcia-Cehic, D.; Ibáñez, M.; Pumarola, T.; et al. Next-Generation Sequencing for Confronting Virus Pandemics. *Viruses* **2022**, *14*, 600. [CrossRef]
5. Whitworth, K.M.; Green, J.A.; Redel, B.K.; Geisert, R.D.; Lee, K.; Telugu, B.P.; Wells, K.D.; Prather, R.S. Improvements in pig agriculture through gene editing. *Cabi Agric. Biosci.* **2022**, *3*, 41. [CrossRef] [PubMed]
6. Ouma, E.; Dione, M.; Birungi, R.; Lule, P.; Mayega, L.; Dizyee, K. African swine fever control and market integration in Ugandan peri-urban smallholder pig value chains: An ex-ante impact assessment of interventions and their interaction. *Prev. Vet. Med.* **2018**, *151*, 29–39. [CrossRef]
7. Pavone, S.; Iscaro, C.; Dettori, A.; Feliziani, F. African Swine Fever: The State of the Art in Italy. *Animals* **2023**, *13*, 2998. [CrossRef]
8. Wang, Y.; Gou, Y.; Yuan, R.; Zou, Q.; Zhang, X.; Zheng, T.; Fei, K.; Shi, R.; Zhang, M.; Li, Y.; et al. A chromosome-level genome of Chenghua pig provides new insights into the domestication and local adaptation of pigs. *Int. J. Biol. Macromol.* **2024**, *270*, 131796. [CrossRef]
9. Li, Z.; Liu, X.; Wang, C.; Li, Z.; Jiang, B.; Zhang, R.; Tong, L.; Qu, Y.; He, S.; Chen, H.; et al. The pig pangenome provides insights into the roles of coding structural variations in genetic diversity and adaptation. *Genome Res.* **2023**, *33*, 1833–1847. [CrossRef] [PubMed]
10. Bilinska, A.; Pszczola, M.; Stachowiak, M.; Stachecka, J.; Garbacz, F.; Aksoy, M.O.; Szczerbal, I. Droplet Digital PCR Quantification of Selected Intracellular and Extracellular microRNAs Reveals Changes in Their Expression Pattern during Porcine In Vitro Adipogenesis. *Genes* **2023**, *14*, 683. [CrossRef] [PubMed]
11. Wang, Z.; Chai, J.; Wang, Y.; Gu, Y.; Long, K.; Li, M.; Jin, L. LncPLAAT3-AS Regulates PLAAT3-Mediated Adipocyte Differentiation and Lipogenesis in Pigs through miR-503-5p. *Genes* **2023**, *14*, 161. [CrossRef] [PubMed]
12. Zhu, J.; Yang, Z.; Hao, W.; Li, J.; Wang, L.; Xia, J.; Zhang, D.; Liu, D.; Yang, X. Characterization of a Read-through Fusion Transcript, BCL2L2-PABPN1, Involved in Porcine Adipogenesis. *Genes* **2022**, *13*, 445. [CrossRef]
13. Lin, W.; Zhao, J.; Yan, M.; Li, X.; Yang, K.; Wei, W.; Zhang, L.; Chen, J. SESN3 Inhibited SMAD3 to Relieve Its Suppression for MiR-124, Thus Regulating Pre-Adipocyte Adipogenesis. *Genes* **2021**, *12*, 1852. [CrossRef]
14. Piórkowska, K.; Zukowski, K.; Ropka-Molik, K.; Tyra, M. Variations in Fibrinogen-like 1 (FGL1) Gene Locus as a Genetic Marker Related to Fat Deposition Based on Pig Model and Liver RNA-Seq Data. *Genes* **2022**, *13*, 1419. [CrossRef]
15. Srihi, H.; Noguera, J.L.; Topayan, V.; de Hijas, M.M.; Ibañez-Escriche, N.; Casellas, J.; Vázquez-Gómez, M.; Martínez-Castillero, M.; Rosas, J.P.; Varona, L. Additive and dominance genomic analysis for litter size in purebred and crossbred iberian pigs. *Genes* **2022**, *13*, 12. [CrossRef]
16. Sell-Kubiak, E.; Dobrzanski, J.; Derks, M.F.L.; Lopes, M.S.; Szwaczkowski, T. Meta-Analysis of SNPs Determining Litter Traits in Pigs. *Genes* **2022**, *13*, 1730. [CrossRef]

17. Nawrocki, M.J.; Jopek, K.; Kaczmarek, M.; Zdun, M.; Mozdziak, P.; Jemielity, M.; Perek, B.; Bukowska, D.; Kempisty, B. Transcriptomic Profile of Genes Regulating the Structural Organization of Porcine Atrial Cardiomyocytes during Primary In Vitro Culture. *Genes* **2022**, *13*, 1205. [CrossRef]
18. Bryl, R.; Nawrocki, M.J.; Jopek, K.; Kaczmarek, M.; Bukowska, D.; Antosik, P.; Mozdziak, P.; Zabel, M.; Dzięgiel, P.; Kempisty, B. Transcriptomic Characterization of Genes Regulating the Stemness in Porcine Atrial Cardiomyocytes during Primary In Vitro Culture. *Genes* **2023**, *14*, 1223. [CrossRef] [PubMed]

genes

Article

Altered Transcript Levels of *MMP13* and *VIT* Genes in the Muscle and Connective Tissue of Pigs with Umbilical Hernia

Jakub Wozniak [1], Weronika Loba [1], Alicja Wysocka [1,†], Stanislaw Dzimira [2], Przemyslaw Przadka [3], Marek Switonski [1] and Joanna Nowacka-Woszuk [1,*]

[1] Department of Genetics and Animal Breeding, Poznan University of Life Sciences, Wolynska 33, 60-637 Poznan, Poland; jakub.wozniak@up.poznan.pl (J.W.); weronika.loba@up.poznan.pl (W.L.); alicja.wysocka@igcz.poznan.pl (A.W.); marek.switonski@up.poznan.pl (M.S.)
[2] Department of Pathology, Wroclaw University of Environmental and Life Sciences, C.K. Norwida 31, 50-375 Wroclaw, Poland; stanislaw.dzimira@upwr.edu.pl
[3] Department of Surgery, Wroclaw University of Environmental and Life Sciences, Plac Grunwaldzki 51, 50-366 Wroclaw, Poland; przemyslaw.przadka@upwr.edu.pl
* Correspondence: joanna.nowacka-woszuk@up.poznan.pl; Tel.: +48-61-8487242
† Current address: Institute of Human Genetics, Polish Academy of Sciences, Strzeszynska 32, 60-479 Poznan, Poland.

Citation: Wozniak, J.; Loba, W.; Wysocka, A.; Dzimira, S.; Przadka, P.; Switonski, M.; Nowacka-Woszuk, J. Altered Transcript Levels of *MMP13* and *VIT* Genes in the Muscle and Connective Tissue of Pigs with Umbilical Hernia. *Genes* **2023**, *14*, 1903. https://doi.org/10.3390/genes14101903

Academic Editors: Katarzyna Piórkowska and Katarzyna Ropka-Molik

Received: 1 September 2023
Revised: 27 September 2023
Accepted: 29 September 2023
Published: 1 October 2023

Abstract: Umbilical hernia (UH) and inguinal hernia (IH) are among the most common defects in pigs, affecting their welfare and resulting in economic losses. In this study, we aimed to verify the association of previously reported differences in transcript levels of the *ACAN*, *COL6A5*, *MMP13*, and *VIT* genes with the occurrence of UH and IH. We examined mRNA levels in muscle and connective tissue from 68 animals—34 affected by UH and 34 controls. In a second cohort, we examined inguinal channel samples from 46 pigs (in four groups). We determined DNA methylation levels in muscle tissue for the UH and control animals. The transcript level of *MMP13* changed in the UH cases, being upregulated and downregulated in muscle and connective tissue, respectively, and the *VIT* gene also showed an increased muscular mRNA level. The transcript of the *ACAN* gene significantly decreased in old pigs with IH. We further observed an increased DNA methylation level for one CpG site within the *MMP13* gene in UH individuals. We conclude that these alterations in gene mRNA levels in the UH animals depend on the tissue and can sometimes be a consequence of, not a cause of, the affected phenotype.

Keywords: umbilical hernia; inguinal hernia; pig; DNA methylation; transcript level

1. Introduction

Global pig production continues to grow with the increasing human population. Livestock provide many benefits such as employment growth, protein-rich food, and carbon storage [1], and the pig has become one of the most important livestock species [2]. The expanding pig industry needs to balance meat quality and animal welfare [3,4]. An important health issue in the pig industry is disorders like umbilical and inguinal hernia [5–7], which can result in economic losses.

Hernia is one of the most widespread condition in pigs; it involves the abdominal contents protruding and forming a subcutaneous bladder beneath the navel (umbilicus) [8], in the case of umbilical hernia (UH), or protruding via the inguinal canal in the case of inguinal hernia (IH) [9]. It can be affected by both genetic and environmental factors, such as incorrect cutting of the umbilical cord, high pressure in the abdomen, and infections and lesions of the body layers [7,10]. It is know that the occurrence of hernia ranges between 0.05% and 6.6%, depending on the breed and size of the studied cohorts. For example, the incidence of umbilical hernia in the Landrace and Pietrain breeds was reported to be 0.05% and 0.08%, respectively. For scrotal hernia, these estimations reached 2.4% for the Landrace and 6.6% for the Yorkshire breed [11–13]. The heritability of this disorder is rather low at approximately 0.3 [14]. Affected animals often show retarded growth and reduced feed efficiency [5]. They also suffer from pain, which increases morbidity and mortality [15]. Hernia is thus a severe problem for both breeders and veterinarians [16].

Although several studies have focused on identifying the genetic markers associated with hernia (reviewed, e.g., by [16,17]), the molecular mechanism responsible for this disorder in pigs remains poorly understood [5,8,15,18,19]. Recent studies have used transcriptome analysis to highlight several genes as strong candidates which correlated with the occurrence of umbilical hernia [6]. The researchers used RNA-seq technology for umbilical ring tissue fragments in five healthy and five umbilical hernia pigs, identifying 230 differentially expressed genes. Some of these genes (*ACAN*, *VIT*, *LGALS3*, *EPYC*, *CCBE1*, as well as genes from the *COL* and *MMP* families) have been taken as key contributors, since they are involved in the remodeling of the extracellular matrix as well as in resistance, production, and integrity of collagen [6]. A similar study concerning scrotal hernia (a form of inguinal hernia) was also performed for the inguinal canal tissue collected from four affected and four normal pigs. This research uncovered 703 differentially expressed genes, and the *MAP1CL3C*, *MYBPC1*, *SLC8A3*, *FGF1*, *BOK*, *DES*, *TPM2*, and *SLC25A4* genes were indicated as crucial in hernia pathogenesis. These genes are involved in important processes such as cell differentiation, muscular system regulation, and myofibril assembly [20].

DNA methylation is one of the regulatory mechanisms which has an influence on gene expression. The effect of this epigenetic modification may vary depending on the location of methylated cytosines and the level of methylation itself. It was already reported that cytosine methylation in gene body is connected with increased expression. On the other hand, the existence of methylated sites in $5'$-flanking regions, rich in CpG dinucleotides (called CpG islands) is associated with gene silencing and repression of transcription [21]. Therefore, it is justified to search for changes in DNA methylation level as a potential mechanism for altered gene expression.

The aim of this study was to determine the transcript level of the *VIT* (Vitrin), *MMP13* (Matrix Metallopeptidase 13), *ACAN* (Aggrecan), and *COL6A5* (Collagen type VI α 5 chain) candidate genes in the muscle and connective tissues dissected from the umbilical ring of pigs with UH and in the muscle tissue of inguinal canal of pigs with IH, and to search for epigenetic mechanisms, such as DNA methylation changes, that are responsible for mRNA level alterations.

2. Materials and Methods

2.1. Ethics Statement

The samples in this project were collected during routine veterinary procedures. This is permitted under Polish law and does not require the approval of the local Bioethical Commission for Animal Care and Use in Poznan, Poland (13 January 2020).

2.2. Groups, Phenotypes, and Sample Collection

2.2.1. Pigs with Umbilical Hernia

Sixty-eight piglets were derived from a cross between a DunBred (Landrace × Yorkshire) maternal hybrid line and Landrace or Duroc paternal lines. The animals all came from the same commercial breeding farm located in Wielkopolska Province, Poland. The animals were of both sexes and were aged 3–4 weeks. They were assigned into two groups: 34 animals were classified by veterinarians as umbilical hernia cases, while the control group (N = 34) was composed of healthy siblings (animals from the same litter as the hernia cases but showing no signs of hernia). This resulted in 34 full or half sibling pairs, so the affected animal always had a healthy sibling. Most of these pairs were born by different sows with the exception of 5 litters (delivered by 5 sows), where samples were collected from two hernia cases and two healthy controls. The animals were sacrificed by headshot, which was followed by collection of muscle and connective tissue samples separately from around the navel. Tissue samples were immediately frozen in dry ice for further analysis.

2.2.2. Pigs with Inguinal Hernia

Forty-six unrelated male pigs were included in this part of the analysis. The animals were a cross between the Polish Large White and Polish Landraces and came from local breeding farms. The pigs were divided into four groups on the basis of veterinarian opinion: group G1 contained ten healthy young animals weighing 30–35 kg without any signs of hernia. The second group (G2) consisted of fifteen animals weighing 30–35 kg and with nonregressive inguinal hernias. G3 group contained eleven pigs weighing 30–35 kg with regressed IH (the hernia was at first clinically observed, but disappeared once the animal's body weight exceeded 25 kg). The fourth group (G4) contained ten adult animals weighing 100–120 kg with IH observed throughout their lives. For the G2 and G3 animals, a fragment of the inguinal canal was collected during the delayed castration procedure, while the samples from G1 (at a weight of 30–35 kg) and G4 (at weight of 100–120 kg) were collected post-mortem from the slaughter house.

2.3. RNA Extraction and cDNA Synthesis for qPCR Assay

Total RNA extraction was performed using an RNeasy Fibrous Tissue Mini Kit (Qiagen, Germantown, MD, USA). The 30 mg of tissue from each sample was homogenized in buffer RTL supplied with the RNeasy Fibrous Tissue Mini Kit (Qiagen) using TissueLyser LT (Qiagen). The following steps were performed in line with the manufacturer's instructions. The quantity and quality of total RNA were assessed by using a NanoDrop ND-2000 spectrophotometer (Thermo Scientific, Waltham, MA, USA) with 260/280 nm wavelength readings.

Around 1 µg of RNA from each sample was reverse-transcribed using a Transcriptor First Strand cDNA Synthesis Kit (Roche, Mannheim, Germany) in a total volume of 20 µL, following the manufacturer's directions. The cDNA was subsequently evaluated with a NanoDrop ND-2000 spectrophotometer (Thermo Scientific, Waltham, MA, USA) and diluted with nuclease-free water to approximately 700 ng/µL.

2.4. Quantitative Real-Time PCR

The cDNA was used in real-time PCR reactions running on a LightCycler® 480 II (Roche, Mannheim, Germany). We analyzed *ACAN*, *COL6A5*, *MMP13*, and *VIT* from both connective and muscle tissue of 34 healthy animals and 34 animals with umbilical hernia. The procedure was additionally performed on muscle tissue from the inguinal canal for healthy and inguinal hernia pigs (46 animals in total). Primers were designed using Primer3Plus software (version 2.4; https://primer3plus.com/cgi-bin/dev/primer3plus.cgi; accessed on 2 June 2021; Supplementary Material Table S1). The reactions for each gene were performed in duplicate using a LightCycler® 480 SYBR Green I Master (Roche, Mannheim, Germany) in a total reaction mix of 10 µL per well, following the manufacturer's instructions. The thermocycler program began with an initial denaturation cycle at 95 °C for 10 min, followed by 45 cycles including denaturation at 95 °C for 10 s, annealing at 60 °C for 5 s, and elongation at 72 °C for 5 s. Melting curve analysis was carried out after each amplification in order to verify product specificity. The relative expression values of each gene were computed on the basis of the standard curve method with a series of ten-fold dilutions of a DNA of known concentration (standards). The relative changes in expression level of genes were normalized against the level of reference (internal control) of the *H3F3A* (H3 histone, family 3A) and *PPIA* (Peptidyl-prolyl cis-trans isomerase A) genes [22], as described by [23].

2.5. Methylation Analysis

An in silico analysis was performed for the two differentially expressed genes, *MMP13* and *VIT*, for which statistically significant results were observed in the transcript level.

The analysis with CpGPlot software (version EMBOSS 6.6.0; http://www.ebi.ac.uk/Tools/seqstats/emboss_cpgplot/; accessed on 17 November 2022) pointed to several CpG sites in the 5′ untranslated region (5′UTR) of the first exon of *MMP13* and in exon 12 of *VIT* (Supplementary Material Table S2).

The PyroMark Assay Design 2.0 software (Qiagen) was used to design primers for the pyrosequencing analysis (Supplementary Material Table S1). Thirty-four animals with umbilical hernia and 34 controls were analyzed in this way. The DNA was isolated from the muscle tissue (the amount of connective tissue was insufficient to perform this analysis) using a Genomic Mini Kit (A&A Biotechnology, Gdańsk, Poland). The quantity and quality of DNA was assessed with the use of a NanoDrop ND-2000 spectrophotometer (Thermo Scientific, Waltham, MA, USA) with 260/280 nm wavelength readings. Five hundred nanograms of DNA was bisulfite-converted using an EZ DNA Methylation-Gold kit (Zymo Research, Irvine, CA, USA). Additionally, nonmethylated controls were prepared using REPLI g Mini Kits (Qiagen), in line with the manufacturer's instructions. The fully methylated controls were prepared from 500 ng of DNA using CpG methyltransferase (*M.SssI*, Thermo Fisher) with incubation for 3 h at 37 °C. Next, PCR was performed with a PyroMark PCR kit (Qiagen), following the manufacturer's instructions. The PCR conditions were: initial denaturation at 95 °C for 15 min; 44 cycles of denaturation at 94 °C for 30 s, primer annealing at 56 °C for both genes, for 30 s and elongation at 72 °C for 30 s, followed by a final extension at 72 °C for 10 min. The negative control samples (without the DNA template) were also prepared for each reaction. PCR amplification results were checked on 1.5% agarose gel electrophoresis. The pyrosequencing reaction for the amplicons was carried out on a PyroMark Q48 Autoprep (Qiagen) system using PyroMark Q48 Advanced CpG Reagents (Qiagen), following the manufacturer's recommended protocol. Finally, PyroMark Q48 Autoprep software (Qiagen) was used to calculate the percentage methylation for each of the CpG sites.

2.6. Statistical Analysis

Statistical analysis was carried out in the R statistical environment (version 4.1.2) [24]. The Shapiro–Wilk test was used to test the normality of the data using the "shapiro.test" function from the "stats" package (version 4.1.2). Subsequently, to compare mRNA results for pigs with UH and controls, a p-value was computed using the nonparametric, two-tailed Mann–Whitney U test. The same test was used to compare different DNA methylation levels between the UH and control animals. Each methylated cytosine was examined independently. The "wilcox.test" function was used to calculate p-values with parameter "paired = FALSE". Moreover, adjustment using the Benjamini and Hochberg (1995) [25] procedure (FDR) was performed for the mRNA results using the "p.adjust" function. Finally, the "cor.test" function was applied to perform the Spearman's rank correlation test to examine the significance of the correlation between the *MMP13* relative mRNA level and the CpG methylation level.

Four groups of animals with inguinal hernia were compared. Initially the Shapiro–Wilk test and the Kruskal–Wallis rank sum test were used ("kruskal.test" function from the "stats" package (version 4.1.2)). Subsequently, Dunn's test for multiple comparisons with the FDR correction was chosen as a post hoc test, in order to identify exactly which groups differed, using the "dunnTest" function from the "FSA" package (version 0.9.3). For the performed tests, the statistical significance cutoff was set at $p < 0.05$.

3. Results

3.1. Relative mRNA Levels of the Studied Genes in the UH Pigs

Quantitative real-time PCR was performed for *MMP13*, *VIT*, *ACAN*, and *COL6A5* in muscle and connective tissue for all 68 animals, revealing differences in the mRNA level of *MMP13* and *VIT* between the control and hernia pigs. The transcript level of *MMP13* was significantly higher in pigs with UH than in the muscle tissue of controls ($p_{adj.} < 0.01$); this was unlike the connective tissue, whose mRNA level was lower in the UH animals

($p_{adj.} < 0.001$). This indicates that the transcript level depends on the analyzed tissue. The only significant result for the *VIT* gene was observed in muscle tissue ($p_{adj.} < 0.05$), where a higher level of mRNA was found in UH pigs than in controls. For *ACAN* and *COL6A5*, the differences proved insignificant in all of the tissues (Figure 1).

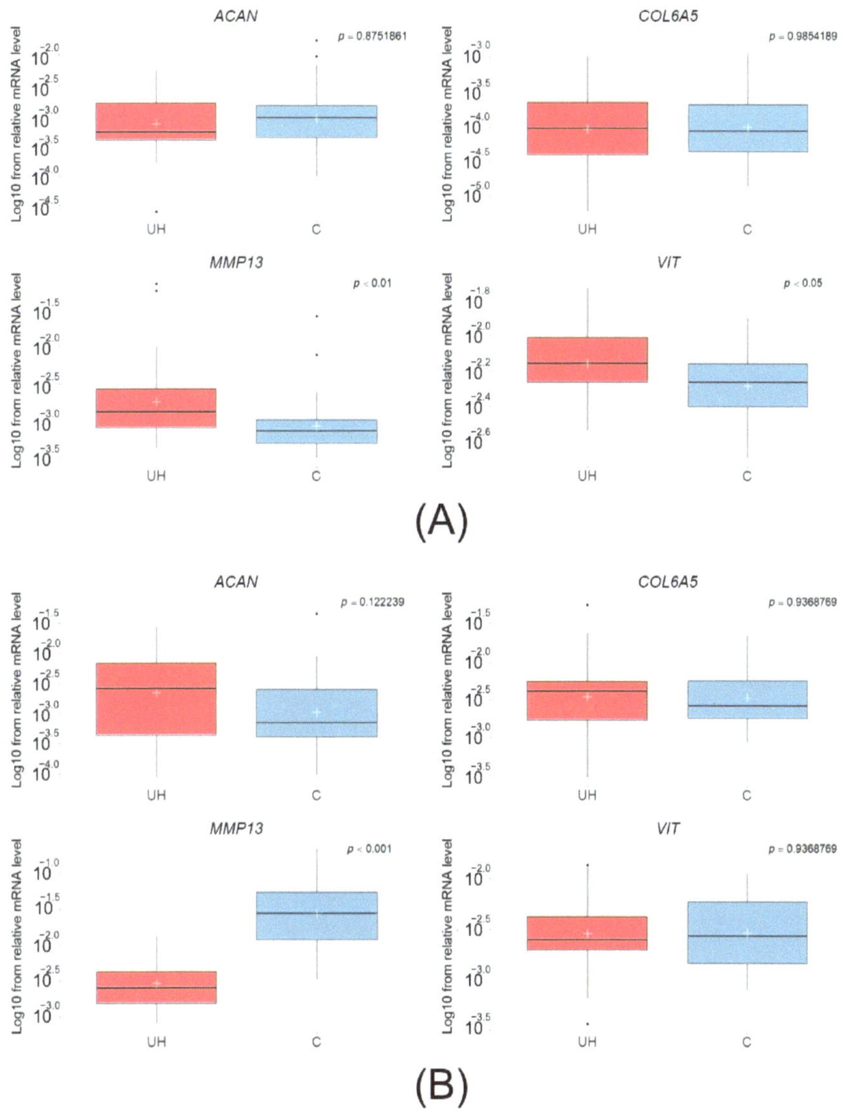

Figure 1. The Log10 of the relative mRNA levels of the *ACAN*, *COL6A5*, *MMP13*, and *VIT* genes in muscle (**A**) and in connective tissue (**B**) obtained from the UH (cases) and C (controls) animals. The white crosses denote mean values and the lines crossing the boxes represent medians. The lines under and over the rectangles denote the highest and lowest values, and the black dots under and over the boxes are outliers.

3.2. Relative mRNA Levels of Studied Genes in IH Pigs

The mRNA level was determined for *MMP13*, *VIT*, *ACAN* and *COL6A5*, but the Kruskal–Wallis test indicated significant results only for the *ACAN* ($p = 0.02$) gene. Subsequently, the post hoc test showed significantly lower transcript levels ($p_{adj.} < 0.05$) in group G4 than in groups G1, G2, and G3 (Figure 2), indicating that the altered *ACAN* expression was not related with the presence of hernia. On the other hand, we found that the expression of this gene was significantly downregulated in group G4 represented by adult IH pigs (>100 kg), so the observed differences can be related to the animal's age.

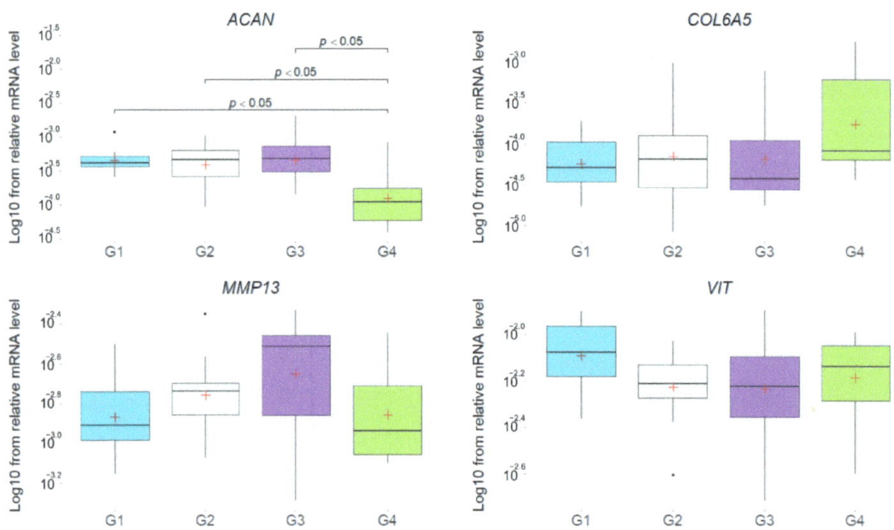

Figure 2. The Log10 of relative mRNA levels of the *ACAN*, *COL6A5*, *MMP13*, and *VIT* genes in groups of pigs without inguinal hernia and with different stages thereof. The white crosses denote mean values and the lines crossing the boxes represent medians. The lines under and over the rectangles denote the highest and lowest values, and the black dots under and over the boxes are outliers. The groups are indicated by G1, G2, G3, and G4.

3.3. CpG Methylation Level

DNA methylation levels were examined in exon 12 for *VIT* and in the 5′UTR for *MMP13*. For both genes, cytosine methylation level was determined in the muscle tissue and seven CpG sites were studied for each gene (Supplementary Material Table S3). Only one statistically significant result was found for the single cytosine marked CpG7 in the *MMP13* gene (chr9: 33616509). In this case, we observed significant higher ($p < 0.001$) cytosine methylation levels in the UH animals (mean = 34%) than in the controls (mean = 29%); see Figure 3. However, there was no significant correlation between the *MMP13* relative mRNA level and the CpG methylation level at this position ($\rho = 0.09342564$; $p = 0.4486$). This observation shows that the DNA methylation of the studied region is probably not responsible for altered transcript level of *MMP13* gene.

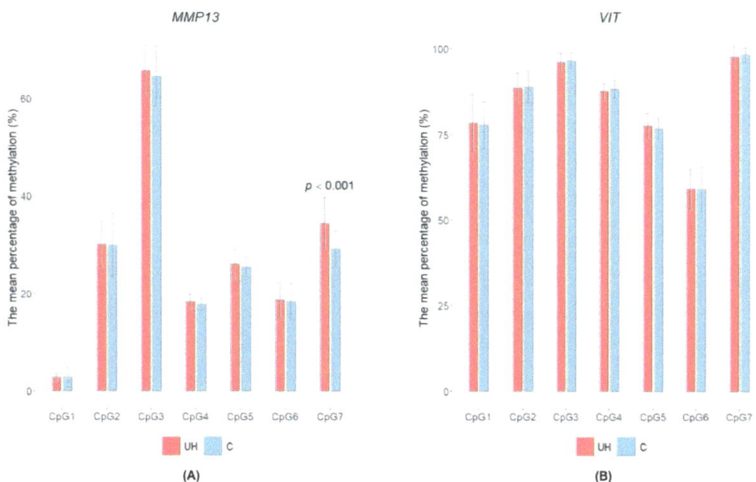

Figure 3. Mean percentage of DNA methylation ± SD in the CpG of two candidate genes: *MMP13* (**A**) and *VIT* (**B**) for UH and control pigs. The cytosines are indicated as CpG1, CpG2, CpG3, CpG4, CpG5, CpG6, and CpG7. Only one significant difference was observed for CpG7 in *MMP13*; the *p*-value refers only to this cytosine.

4. Discussion

Information on the background of porcine hernia remains scarce [16]. Much effort has been put into identifying genetic variants that predispose to hernia development [5,6,8,15,18,19,26], but the polygenic nature of this condition has made such studies challenging. A new approach involving the transcriptome differences between hernia and control animals was thus undertaken for umbilical and inguinal hernia [6,20].

Here we studied the mRNA levels of four genes (*ACAN*, *COL6A5*, *MMP13*, and *VIT*) initially noted by Souza et al. [6] as strong candidates for hernia formation. The novelty of our study is that, in the case of umbilical hernia, we examined the muscle and connective tissues separately (dissected from the umbilical ring), which has not been done previously [6]. We also determined whether altered transcript levels in muscle tissue might be caused by differences in epigenetic mechanisms (DNA methylation). We additionally checked whether the same genes have altered expression in inguinal canal tissue collected from inguinal hernia pigs.

Our most interesting results were for the *MMP13* gene, in which the transcript level was increased in the muscle tissue and decreased in the connective tissue of the pigs with umbilical hernia. Expression of this gene has previously been associated with inguinal hernia in humans [27], while its downregulation has been noted in pigs with umbilical hernia by Souza et al. [6]. Those researchers, however, studied the transcriptome in a mix of different cell types collected from the umbilical ring. Here we were able to distinguish between muscle and connective tissues and found the opposite results. These differences may be caused by the different molecular function of the MMP13 protein depending on the expression site. MMP13 (known also as collagenase-3) is a member of the matrix metalloproteinase (MMP) family, which are proteins that degrade the components of the extracellular matrix and remodel connective tissue [28]. We can thus speculate that the reduced level of the *MMP13* gene in the connective tissue of the affected animals leads to incorrect tissue remodeling and possible lower tissue strength, which can predispose for hernia development. MMP13 also plays a role in wound healing [29], and it is known that cytokines and growth factors formed under pathological condition induce MMP levels [30]. The increased *MMP13* transcript level observed in the muscle tissue in our study might thus be rather a consequence of the hernia condition. It can be suggested that an elevated level of MMP13 is needed for wound healing (e.g., that the disruption of muscle tissue by the hernia induces increased repair processes). This hypothesis is supported by earlier studies, in which the involvement of the *MMP13* gene in tissue healing and repair processes has already been confirmed [31,32].

We also checked whether the differences in the mRNA level of the *MMP13* gene are the results of DNA methylation changes. Due to the limited quantity of connective tissue sample, this analysis was performed only for the muscle tissue. We found that in the studied 5′UTR region (exon 1) the methylation level for one cytosine (CpG7) was significantly increased in the affected animals. It is known that the 5′UTR region contains multiple regulatory elements that are important for the correct initiation of translation [33]. Moreover, a recent broad study of DNA methylation changes during human early embryo development concluded that, in the 5′UTR region (where the high CpG density was observed), the methylation level was low and negatively correlated with gene expression [34]. However, in our study, the significantly methylated cytosine in the 5′UTR of the *MMP13* gene did not correlate with the transcript level of the *MMP13* gene. We are thus unable to determine whether the difference we observed in DNA methylation has a significant relation to the hernia condition.

One interesting result concerned elevated transcript levels of the *VIT* gene in UH pigs. This observation did not confirm the result reported by Souza et al. [6], who examined a mix of tissues from the umbilical ring to compare 5 UH with 5 control pigs; they noted that the level of *VIT* expression was reduced. The *VIT* gene encodes an extracellular matrix (ECM) protein, which contains two von Willebrand A domains. It seems that the VIT protein is associated with migration and cell adhesion, like the majority of von Willebrand A domain-containing proteins [35,36]. Based on this information, we suggest that the observed overexpression of this gene in muscle tissue may be an effect of the repair mechanism acting in response to muscle damage by the hernia. Our study did not show altered transcript levels of the *ACAN* and *COL6A5* genes in the tissues of the UH pigs.

Interestingly, significant differences in the mRNA level of the *ACAN* gene were found in the IH pigs. Differences were observable between young (group G1, G2, and G3) and old (group G4) pigs, but not between those with inguinal hernia and those which were healthy. We thus assume that the reduced mRNA level of *ACAN* is an effect of aging. It is known that this protein (aggrecan) is a proteoglycan and is a component of the extracellular matrix which exists in form of proteoglycan aggregates. It plays an important role in the function of cartilaginous tissues [37], as well as in skeletal development through its involvement in the endochondral ossification process [38]. Our observation could thus be explained by the role of the ACAN protein in collagen production, with young pigs needing higher levels in the intense growth phase, whereas aggrecan in adults is involved in skeletal development. Moreover, ACAN is an important element of cartilage extracellular matrix that is reduced as aging of the organism proceeds [39]. The *ACAN* gene was the only gene that was found to be differently expressed in the pigs with inguinal hernia in this study, and no differences were observed for *MMP13*, *VIT*, or *COL6A5*.

5. Conclusions

Pigs with umbilical hernia showed differences in mRNA levels for the *MMP13* gene in muscle and connective tissue dissected from the umbilical ring as well as changes in the DNA methylation of a single cytosine located in the 5′UTR of *MMP13*. This may be the result of the demand for the MMP13 protein in UH pigs depending on tissue type. The *VIT* gene differed in transcript level in muscle tissues, which may be a result of umbilical hernia, rather than a cause. In the case of inguinal hernia, only the *ACAN* gene showed decreased mRNA levels in the muscle tissue of the inguinal canal in older animals, which may be a consequence of aging rather than pathological status.

Supplementary Materials: The following supporting information can be downloaded from: https://www.mdpi.com/article/10.3390/genes14101903/s1, Table S1: primer sequences for the *ACAN*, *COL6A5*, *MMP13*, *VIT*, *H3F3A*, and *PPIA* genes used in qPCR and for the *MMP13* and *VIT* genes used in pyrosequencing; Table S2: CpG genomic position according to Sscrofa11.1. and their location in the gene context; Table S3: the mean percentage of DNA methylation with standard deviation and *p*-value in each studied cytosine, in two examined genes (*MMP13* and *VIT*), for both control and umbilical hernia groups.

Author Contributions: Conceptualization: J.N.-W. and M.S.; methodology: J.W., W.L., A.W. and J.N.-W.; software: J.W.; validation: J.W., W.L., A.W. and J.N.-W.; formal analysis: J.W., W.L., A.W. and J.N.-W.; investigation: J.W., W.L., A.W. and J.N.-W.; resources: S.D. and P.P.; data curation: J.W. and J.N.-W.; writing—original draft preparation: J.W. and J.N.-W.; writing—review and editing: J.W., M.S., S.D., P.P. and J.N.-W.; visualization: J.W., supervision: J.N.-W.; project administration: J.N.-W.; funding acquisition: J.N.-W. All authors have read and agreed to the published version of the manuscript.

Funding: This research was funded by the National Science Centre, Poland, project No. 2019/33/B/NZ9/00654.

Institutional Review Board Statement: The samples were collected during routine veterinary procedures, which under Polish law do not require the permission of the local Bioethical Commission for Animal Care and Use in Poznan, Poland (13 January 2020).

Informed Consent Statement: Not applicable.

Data Availability Statement: Not applicable.

Acknowledgments: We thank Rafal Niemyjski and Jaroslaw Krzyszton for their help in sample collection.

Conflicts of Interest: The authors declare no conflict of interest.

References

1. Rauw, W.M.; Rydhmer, L.; Kyriazakis, I.; Øverland, M.; Gilbert, H.; Dekkers, J.C.; Hermesch, S.; Bouquet, A.; Gómez Izquierdo, E.; Louveau, I.; et al. Prospects for Sustainability of Pig Production in Relation to Climate Change and Novel Feed Resources. *J. Sci. Food Agric.* **2020**, *100*, 3575–3586. [CrossRef] [PubMed]
2. Gerber, P.J.; Vellinga, T.V.; Steinfeld, H. Issues and Options in Addressing the Environmental Consequences of Livestock Sector's Growth. *Meat Sci.* **2010**, *84*, 244–247. [CrossRef] [PubMed]
3. Davoli, R.; Braglia, S. Molecular Approaches in Pig Breeding to Improve Meat Quality. *Brief. Funct. Genom. Proteom.* **2008**, *6*, 313–321. [CrossRef] [PubMed]
4. Thorslund, C.A.H.; Aaslyng, M.D.; Lassen, J. Perceived Importance and Responsibility for Market-Driven Pig Welfare: Literature Review. *Meat Sci.* **2017**, *125*, 37–45. [CrossRef]
5. Lago, L.; Nery Da Silva, A.; Zanella, E.; Groke Marques, M.; Peixoto, J.; Da Silva, M.; Ledur, M.; Zanella, R. Identification of Genetic Regions Associated with Scrotal Hernias in a Commercial Swine Herd. *Vet. Sci.* **2018**, *5*, 15. [CrossRef]
6. Souza, M.R.; Ibelli, A.M.G.; Savoldi, I.R.; Cantão, M.E.; Peixoto, J.D.O.; Mores, M.A.Z.; Lopes, J.S.; Coutinho, L.L.; Ledur, M.C. Transcriptome Analysis Identifies Genes Involved with the Development of Umbilical Hernias in Pigs. *PLoS ONE* **2020**, *15*, e0232542. [CrossRef]
7. Rodrigues, A.F.G.; Ibelli, A.M.G.; Peixoto, J.D.O.; Cantão, M.E.; Oliveira, H.C.D.; Savoldi, I.R.; Souza, M.R.; Mores, M.A.Z.; Carreño, L.O.D.; Ledur, M.C. Genes and SNPs Involved with Scrotal and Umbilical Hernia in Pigs. *Genes* **2021**, *12*, 166. [CrossRef]
8. Grindflek, E.; Hansen, M.H.S.; Lien, S.; Van Son, M. Genome-Wide Association Study Reveals a QTL and Strong Candidate Genes for Umbilical Hernia in Pigs on SSC14. *BMC Genom.* **2018**, *19*, 412. [CrossRef]
9. Hammoud, M.; Gerken, J. Inguinal Hernia. In *StatPearls*; StatPearls Publishing: Treasure Island, FL, USA, 2023.
10. Liao, Y.; Smyth, G.K.; Shi, W. The R Package Rsubread Is Easier, Faster, Cheaper and Better for Alignment and Quantification of RNA Sequencing Reads. *Nucleic Acids Res.* **2019**, *47*, e47. [CrossRef]
11. Mikami, H.; Fredeen, H.T. A GENETIC STUDY OF CRYPTORCHIDISM AND SCROTAL HERNIA IN PIGS. *Can. J. Genet. Cytol.* **1979**, *21*, 9–19. [CrossRef]
12. Searcy-Bernal, R.; Gardner, I.A.; Hird, D.W. Effects of and Factors Associated with Umbilical Hernias in a Swine Herd. *J. Am. Vet. Med. Assoc.* **1994**, *204*, 1660–1664. [PubMed]
13. Thaller, G.; Dempfle, L.; Hoeschele, I. Investigation of the Inheritance of Birth Defects in Swine by Complex Segregation Analysis. *J. Anim. Breed. Genet.* **1996**, *113*, 77–92. [CrossRef]
14. Sevillano, C.A.; Lopes, M.S.; Harlizius, B.; Hanenberg, E.; Knol, E.F.; Bastiaansen, J. Genome-Wide Association Study Using Deregressed Breeding Values for Cryptorchidism and Scrotal/Inguinal Hernia in Two Pig Lines. *Genet. Sel. Evol.* **2015**, *47*, 18. [CrossRef] [PubMed]

15. Ding, N.S.; Mao, H.R.; Guo, Y.M.; Ren, J.; Xiao, S.J.; Wu, G.Z.; Shen, H.Q.; Wu, L.H.; Ruan, G.F.; Brenig, B.; et al. A Genome-Wide Scan Reveals Candidate Susceptibility Loci for Pig Hernias in an Intercross between White Duroc and Erhualian1. *J. Anim. Sci.* **2009**, *87*, 2469–2474. [CrossRef]
16. Nowacka-Woszuk, J. The Genetic Background of Hernia in Pigs: A Review. *Livest. Sci.* **2021**, *244*, 104317. [CrossRef]
17. Krupova, Z. Candidate Genes for Congenital Malformations in Pigs. *Acta Fytotechn Zootech.* **2021**, *24*, 309–314. [CrossRef]
18. Long, Y.; Su, Y.; Ai, H.; Zhang, Z.; Yang, B.; Ruan, G.; Xiao, S.; Liao, X.; Ren, J.; Huang, L.; et al. A Genome-Wide Association Study of Copy Number Variations with Umbilical Hernia in Swine. *Anim. Genet.* **2016**, *47*, 298–305. [CrossRef]
19. Li, X.; Xu, P.; Zhang, C.; Sun, C.; Li, X.; Han, X.; Li, M.; Qiao, R. Genome-wide Association Study Identifies Variants in the *CAPN 9* Gene Associated with Umbilical Hernia in Pigs. *Anim. Genet.* **2019**, *50*, 162–165. [CrossRef]
20. Romano, G.D.S.; Ibelli, A.M.G.; Lorenzetti, W.R.; Weber, T.; Peixoto, J.D.O.; Cantão, M.E.; Mores, M.A.Z.; Morés, N.; Pedrosa, V.B.; Coutinho, L.L.; et al. Inguinal Ring RNA Sequencing Reveals Downregulation of Muscular Genes Related to Scrotal Hernia in Pigs. *Genes* **2020**, *11*, 117. [CrossRef]
21. Greenberg, M.V.C.; Bourc'his, D. The Diverse Roles of DNA Methylation in Mammalian Development and Disease. *Nat. Rev. Mol. Cell Biol.* **2019**, *20*, 590–607. [CrossRef]
22. Lorenzetti, W.R.; Ibelli, A.M.G.; Peixoto, J.D.O.; Mores, M.A.Z.; Savoldi, I.R.; Carmo, K.B.D.; Oliveira, H.C.D.; Ledur, M.C. Identification of Endogenous Normalizing Genes for Expression Studies in Inguinal Ring Tissue for Scrotal Hernias in Pigs. *PLoS ONE* **2018**, *13*, e0204348. [CrossRef] [PubMed]
23. Vandesompele, J.; De Preter, K.; Pattyn, F.; Poppe, B.; Van Roy, N.; De Paepe, A.; Speleman, F. Accurate Normalization of Real-Time Quantitative RT-PCR Data by Geometric Averaging of Multiple Internal Control Genes. *Genome Biol.* **2002**, *3*, research0034.1. [CrossRef] [PubMed]
24. R Core Team. *R: A Language and Environment for Statistical Computing*; R Foundation for Statistical Computing: Vienna, Austria, 2021. Available online: https://www.R-Project.Org (accessed on 10 February 2022).
25. Benjamini, Y.; Hochberg, Y. Controlling the False Discovery Rate: A Practical and Powerful Approach to Multiple Testing. *J. R. Stat. Soc. B Stat. Methodol.* **1995**, *57*, 289–300. [CrossRef]
26. Wozniak, J.; Loba, W.; Iskrzak, P.; Pszczola, M.; Wojtczak, J.; Switonski, M.; Nowacka-Woszuk, J. A Confirmed Association between DNA Variants in *CAPN9*, *OSM*, and *ITGAM* Candidate Genes and the Risk of Umbilical Hernia in Pigs. *Anim. Genet.* **2023**, *54*, 307–314. [CrossRef]
27. Zheng, Z.S.; Reiner Kasperk, H. Recurrent Inguinal Hernia: Disease of the Collagen Matrix? *World J. Surg.* **2002**, *26*, 401–408. [CrossRef]
28. Bracale, U.; Peltrini, R.; Iacone, B.; Martirani, M.; Sannino, D.; Gargiulo, A.; Corcione, F.; Serra, R.; Bracale, U.M. A Systematic Review on the Role of Matrix Metalloproteinases in the Pathogenesis of Inguinal Hernias. *Biomolecules* **2023**, *13*, 1123. [CrossRef]
29. Amar, S.; Smith, L.; Fields, G.B. Matrix Metalloproteinase Collagenolysis in Health and Disease. *Biochim. Biophys. Acta (BBA)-Mol. Cell Res.* **2017**, *1864*, 1940–1951. [CrossRef]
30. Vincenti, M.P.; Brinckerhoff, C.E. Transcriptional Regulation of Collagenase (MMP-1, MMP-13) Genes in Arthritis: Integration of Complex Signaling Pathways for the Recruitment of Gene-Specific Transcription Factors. *Arthritis Res. Ther.* **2002**, *4*, 157. [CrossRef]
31. Toriseva, M.; Laato, M.; Carpén, O.; Ruohonen, S.T.; Savontaus, E.; Inada, M.; Krane, S.M.; Kähäri, V.-M. MMP-13 Regulates Growth of Wound Granulation Tissue and Modulates Gene Expression Signatures Involved in Inflammation, Proteolysis, and Cell Viability. *PLoS ONE* **2012**, *7*, e42596. [CrossRef]
32. Lei, H.; Leong, D.; Smith, L.R.; Barton, E.R. Matrix Metalloproteinase 13 Is a New Contributor to Skeletal Muscle Regeneration and Critical for Myoblast Migration. *Am. J. Physiol. Cell Physiol.* **2013**, *305*, C529–C538. [CrossRef]
33. Barrett, L.W.; Fletcher, S.; Wilton, S.D. Regulation of Eukaryotic Gene Expression by the Untranslated Gene Regions and Other Non-Coding Elements. *Cell Mol. Life Sci.* **2012**, *69*, 3613–3634. [CrossRef] [PubMed]
34. Luo, R.; Bai, C.; Yang, L.; Zheng, Z.; Su, G.; Gao, G.; Wei, Z.; Zuo, Y.; Li, G. DNA Methylation Subpatterns at Distinct Regulatory Regions in Human Early Embryos. *Open Biol.* **2018**, *8*, 180131. [CrossRef]
35. Colombatti, A.; Bonaldo, P.; Doliana, R. Type A Modules: Interacting Domains Found in Several Non-Fibrillar Collagens and in Other Extracellular Matrix Proteins. *Matrix* **1993**, *13*, 297–306. [CrossRef] [PubMed]
36. Whittaker, C.A.; Hynes, R.O. Distribution and Evolution of von Willebrand/Integrin A Domains: Widely Dispersed Domains with Roles in Cell Adhesion and Elsewhere. *MBoC* **2002**, *13*, 3369–3387. [CrossRef]
37. Roughley, P.J.; Mort, J.S. The Role of Aggrecan in Normal and Osteoarthritic Cartilage. *J. Exp. Ortop.* **2014**, *1*, 8. [CrossRef] [PubMed]
38. Aspberg, A. The Different Roles of Aggrecan Interaction Domains. *J. Histochem. Cytochem.* **2012**, *60*, 987–996. [CrossRef]
39. Miranda-Duarte, A. DNA Methylation in Osteoarthritis: Current Status and Therapeutic Implications. *TORJ* **2018**, *12*, 37–49. [CrossRef]

 *genes*

Communication

Identification of ROH Islands Conserved through Generations in Pigs Belonging to the Nero Lucano Breed

Paola Di Gregorio *, Annamaria Perna, Adriana Di Trana and Andrea Rando

Scuola di Scienze Agrarie, Forestali, Alimentari ed Ambientali, University of Basilicata,
Via dell'Ateneo Lucano 10, 85100 Potenza, Italy; anna.perna@unibas.it (A.P.); adriana.ditrana@unibas.it (A.D.T.);
andrea.rando@unibas.it (A.R.)
* Correspondence: paola.digregorio@unibas.it

Abstract: The recovery of Nero Lucano (NL) pigs in the Basilicata region (Southern Italy) started in 2001 with the collaboration of several public authorities in order to preserve native breeds that can play a significant economic role both due to their remarkable ability to adapt to difficult environments and the value of typical products from their area of origin. In this study, by using the Illumina Porcine SNP60 BeadChip, we compared the genetic structures of NL pigs reared in a single farm in two different periods separated by a time interval corresponding to at least three generations. The results showed an increase in the percentage of polymorphic loci, a decrease in the inbreeding coefficient calculated according to ROH genome coverage (F_{ROH}), a reduction in the number of ROH longer than 16 Mb and an increase in ROH with a length between 2 and 4 Mb, highlighting a picture of improved genetic variability. In addition, ROH island analysis in the two groups allowed us to identify five conserved regions, located on chromosomes 1, 4, 8, 14 and 15, containing genes involved in biological processes affecting immune response, reproduction and production traits. Only the conserved ROH island on chromosome 14 contains markers which, according to the literature, are associated with QTLs affecting thoracic vertebra number, teat number, gestation length, age at puberty and mean platelet volume.

Keywords: Nero Lucano pig; Southern Italy; inbreeding coefficient (F_{ROH}); runs of homozygosity (ROH) islands

Citation: Di Gregorio, P.; Perna, A.; Di Trana, A.; Rando, A. Identification of ROH Islands Conserved through Generations in Pigs Belonging to the Nero Lucano Breed. *Genes* **2023**, *14*, 1503. https://doi.org/10.3390/genes14071503

Academic Editors: Katarzyna Piórkowska and Katarzyna Ropka-Molik

Received: 30 June 2023
Revised: 18 July 2023
Accepted: 21 July 2023
Published: 23 July 2023

1. Introduction

Farms rearing Nero Lucano (NL) pigs, a typical autochthonous black pig, are characterized by a very low level of environmental impact [1], since animals are raised outdoors in the Basilicata region (Southern Italy) and are able to exploit marginal areas and feed resources available in the environment. The sustainability of the production chain and the specific organoleptic properties of NL pig cured products are particularly appreciated by customers and, therefore, economically advantageous for breeders. In addition, the enhancement and protection of this breed is useful for the conservation of biodiversity and for the reduction in the environmental impact and climate footprint. These matters are goals of the "Farm to Fork Strategy" of the European Union [2], aiming to obtain a sustainable and affordable food chain effective for consumers, producers, climate and environment.

Starting from 2001, the NL pig breed was subject to an intervention for its recovery from extinction thanks to public authorities (the Basilicata Region, the University of Basilicata, the Regional Breeders Association, the Comunità Montana Medio Basento). However, it still suffers all the problems (for example, slow growth speed and a low number of newborns per delivery) determined by the inbreeding associated with the low number of sires and dams typical of "small" populations [3,4]. In addition, this breed is characterized by very low frequencies of alleles at the IGF2, MC4R and VRTN loci that, according to the literature, are associated with positive effects on some meat production traits in cosmopolitan breeds, and is free from malignant hyperthermia (MH) [5]. Recently, we analyzed the genotypes using the 61,565 SNPs of the Illumina Porcine SNP60 BeadChip of about 70% of sires and dams born from 2004 to 2014 of the NL pig population, in order to obtain a first picture of its genetic structure [6]. As expected for a recovering population starting from only six subjects, the analyzed individuals were characterized by high levels of inbreeding coefficients (both F_{MOL}, calculated by referring to allelic frequencies, and F_{ROH}, calculated according to the ROH extension), low effective population size and long generation intervals. These results depict a population still at risk and the need for actions to avoid an excessive inbreeding coefficient increase.

At the end of 2021, we received from the owners of a farm, whose NL pigs were analyzed ten years ago, blood samples collected by the veterinary service during the normal activities of animal controls. As a consequence, we had the opportunity to make comparisons between two groups of NL pigs reared in the same farm and separated by a time period of about ten years, corresponding to at least three generations, in order to obtain a picture of the evolution of their genetic structure by using the ROH approach and to identify ROH islands characterized by a conserved structure through generations.

2. Materials and Methods

2.1. Animals

Animal blood samples were obtained from 76 Nero Lucano pigs reared in a single farm where, about ten years ago, 66 samples of the same breed had already been collected.

2.2. DNA Analyses

DNA samples were genotyped with the Illumina Porcine SNP60 BeadChip v2. The data quality control, accomplished by using PLINK v.1.9 [7], determined the removal of 3 samples due to the genotyping rate being lower than 95% and 2548 SNPs due to a call rate of lower than 95%. Hardy–Weinberg equilibrium was calculated by considering only polymorphic loci located on the 18 autosomal chromosomes. The runs of homozygosity (ROH) were obtained by defining a sliding 'window' of 50 SNPs, allowing a maximum of one heterozygote and one missing call in the 'window' and at least 50 SNPs per 'window'. Individual inbreeding values based on ROH (F_{ROH}) were calculated as $F_{ROH} = \Sigma L_{ROH}/L$, where ΣL_{ROH} is the total ROH length per individual and L is the autosomal genome length (2265.77 Mb, according to Sscrofa 11.1 Genome Assembly).

Gene location was accomplished by referring to NCBI Sus scrofa Annotation Release 106 (https://www.ncbi.nlm.nih.gov/genome/annotation_euk/Sus_scrofa/106/, accessed on 30 December 2022).

Gene Ontology (GO) enrichment analysis was performed by using DAVID Knowledgebase v2022q4 [8,9] (https://david.ncifcrf.gov/home.jsp, accessed on 15 February 2023) and protein–protein interaction by using STRING 11.5 software [10] (https://string-db.org/, accessed on 15 February 2023). REVIGO 1.8.1 software was used to reduce and visualize GO terms [11] (http://revigo.irb.hr/, accessed on 15 February 2023).

Significance tests and graphic plots were obtained by using R-4.1.2 software [12].

3. Results

The variation of the genetic structure of Nero Lucano pigs in a time period corresponding to at least three generations was analyzed by comparing 66 samples, NL-A, collected about 10 years ago and 73 samples, NL-B, collected in 2021 from the same farm. Samples belonging to the two groups were compared without considering the separation of individuals according to their generation since. Unfortunately, pedigree data were not available for samples collected in 2021.

The comparison of the minimum allele frequency (MAF) of the SNPs in the two groups showed that the percentage of the polymorphic SNPs increased from 67% to 83% in about ten years (Figure 1) (see also Figure S1 for MAF chromosomal distribution).

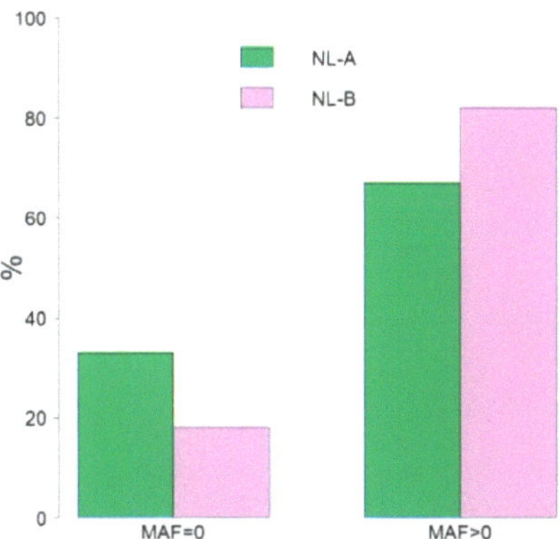

Figure 1. Comparison of the SNPs' distribution according to the minimum allele frequencies (MAF) in the two Nero Lucano pig groups ($z = 61.22$, $p < 0.00001$).

The analysis of genotype distributions accomplished only for the SNPs located in the 18 autosomal chromosomes showed that in NL-B pigs, 6.32% of SNPs were not in Hardy–Weinberg equilibrium, and in 65.48% of cases, the disequilibrium was determined by an excess of homozygotes. The corresponding values in NL-A pigs were 3.43% and 46.61%, respectively.

The ROH analysis of the 73 NL-B pigs allowed us to identify 3690 total ROH, covering 28.11% of the 18 autosomal chromosomes, whereas in the 66 NL-A pigs, the number of total ROH was 3626, covering 38.63% of the autosomal genome (Table 1).

Table 1. Features of ROH clustered according to length in NL-A and NL-B pigs.

ROH Class	NL-A			NL-B		
	ROH Number	SNPs/ROH (Mean ± SD)	% Genome Coverage	ROH Number	SNPs/ROH (Mean ± SD)	% Genome Coverage
<2 Mb	260	57.30 ± 7.67	0.29	236	59.82 ± 8.71	0.25
2–4 Mb	897	85.67 ± 22.93	1.73	1052	84.14 ± 20.18	1.85
4–8 Mb	859	169.26 ± 46.37	3.24	913	161.81 ± 41.76	3.16
8–16 Mb	707	329.64 ± 76.12	5.40	741	321.53 ± 72.55	5.07
>16 Mb	903	1180.92 ± 783.82	27.97	748	1002.98 ± 636.18	17.78
Total	3626	-	38.63	3690	-	28.11

In addition, the mean ROH number per pig was 50.55 ± 10.97 in NL-B and 54.94 ± 7.66 in NL-A ($t(137) = 2.7$, $p = 0.008$), with a total ROH length per pig spanning from a minimum of 38.09 Mb to a maximum of 1260.11 Mb (mean 636.80 ± 219.04 Mb) in NL-B and from a minimum of 236.82 Mb to a maximum of 1366.99 Mb (mean 875.1 ± 198.49 Mb) in NL-A ($t(137) = 6.7$, $p < 0.00001$). As a consequence, the mean ROH length per pig was 12.47 ± 3.69 Mb in NL-B and 15.90 ± 3.10 Mb in NL-A. In both groups, the distribution of ROH among the size classes was balanced, with the exception of those with a length lower than 2 Mb. As shown in Table 1, the ROH number and genome coverage percentage showed the greatest increase for the class 2–4 Mb and the greatest decrease for the class >16 Mb in NL-B pigs.

Individual inbreeding values based on ROH extension (F_{ROH}) ranged from a minimum of 0.02 to a maximum of 0.56, with a mean value of 0.28 ± 0.10 in NL-B, whereas in NL-A, the corresponding values were 0.10, 0.60 and 0.39 ± 0.09, respectively (Figure 2). As a consequence, the F_{ROH} mean value decreased by 28% in about ten years.

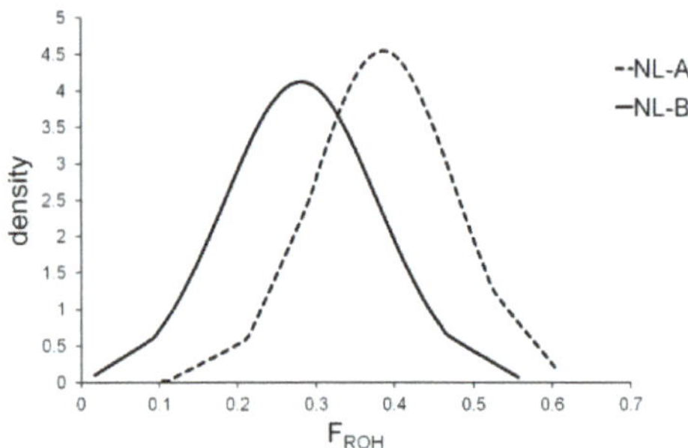

Figure 2. Distribution of the individual F_{ROH} (inbreeding coefficient based on runs of homozygosity) values in two groups of Nero Lucano pigs (t(137) = 6.7, $p < 0.00001$).

The number of core ROH, defined as the consensus regions determined by the overlapping of individual ROH [7], was 587 in NL-B and 494 in NL-A (Figure S2, Table S1). ROH islands were obtained by considering an uninterrupted stretch of at least three SNPs both located in a core ROH and exceeding 99% of the standardized distribution [13] (Figure 3). By using this approach, 24 and 19 ROH islands were identified in NL-B and NL-A, respectively (Table S1). The higher number of ROH islands observed in NL-B was, however, associated with a lower total extension (NL-B 22.79 Mb versus NL-A 29.94 Mb) (Table S2).

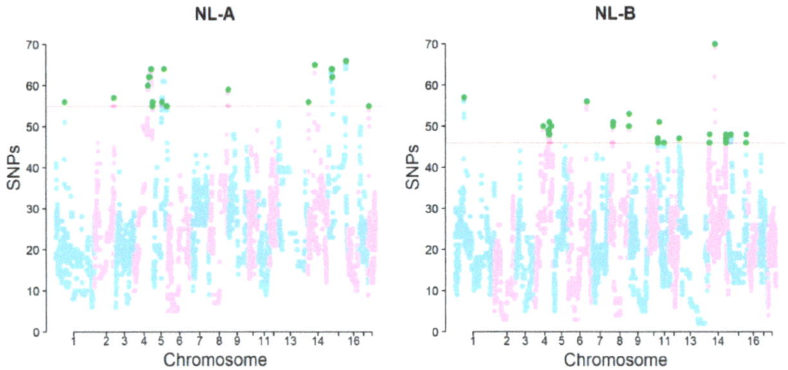

Figure 3. Manhattan plot of SNPs' distribution in runs of homozygosity in NL-A and NL-B pig autosomal chromosomes. The red lines, different for each group, indicate the threshold of the top 1% of the observations. Green dots represent SNPs located in ROH islands.

As shown in Table 2, only in five ROH islands, located in chromosomes 1, 4, 8, 14, and 15, can a partial or total overlap be observed between the two groups.

Table 2. Features (extension, number of SNPs and genes) of ROH islands conserved between NL-A and NL-B pigs.

SSC	NL-A		NL-B		N. Genes
	from-to	N. SNPs	from-to	N. SNPs	
1	61,286,411-63,123,772	52	61,286,411-63,012,075	47	1
4	96,094,167-96,682,751	19	96,094,167-96,243,742	8	7
8	135,852,700-136,018,155	4	135,852,700-136,018,155	4	2
14	46,176,964-48,062,822	60	46,176,964-47,999,414	58	39
15	126,097,384-126,642,052	17	126,097,384-126,642,052	17	3

According to Sus scrofa 11.1 Genome Assembly, the five overlapping regions contain 52 genes (Table S3). Gene Ontology (GO) analysis using DAVID Knowledgebase v2022q4 showed that 26 genes were significantly involved in 15 biological processes (Table 3) whose GO terms were grouped into four superclusters (inflammatory response, system development, regulation of cellular component organization and glycerolipid metabolism) by using REVIGO software (Figures 4 and S3).

Table 3. Genes associated in biological processes after Gene Ontology (GO) analysis by using DAVID v2022q4 software.

GO	Biological Process	Genes
GO:0006954	inflammatory response	S100A12, S100A8, S100A9, OSM, PLA2G3, RHBDD3,
GO:0048731	system development	LIF, LIMK2, NF2, S100A8, S100A9, THOC5, AP1B1, CUL3, GAL3ST1, GAS2L1, INPP5J, KREMEN1, NEFH, PLA2G3, RHBDD3, SELENOM, ZNRF3
GO:0051128	regulation of cellular component organization	LIF, LIMK2, MORC2, NF2, S100A9, CUL3, GAS2L1, HNRNPD, INPP5J, KREMEN1, MTMR3, PLA2G3,
GO:0033043	regulation of organelle organization	LIF, LIMK2, MORC2, NF2, CUL3, GAS2L1, HNRNPD, MTMR3,
GO:0009914	hormone transport	LIF, OSM, PLA2G3, SELENOM,
GO:0051046	regulation of secretion	LIF, S100A8, OSM, PLA2G3, RHBDD3,
GO:0046903	secretion	LIF, S100A8, OSM, PLA2G3, RHBDD3, SELENOM,
GO:0050729	positive regulation of inflammatory response	S100A8, OSM, PLA2G3
GO:0031345	negative regulation of cell projection organization	LIMK2, INPP5J, KREMEN1,
GO:0023061	signal release	LIF, OSM, PLA2G3, SELENOM,
GO:0051247	positive regulation of protein metabolic process	LIF, LIMK2, S100A8, TBC1D10A, CUL3, HNRNPD, OSM, RHBDD3,
GO:0010648	negative regulation of cell communication	LIF, NF2, CUL3, CASTOR1, KREMEN1, OSM, ZNRF3,
GO:0023057	negative regulation of signaling	LIF, NF2, CUL3, CASTOR1, KREMEN1, OSM, ZNRF3,
GO:0048232	male gamete generation	LIMK2, GAL3ST1, OSBP2, PLA2G3,
GO:0046486	glycerolipid metabolic process	GAL3ST1, INPP5J, MTMR3, PLA2G3,

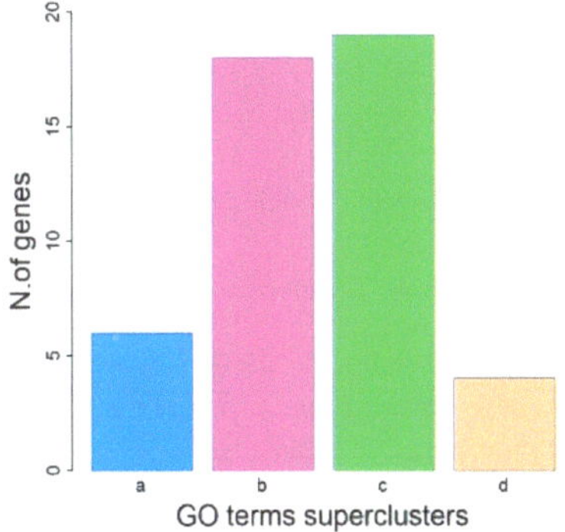

Figure 4. GO terms grouped into four superclusters by using REVIGO software: (a) inflammatory response, (b) system development, (c) regulation of cellular component organization (d) glycerolipid metabolism.

Furthermore, three genes (S100A7, S100A8, S100A9) were involved in the IL-17 signaling pathway, which is engaged in several immune regulatory functions such as the response to injury, physiological stress, and infection [14].

Analysis of the 52 genes with STRING 11.5 software evidenced more protein–protein interactions than expected (26 edges rather than 4, PPI enrichment p-value 8.5×10^{-13})

(Figure S4). As a consequence, some of the proteins coded by these genes are at least partially biologically connected. However, only part of the biological processes identified by DAVID software found a correspondence with protein–protein interactions highlighted by STRING software.

According to the Pig QTL database (version Sus scrofa 11.1 Genome Assembly), only the conserved ROH island on chromosome 14 contained six QTLs affecting thoracic vertebra number (#64660 and #64771) [15], teat number (#126628) [16], gestation length (#173177) [17], mean platelet volume (#37844) [18], and age at puberty (#22109) [19] (Table S3). Each of the six SNPs located at the peak of association with the above-mentioned QTLs were characterized by a MAF = 0 in both NL-A and NL-B pigs.

4. Discussion

In this work, we compared the results of Illumina PorcineSNP60 BeadChip genotyping accomplished on two groups of Nero Lucano pigs, NL-A and NL-B, reared in the same herd and sampled in two time periods separated by about ten years, corresponding to at least three generations.

Analysis of the results showed that in this period the number of polymorphic loci increased by about 24%. In addition, the ROH analyses in the two groups showed that the mean ROH number, mean total ROH length and mean ROH length per pig were characterized by decreased values in NL-B. The distribution of the ROH per class was consistent with these results, showing a reduction in the number of ROH longer than 16 Mb and an increase in the number of ROH with a length between 2 and 4 Mb in NL-B. All the above-mentioned results were responsible for the strong decrease in the inbreeding coefficient over ten years, from 0.39 to 0.28, based on ROH genome coverage (F_{ROH}). The origin of this decrease could be explained either by introgression events or by the use of sires and dams belonging to other NL herds. However, according to what was stated by the breeder and to the typical morphological traits characterizing the NL-B individuals, the former hypothesis can be excluded.

According to several authors, the frequency and extension of ROH are affected by ancient or recent inbreeding events in the population and/or by the presence of genes under strong selection [20,21]. ROH islands are stretches of homozygous genomic sequences characterized by a very high frequency in a population. As a consequence, the analysis of ROH islands has been the object of several studies in different species in order to identify genes associated with domestication or useful for the improvement of animal production and reproduction. We identified five ROH islands in common between NL-A and NL-B groups—that is, genomic regions constantly maintained through generations as homozygous milestones. Of course, genes contained in these homozygous milestones should be strong candidates to have effects on the adaptive selection of this breed.

Some of the genes identified in these five regions are involved in biological processes affecting immune response, reproduction and production traits. For example, three of the calcium ion binding proteins belonging to the S100 family (S100A7, S100A8, S100A9) are involved in the IL-17 signaling pathway, which is important in the immunity to pathogens, or contribute to the pathogenesis of inflammatory diseases [14]. Furthermore, S100A6 is involved in the response to the viral infection causing porcine reproductive and respiratory syndrome (PRRS), one of the most economically significant swine infectious diseases [22,23]; PPGRP-S has a role in the immunity against intestinal microorganisms [24]; RHBDD3 suppresses, in mice, the production of IL6, preventing the development of autoimmune diseases [25]; and PLA2G3 affects the maturation and function of mast cells, key players in the inflammatory response [26]. The conservation of regions containing genes affecting immune response is probably due to the need for an efficient immunological system in animals living outdoors in semi-wild conditions, such as NL pigs.

As far as pig reproduction traits are concerned, the MORC2 gene prevents apoptosis of granulosa cells [27], whereas SMTN, coding for a structural protein found exclusively in contractile smooth muscle cells, is associated with ovarian follicle growth and development [28]. Furthermore, both OSBP2 and LIMK2 are involved in spermatogenesis [29]. In mice, the defective gene MORC2b is responsible for male and female sterility [30], and mice deficient in OSBP2 produce a severely reduced number of spermatozoa that show very low motility and no fertilizing ability in vitro [31]. Similar results are observed for mice with a disrupted LIMK2 gene in which the progression of spermatogenesis is strongly affected [32] and for humans where mutations in the heterozygous state in LIMK2 are present in infertile males [33]. Finally, an LIF gene polymorphism (rs3463076786: C/T) is associated with the number of stillbirths in pigs [34].

In Chinese Anqing six-end-white pig, SELENOM has been identified as a candidate gene affecting body weight [29]. This gene is highly expressed in the brain and may be involved in neurodegenerative disorders. Transgenic mice with targeted deletion of this gene are characterized by obesity, suggesting a possible role of SELENOM in the regulation of body weight and energy metabolism [35]. In addition, DOCK10 is a candidate gene associated with intramuscular fat [36], and EWSR1 is associated with meat quality traits, in particular meat colour [37]. The PLA2G3 gene belongs to the phospholipase A2 family that in pigs, as in other vertebrates, is involved in lipid metabolism [38]. In humans, KREMEN1 and ZNRF3 are involved in body fat distribution [39].

The conserved five ROH islands were also analyzed for the presence of QTLs, and only the one on chromosome 14 contained markers associated with effects on thoracic vertebra number, teat number, gestation length, mean platelet volume and age at puberty. Unfortunately, these markers were monomorphic in all of the genotyped NL pigs, and therefore cannot be used to analyze the variation of these traits in this breed. In any case, the conserved ROH island located on chromosome 14 could have had a key role in pig domestication since it was also identified in other breeds. In fact, the upstream region containing the AP1B1, EWSR1, KREMEN1, NEFH, THOC5, and ZNRF3 genes (see Table S3) was identified in Russian Large White pigs as a signature of selection associated with QTLs affecting reproduction and production traits [40], whereas the downstream region containing MORC2, SMTN, INPP5J, PLA2G3 and RNF185 genes (see Table S3) was identified by Li et al. [41] as a signature of selection in Chinese pigs. This region is associated with a QTL affecting intramuscular fat linoleic acid content [42,43] that is positively correlated with pork flavor [44]. As a consequence, it is plausible that genes contained in this ROH island could also play an important role in the quality of NL pig cured meat products which are particularly appreciated for their organoleptic properties.

In conclusion, the search for ROH islands conserved through generations can be used to mark boundaries of genomic regions to be analyzed for the identification of mutations affecting the expression of economically important genes and/or differentiating cosmopolitan from small autochthonous populations.

Supplementary Materials: The following supporting information can be downloaded at: https://www.mdpi.com/article/10.3390/genes14071503/s1, Figure S1. Chromosomal distribution according to the minimum allele frequencies (MAF) of the SNPs in the two groups, NL-A and NL-B, of Nero Lucano pigs (0 = non-defined chromosome position, 23 = X chromosome, 24 = Y chromosome, 25 = XY ψ-autosomal region). Figure S2. Chromosomal distribution of core ROH (bars), and minimum and maximum frequencies of core ROH (lines) in the two groups, NL-A and NL-B, of Nero Lucano pigs. Dots indicate ROH islands. Figure S3. "Revigo Treemap" of Gene Ontology terms for genes located in the five ROH islands conserved between NL-A and NL-B pigs. Figure S4. Network of protein–protein interaction (PPI) analysis carried out on genes located in the five ROH islands conserved between NL-A and NL-B pigs (nodes represent proteins). Table S1. Chromosomal distribution of the number of core ROH and ROH islands in the two groups, NL-A and NL-B, of Nero Lucano pigs. Table S2. Features of the ROH islands in the two groups, NL-A and NL-B, of Nero Lucano pigs (total overlaps are highlighted in yellow, partial overlaps are in red). Table S3. Genes and QTLs mapped in the five ROH islands conserved between NL-A and NL-B pigs.

Author Contributions: Conceptualization and Methodology, P.D.G. and A.R.; Investigation, P.D.G.; Project administration, A.P.; Funding acquisition, A.R. and A.P.; Formal analysis, P.D.G. and A.D.T.; Writing—Original Draft Preparation, P.D.G.; Writing—Review and Editing, P.D.G. and A.R. All authors have read and agreed to the published version of the manuscript.

Funding: This work was supported by PSR of the Basilicata Region 2014–2020, Misura 10, sottomisura 10.2. "Conservazione e uso sostenibile delle risorse genetiche in agricoltura" (Conservation and sustainable use of genetic resources in agriculture), project "(TGA)—Standardizzazione, stabilizzazione e valorizzazione dei tipi genetici autoctoni suini, ovicaprini ed equini" (TGA—Standardization, stabilization and enhancement of native pig, sheep, goat and equine genetic types); CUP C36C18000160006.

Institutional Review Board Statement: This research does not fall within Directive 63/210 of the European Parliament and of the Council on the protection of animals used for experimental purposes (transposed into Italian law by Legislative Decree 26/2014) and, thus, it does not require any authorization from the National competent Authorities (Protocol code: OpBA 07_2023_UNIBAS, Organismo Preposto al Benessere degli Animali, University of Basilicata).

Data Availability Statement: The data analyzed during the current study are available from the corresponding author on reasonable request.

Acknowledgments: We thank the Associazione Regionale Allevatori, Basilicata (Basilicata Region Breeders Association) for their kind collaboration.

Conflicts of Interest: The authors declare no conflict of interest.

References

1. Hilborn, R.; Banobi, J.; Hall, S.J.; Pucylowski, T.; Walsworth, T.E. The environmental cost of animal source foods. *Front. Ecol. Environ.* **2018**, *16*, 329–335. [CrossRef]
2. European Commission. Farm to Fork Strategy: For a Fair, Healthy and Environmentally-Friendly Food System. COM/2020/381 Final. Date of Document: 20/05/2020. Available online: https://ec.europa.eu/commission/presscorner/detail/en/fs_20_908 (accessed on 15 February 2023).
3. Farkas, J.; Curik, I.; Csato, L.; Csornyei, Z.; Baumung, R.; Nagy, I. Bayesian inference of inbreeding effects on litter size and gestation length in Hungarian Landrace and Hungarian Large White pigs. *Livest. Sci.* **2007**, *112*, 109–114. [CrossRef]
4. Zhang, Y.; Zhuo, Y.; Ning, C.; Zhou, L.; Liu, J.F. Estimate of inbreeding depression on growth and reproductive traits in a Large White pig population. *G3 Genes Genomes Genet.* **2022**, *12*, jkac118. [CrossRef]
5. Valluzzi, C.; Rando, A.; Di Gregorio, P. Genetic variability of Nero Lucano pig breed at IGF2, LEP, MC4R, PIK3C3, RYR1 and VRTN loci. *Ital. J. Anim. Sci.* **2019**, *18*, 1321–1326. [CrossRef]
6. Valluzzi, C.; Rando, A.; Macciotta, N.P.P.; Gaspa, G.; Di Gregorio, P. The Nero Lucano Pig Breed: Recovery and Variability. *Animals* **2021**, *11*, 1331. [CrossRef]
7. Purcell, S.; Neale, B.; Todd-Brown, K.; Thomas, L.; Ferreira, M.A.R.; Bender, D.; Maller, J.; Sklar, P.; de Bakker, P.I.W.; Daly, M.J.; et al. PLINK: A toolset for whole-genome association and population-based linkage analysis. *Am. J. Hum. Genet.* **2007**, *81*, 559–575. [CrossRef]
8. Huang, D.W.; Sherman, B.T.; Lempicki, R.A. Systematic and integrative analysis of large gene lists using DAVID Bioinformatics Resources. *Nat. Protoc.* **2009**, *4*, 44–57. [CrossRef]
9. Sherman, B.T.; Hao, M.; Qiu, J.; Jiao, X.; Baseler, M.W.; Lane, H.C.; Imamichi, T.; Chang, W. DAVID: A web server for functional enrichment analysis and functional annotation of gene lists (2021 update). *Nucleic Acids Res.* **2022**, *50*, W216–W221. [CrossRef] [PubMed]
10. Szklarczyk, D.; Gable, A.L.; Nastou, K.C.; Lyon, D.; Kirsch, R.; Pyysalo, S.; Doncheva, N.T.; Legeay, M.; Fang, T.; Bork, P.; et al. The STRING database in 2021: Customizable protein–protein networks, and functional characterization of user-uploaded gene/measurement sets. *Nucleic Acids Res.* **2021**, *49*, D605–D612. [CrossRef]
11. Supek, F.; Bosnjak, M.; Skunca, N.; Smuc, T. REVIGO summarizes and visualizes long lists of Gene Ontology terms. *PLoS ONE* **2011**, *6*, e21800. [CrossRef]
12. R Core Team. *R: A Language and Environment for Statistical Computing*; R Foundation for Statistical Computing: Vienna, Austria, 2022. Available online: https://www.R-project.org/ (accessed on 15 February 2023).
13. Szmatoła, T.; Gurgul, A.; Ropka-Molik, K.; Jasielczuk, I.; Ząbek, T.; Bugno-Poniewierska, M. Characteristics of runs of homozygosity in selected cattle breeds maintained in Poland. *Livest. Sci.* **2016**, *188*, 72–80. [CrossRef]
14. McGeachy, M.J.; Cua, D.J.; Gaffen, S.L. The IL-17 family of cytokines in health and disease. *Immunity* **2019**, *50*, 892–906. [CrossRef] [PubMed]
15. Rohrer, G.A.; Nonneman, D.J.; Wiedmann, R.T.; Schneider, J.F. A study of vertebra number in pigs confirms the association of vertnin and reveals additional QTL. *BMC Genet.* **2015**, *16*, 129. [CrossRef]

16. Rohrer, G.A.; Nonneman, D.J. Genetic analysis of teat number in pigs reveals some developmental pathways independent of vertebra number and several loci which only affect a specific side. *Genet. Sel. Evol.* **2017**, *49*, 4. [CrossRef] [PubMed]
17. See, G.M.; Trenhaile-Grannemann, M.D.; Spangler, M.L.; Ciobanu, D.C.; Mote, B.E. A genome-wide association study for gestation length in swine. *Anim. Genet.* **2019**, *50*, 539–542. [CrossRef]
18. Wang, J.Y.; Luo, Y.R.; Fu, W.X.; Lu, X.; Zhou, J.P.; Ding, X.D.; Liu, J.F.; Zhang, Q. Genome-wide association studies for hematological traits in swine. *Anim. Genet.* **2013**, *44*, 34–43. [CrossRef]
19. Nonneman, D.; Lents, C.; Rohrer, G.; Rempel, L.; Vallet, J. Genome-wide association with delayed puberty in swine. *Anim. Genet.* **2014**, *45*, 130–132. [CrossRef]
20. Curik, I.; Ferencakovic, M.; Sölkner, J. Inbreeding and runs of homozygosity: A possible solution to an old problem. *Livest. Sci.* **2014**, *166*, 26–34. [CrossRef]
21. Toro-Ospina, A.M.; Herrera Rios, A.C.; Pimenta Schettini, G.; Vallejo Aristizabal, V.H.; Bizarria dos Santos, W.; Zapata, C.A.; Ortiz Morea, E.G. Identification of Runs of Homozygosity Islands and Genomic Estimated Inbreeding Values in Caqueteño Creole Cattle (Colombia). *Genes* **2022**, *13*, 1232. [CrossRef]
22. Zhou, X.; Wang, P.; Michal, J.J.; Wang, Y.; Zhao, J.; Jiang, Z.; Liu, B. Molecular characterization of the porcine S100A6 gene and analysis of its expression in pigs infected with highly pathogenic porcine reproductive and respiratory syndrome virus (HP-PRRSV). *J. Appl. Genet.* **2015**, *56*, 355–363. [CrossRef]
23. Raymond, P.; Bellehumeur, C.; Nagarajan, M.; Longtin, D.; Ferland, A.; Muller, P.; Bissonnette, R.; Simard, C. Porcine reproductive and respiratory syndrome virus (PRRSV) in pig meat. *Can. J. Vet. Res.* **2017**, *81*, 162–170. [PubMed]
24. Ueda, W.; Tohno, M.; Shimazu, T.; Fujie, H.; Aso, H.; Kawai, Y.; Numasaki, M.; Saito, T.; Kitazawa, H. Molecular cloning, tissue expression, and subcellular localization of porcine peptidoglycan recognition proteins 3 and 4. *Vet. Immunol. Immunopathol.* **2011**, *143*, 148–154. [CrossRef] [PubMed]
25. Liu, J.; Han, C.; Xie, B.; Wu, Y.; Liu, S.; Chen, K.; Xia, M.; Zhang, Y.; Song, L.; Li, Z.; et al. Rhbdd3 controls autoimmunity by suppressing the production of IL-6 by dendritic cells via K27-linked ubiquitination of the regulator NEMO. *Nat. Immunol.* **2014**, *15*, 612–622. [CrossRef]
26. Starkl, P.; Marichal, T.; Galli, S.J. PLA2G3 promotes mast cell maturation and function. *Nat. Immunol.* **2013**, *14*, 527–529. [CrossRef]
27. Liu, J.; Qi, N.; Xing, W.; Li, M.; Qian, Y.; Luo, G.; Yu, S. The TGF-β/SMAD signaling pathway prevents follicular atresia by upregulating MORC2. *Int. J. Mol. Sci.* **2022**, *23*, 10657. [CrossRef]
28. Bonnet, A.; Le Cao, K.A.; Sancristobal, M.; Benne, F.; Robert-Granie, C.; Law-So, G.; Fabre, S.; Besse, P.; De Billy, E.; Quesnel, H.; et al. In vivo gene expression in granulosa cells during pig terminal follicular development. *Reproduction* **2008**, *136*, 211–224. [CrossRef]
29. Zhang, W.; Yang, M.; Zhou, M.; Wang, Y.; Wu, X.; Zhang, X.; Ding, Y.; Zhao, G.; Yin, Z.; Wang, C. Identification of signatures of selection by whole-genome resequencing of a Chinese native pig. *Front. Genet.* **2020**, *11*, 566255. [CrossRef]
30. Shi, B.; Xue, J.; Zhou, J.; Kasowitz, S.D.; Zhang, Y.; Liang, G.; Guan, Y.; Shi, Q.; Liu, M.; Sha, J.; et al. MORC2B is essential for meiotic progression and fertility. *PLoS Genet.* **2018**, *14*, e1007175. [CrossRef] [PubMed]
31. Udagawa, O.; Ito, C.; Ogonuki, N.; Sato, H.; Lee, S.; Tripvanuntakul, P.; Ichi, I.; Uchida, Y.; Nishimura, T.; Murakami, M.; et al. Oligo-astheno-teratozoospermia in mice lacking ORP4, a sterol-binding protein in the OSBP-related protein family. *Genes Cells* **2014**, *19*, 13–27. [CrossRef]
32. Takahashi, H.; Koshimizu, U.; Miyazaki, J.; Nakamura, T. Impaired spermatogenic ability of testicular germ cells in mice deficient in the LIM-kinase 2 gene. *Dev. Biol.* **2002**, *241*, 259–272. [CrossRef]
33. Kuzmin, A.; Jarvi, K.; Lo, K.; Spencer, L.; Chow, G.Y.; Macleod, G.; Wang, Q.; Varmuza, S. Identification of potentially damaging amino acid substitutions leading to human male infertility. *Biol. Reprod.* **2009**, *81*, 319–326. [CrossRef] [PubMed]
34. Leonova, M.A.; Getmantseva, L.V.; Vasilenko, V.N.; Klimenko, A.I.; Usatov, A.V.; Yu, B.S.; Yu, K.A.; Shirockova, N.V. Leukemia Inhibitory Factor (LIF) Gene Polymorphism and its Impact on Reproductive Traits of Pigs. *Am. J. Anim. Vet. Sci.* **2015**, *10*, 212–216. [CrossRef]
35. Pitts, M.W.; Reeves, M.A.; Hashimoto, A.C.; Ogawa, A.; Kremer, P.; Seale, L.A.; Berry, M.J. Deletion of selenoprotein M leads to obesity without cognitive deficits. *J. Biol. Chem.* **2013**, *288*, 26121–26134. [CrossRef] [PubMed]
36. Li, H.; Xu, C.; Meng, F.; Yao, Z.; Fan, Z.; Yang, Y.; Meng, X.; Zhan, Y.; Sun, Y.; Ma, F.; et al. Genome-Wide Association studies for flesh color and intramuscular fat in (Duroc Landrace Large White) crossbred commercial pigs. *Genes* **2022**, *13*, 2131. [CrossRef] [PubMed]
37. Li, X.; Kim, S.W.; Do, K.T.; Ha, Y.K.; Lee, Y.M.; Yoon, S.H.; Kim, H.B.; Kim, J.J.; Choi, B.H.; Kim, K.S. Analyses of porcine public SNPs in coding-gene regions by re-sequencing and phenotypic association studies. *Mol. Biol. Rep.* **2011**, *38*, 3805–3820. [CrossRef] [PubMed]
38. Huang, Q.; Wu, Y.; Qin, C.; He, W.; Wei, X. Phylogenetic and structural analysis of the phospholipase A2 gene family in vertebrates. *Int. J. Mol. Med.* **2015**, *35*, 587–596. [CrossRef]
39. Heid, I.M.; Jackson, A.U.; Randall, J.C.; Winkler, T.W.; Qi, L.; Steinthorsdottir, V.; Thorleifsson, G.; Zillikens, M.C.; Speliotes, E.K.; Mägi, R.; et al. Meta-analysis identifies 13 new loci associated with waist-hip ratio and reveals sexual dimorphism in the genetic basis of fat distribution. *Nat. Genet.* **2010**, *42*, 949–960, Erratum in *Nat Genet.* **2011**, *43*, 1164. [CrossRef]
40. Bakoev, S.; Getmantseva, L.; Kostyunina, O.; Bakoev, N.; Prytkov, Y.; Usatov, A.; Tatarinova, T.V. Genome-wide analysis of genetic diversity and artificial selection in Large White pigs in Russia. *PeerJ* **2021**, *9*, e11595. [CrossRef]

41. Li, X.; Yang, S.; Dong, K.; Tang, Z.; Li, K.; Fan, B.; Wang, Z.; Liu, B. Identification of positive selection signatures in pigs by comparing linkage disequilibrium variances. *Anim. Genet.* **2017**, *48*, 600–605. [CrossRef] [PubMed]
42. Uemoto, Y.; Nakano, H.; Kikuchi, T.; Sato, S.; Ishida, M.; Shibata, T.; Kadowaki, H.; Kobayashi, E.; Suzuki, K. Fine mapping of porcine SSC14 QTL and SCD gene effects on fatty acid composition and melting point of fat in a Duroc purebred population. *Anim. Genet.* **2012**, *43*, 225–228. [CrossRef]
43. Munoz, M.; Rodríguez, M.C.; Alves, E.; Folch, J.M.; Ibanez-Escriche, N.; Silio, L.; Fernández, A.I. Genome-wide analysis of porcine backfat and intramuscular fat fatty acid composition using high-density genotyping and expression data. *BMC Genom.* **2013**, *14*, 845. [CrossRef]
44. Cameron, N.D.; Enser, M.; Nute, G.R.; Whittington, F.M.; Penman, J.C.; Fisken, A.C.; Perry, A.M.; Wood, J.D. Genotype with nutrition interaction on fatty acid composition of intramuscular fat and the relationship with flavour of pig meat. *Meat Sci.* **2000**, *55*, 187–195. [CrossRef] [PubMed]

 *genes*

Article

Transcriptomic Characterization of Genes Regulating the Stemness in Porcine Atrial Cardiomyocytes during Primary In Vitro Culture

Rut Bryl [1], Mariusz J. Nawrocki [2], Karol Jopek [3], Mariusz Kaczmarek [4,5], Dorota Bukowska [6], Paweł Antosik [7], Paul Mozdziak [8,9], Maciej Zabel [10,11], Piotr Dzięgiel [10] and Bartosz Kempisty [7,9,12,13,*]

1. Section of Regenerative Medicine and Cancer Research, Natural Sciences Club, Faculty of Biology, Adam Mickiewicz University, Poznań, 61-614 Poznan, Poland; rutbryl@gmail.com
2. Department of Anatomy, Poznan University of Medical Sciences, 60-781 Poznan, Poland; mjnawrocki@ump.edu.pl
3. Department of Histology and Embryology, Poznan University of Medical Sciences, 60-781 Poznan, Poland; karoljopek@ump.edu.pl
4. Department of Cancer Immunology, Chair of Medical Biotechnology, Poznan University of Medical Sciences, 61-866 Poznan, Poland; markacz@ump.edu.pl
5. Gene Therapy Laboratory, Department of Cancer Diagnostics and Immunology, Greater Poland Cancer Centre, 61-866 Poznan, Poland
6. Department of Diagnostics and Clinical Sciences, Institute of Veterinary Medicine, Nicolaus Copernicus University in Torun, 87-100 Torun, Poland; dbukowska@umk.pl
7. Department of Veterinary Surgery, Institute of Veterinary Medicine, Nicolaus Copernicus University in Torun, 87-100 Torun, Poland; pantosik@umk.pl
8. Prestage Department of Poultry Science, North Carolina State University, Raleigh, NC 27695, USA; pemozdzi@ncsu.edu
9. Physiology Graduate Faculty, North Carolina State University, Raleigh, NC 27695, USA
10. Department of Human Morphology and Embryology, Division of Histology and Embryology, Wroclaw Medical University, 50-368 Wroclaw, Poland; maciej.zabel@umw.edu.pl (M.Z.); piotr.dziegiel@umw.edu.pl (P.D.)
11. Division of Anatomy and Histology, University of Zielona Góra, 65-046 Zielona Góra, Poland
12. Department of Human Morphology and Embryology, Division of Anatomy, Wroclaw Medical University, 50-367 Wroclaw, Poland
13. Department of Obstetrics and Gynaecology, University Hospital and Masaryk University, 62500 Brno, Czech Republic
* Correspondence: bartosz.kempisty@umw.edu.pl or kempistybartosz@gmail.com

Citation: Bryl, R.; Nawrocki, M.J.; Jopek, K.; Kaczmarek, M.; Bukowska, D.; Antosik, P.; Mozdziak, P.; Zabel, M.; Dzięgiel, P.; Kempisty, B. Transcriptomic Characterization of Genes Regulating the Stemness in Porcine Atrial Cardiomyocytes during Primary In Vitro Culture. *Genes* 2023, 14, 1223. https://doi.org/10.3390/genes14061223

Academic Editors: Katarzyna Piórkowska and Katarzyna Ropka-Molik

Received: 12 May 2023
Revised: 1 June 2023
Accepted: 2 June 2023
Published: 4 June 2023

Abstract: Heart failure remains a major cause of death worldwide. There is a need to establish new management options as current treatment is frequently suboptimal. Clinical approaches based on autologous stem cell transplant is potentially a good alternative. The heart was long considered an organ unable to regenerate and renew. However, several reports imply that it may possess modest intrinsic regenerative potential. To allow for detailed characterization of cell cultures, whole transcriptome profiling was performed after 0, 7, 15, and 30 days of in vitro cell cultures (IVC) from the right atrial appendage and right atrial wall utilizing microarray technology. In total, 4239 differentially expressed genes (DEGs) with ratio > abs |2| and adjusted p-value ≤ 0.05 for the right atrial wall and 4662 DEGs for the right atrial appendage were identified. It was shown that a subset of DEGs, which have demonstrated some regulation of expression levels with the duration of the cell culture, were enriched in the following GO BP (Gene Ontology Biological Process) terms: "stem cell population maintenance" and "stem cell proliferation". The results were validated by RT-qPCR. The establishment and detailed characterization of in vitro culture of myocardial cells may be important for future applications of these cells in heart regeneration processes.

Keywords: cardiomyocytes; porcine cardiac muscle; long-term in vitro cell culture; stemness markers; transcriptomic analysis

1. Introduction

The past three decades have brought progress in cardiovascular medicine leading to a significant decrease in mortality due to acute cardiovascular syndromes in developed countries [1]. Still, in many cases, myocardial injury leads to the development of heart failure, "a complex clinical syndrome that results from any structural or functional impairment of ventricular filling or ejection of blood" [2]. Heart failure (HF) has an annual prevalence of approximately 37.7 million people [3]. Survival of HF patients varies from 3 to 5 years, on average, making their prognosis poorer than for most cancers [4].

There is a need to develop more optimal methods for the treatment of heart failure. The heart was considered an organ unable to regenerate and renew. However, there is now evidence suggesting that the heart may have some ability for self-repair, especially in response to exploitation/physiological stress [5–10]. Therapies based on cells driving heart regeneration may become a promising new strategy for current disease management.

Cardiac progenitor cells (CPCs) are potentially a heterogenous group of cells which activate in response to an injury and contribute to cardiomyocytes' replenishment. They are localized in different heart regions (atria, ventricles, epicardium, or pericardium) [11]. Many studies in the past two decades have focused on the identification and isolation of this cell population in adult mammalian hearts. Their characteristic properties include clonogenicity, self-renewal, differentiation into several cell types of cardiac lineage, expression of specific transcription factors—Isl-1, Nkx2.5, MEF2C, and GATA-4—and stemness markers such as Oct3/4, Bmi-1, and Nanog. Intramyocardial transplantation of such cells leads to a reduction of scar size and the preservation of left ventricular function in preclinical myocardial infarction models [11–13]. These subpopulations include c-kit^{pos} CPCs [14–16], cardiosphere-derived cells (CDCs) [17–20], cardiac side population cells (SP) [21–23], Sca-1^{pos} CPCs (stem cell antigen-1) [24,25], epicardium-derived progenitor cells (EPDCs) [26,27], cardiac colony-forming unit fibroblasts (cCFU-Fs) [28,29], and Islet-1^{pos} cells [30–32].

Two types of adult cardiac progenitor cells have reached clinical testing [33,34]. It is noteworthy that results of CADUCEUS, a clinical study of cardiosphere-derived cells, showed a reduction of infarct size, increased viable heart mass, and regional contractility in the group administered with autologous CDCs compared to the group that received standard treatment. Nevertheless, there was no significant difference in left ventricular (LV) ejection fraction between the groups [34].

A significant challenge to employ cardiac stem cells is the small pool of these cells. Furthermore, the precise role of the majority of CPCs in the adult heart is unknown [11]. As a consequence, molecular characterization of these cell populations and the elucidation of genetic and epigenetic mechanisms which dictate the cells' phenotype and fate is crucial for further progress.

The aim of this work was to establish a methodology for the identification of presumed cardiac stem cells, as well as for primary in vitro cultures' conditions of cardiomyocytes isolated from two different fragments of porcine (*Sus scrofa f. domestica*) heart, namely, the right atrial appendage and right atrial wall. It is suggested that there may reside a specific cell population with regenerative potential within porcine cardiac muscle. The establishment and detailed characterization of in vitro culture of cardiac muscle cells may be important for future applications of these cells in heart regeneration processes.

2. Materials and Methods

2.1. Animals

For this study, a total of 25 pubertal crossbred Landrace gilts bred on a commercial local farm were used. The gilts had a mean age of 155 days (range 140–170 days), and the mean weight was 100 kg (95–120 kg). All of the animals were housed under identical conditions and fed the same forage (depending on age and reproductive status).

This research related to animal use has complied with all the relevant national regulations and institutional policies for the care and use of animals. As the research material is usually disposed of after slaughter, being a remnant by-product, no Ethical Committee approval was needed for this study.

2.2. Cell Isolation

After slaughter, porcine hearts were transported to the laboratory on ice within 40 min of harvest. Next, two heart fragments were excised from the right atrial appendage and right atrial wall. The excised tissue was washed in 10% povidone–iodine solution (Betadine; EGIS, Warsaw, Poland) at room temperature, twice in 0.9% sodium chloride solution (NATRIUM CHLORATUM 0,9% FRESENIUS, Fresenius Kabi, Warsaw, Poland) at room temperature, and eventually in cold (kept at 4 °C) Dulbecco's Phosphate Buffered Saline (D-PBS; SIGMA-ALDRICH, St Louis, MO, USA, Merck KGaA, Darmstadt, Germany) (137 mM NaCl, 27 mM KCl, 10 mM Na_2HPO_4, 2 mM KH_2PO_4, pH 7.4) supplemented with 2% antibiotics/antimycotics (100 U/mL penicillin G sodium, 100 µg/mL streptomycin sulfate, and 0.25 µg/mL amphotericin B) (Antibiotic Antimycotic Solution ($100\times$); SIGMA-ALDRICH, St. Louis, MO, USA, Merck KGaA, Darmstadt, Germany). Next, tissue was subjected to two-step mincing in Petri dishes using sterile tools. Firstly, the heart fragments were placed in a droplet of cold Dulbecco's Modified Eagle Medium (DMEM; SIGMA-ALDRICH, St. Louis, MO, USA, Merck KGaA, Darmstadt, Germany), cut into cubes with sides of ~5 mm, and transferred to 50 mL conical tubes for intensive washing with cold D-PBS. Washed fragments were then minced into ~1 mm^3 fragments, also in a droplet of cold DMEM. Each tissue fragment was incubated in 10 mL of collagenase type II solution (collagenase type II; SIGMA-ALDRICH, St. Louis, MO, USA, Merck KGaA, Darmstadt, Germany) in DMEM (c = 1 mg/mL) with shaking (100 rpm) for 40 min at 37 °C (Orbital Shaker with a Heating Module and Incubator Hood, Incubator 1000/Polymax 1040 model; Heidolph Instruments GmbH & Co. KG, Schwabach, Germany). After digestion, the solution was filtered through an autoclavable stainless steel cell sieve with steel mesh and subsequently through 70 µm nylon cell strainers. The filtrate was transferred to 15 mL conical tubes and centrifuged (5 min, $200\times g$, room temperature) (Centrifuge 5810R, Eppendorf AG, Hamburg, Germany). After centrifugation, the supernatant was discarded and the pellet was resuspended in 5 mL of cold D-PBS. The centrifugation was repeated and the pellet resuspended in 1 mL of DMEM/F12 supplemented (cardiomyocytes culture medium) at 37 °C consisting of DMEM/Nutrient Mixture F-12 Ham (DMEM/F12; SIGMA-ALDRICH, St. Louis, MO, USA, Merck KGaA, Darmstadt, Germany), 20% Fetal Bovine Serum (FBS; SIGMA-ALDRICH, St. Louis, MO, USA, Merck KGaA, Darmstadt, Germany), 10% Horse Serum (HS; SIGMA-ALDRICH, St. Louis, MO, USA, Merck KGaA, Darmstadt, Germany) and 1% antibiotics/antimycotics (100 U/mL penicillin G sodium, 100 µg/mL streptomycin sulfate, and 0.25 µg/mL amphotericin B) (Antibiotic Antimycotic Solution ($100\times$); SIGMA-ALDRICH, St. Louis, MO, USA, Merck KGaA, Darmstadt, Germany). Cells were then seeded in 25 mL bottles (at a final volume of 4 mL of medium).

2.3. Long-Term Primary Cell Culture

Cells were cultured at 37 °C in a humidified atmosphere of 5% CO_2 in 25 mL bottles for a maximum of 30 days.

Once the cultures reached 70–80% confluency, they were passaged by washing with 3 mL of D-PBS, digested with 3 mL of 0.025% trypsin/EDTA (Trypsin-EDTA solution (0.25%), SIGMA-ALDRICH, St. Louis, MO, USA, Merck KGaA, Darmstadt, Germany) at 37 °C for 3–5 min, neutralized by the addition of 6 mL of DMEM supplemented with 10% FBS, 4 mM L-glutamine solution (L-glutamine solution (200 mM), SIGMA-ALDRICH, St. Louis, MO, USA, Merck KGaA, Darmstadt, Germany) and 1% antibiotics/antimycotics (100 U/mL penicillin G sodium, 100 µg/mL streptomycin sulfate and 0.25 µg/mL amphotericin B) (Antibiotic Antimycotic Solution ($100\times$); SIGMA-ALDRICH, St. Louis, MO, USA, Merck KGaA, Darmstadt, Germany), centrifuged (5 min, $200\times g$, room temperature), and resuspended in DMEM/F12 supplemented (cardiomyocytes culture medium).

The medium was changed every 3 days and the culture was observed under an inverted microscope employing relief contrast (IX73, Olympus, Tokyo, Japan) every day. Before the harvesting of cells for further experiments, photos of the culture at three time points, 7 days, 15 days, and 30 days, were also taken to observe possible morphological changes.

2.4. Detection of Chosen Markers Using Flow Cytometry (FC)

After reaching a proper degree of confluence, the cells were digested from the culture plate, using 0.025% trypsin/EDTA solution, and subjected to FC analysis through staining. To identify characteristic markers, the following antibodies were used: Anti-CD44 FITC (Abcam, Cambridge, UK), Anti-CD90 APC, and Anti-CD105 PE (R&D systems). For the staining of cytoplasmic proteins, firstly, cells were permeabilized and fixed in BD Perm/Wash™ buffer on ice. Next, 300 μL of BD Perm/Wash™ buffer was added to 100 μL of cells in solution. The samples were subsequently centrifuged ($250\times g$, 5 min), supernatant was discarded, and cells were incubated with antibodies for 30 min on ice. Following incubation, a wash with BD Perm/Wash™ buffer was performed. Such prepared cells were analyzed with a BD FACSAria™ cytometer and FACSDiva™ software (Becton Dickinson, Franklin Lanes, NJ, USA). Cell surface markers (CD44, CD105, and CD90) were identified, then 5 μL of antibodies were added to 100 μL of cell suspension. The mix was washed twice in PBS and centrifuged ($250\times g$, 5 min). The pellet, remaining after supernatant removal, was resuspended in 100 μL of PBS and subjected to acquisition in a flow cytometer.

2.5. RNA Extraction and Isolation

Total RNA was extracted after 7, 15, and 30 days of the cell culture, respectively. The cells were treated as if they were passaged (see Long-term primary cell culture for details) but the pellet was resuspended in 500 μL of TRI Reagent Solution (TRI Reagent®, SIGMA-ALDRICH, St. Louis, MO, USA, Merck KGaA, Darmstadt, Germany). The aliquots were stored in −80 °C until RNA isolation.

Total RNA was isolated according to the Chomczyński and Sacchi method [35,36]. Firstly, 100 μL of chloroform (SIGMA-ALDRICH, St. Louis, MO, USA, Merck KGaA, Darmstadt, Germany) was added, the samples were mixed by inversion and shaken for 15 sec, and then incubated for 10 min at room temperature. Subsequently, the biphasic emulsion was separated by centrifugation at $12,000\times g$ for 15 min at 4 °C. The upper, aqueous phase which contained RNA was transferred to new Eppendorf tubes. Next, 250 μL of isopropanol (SIGMA-ALDRICH, St. Louis, MO, USA, Merck KGaA, Darmstadt, Germany) was added, the samples were mixed by inversion, shaken for 15 s, and incubated for 15 min at room temperature. The samples were centrifuged at $12,000\times g$ for 10 min at 4 °C. Then, 1 ml of 75% ethanol solution (SIGMA-ALDRICH, St. Louis, MO, USA, Merck KGaA, Darmstadt, Germany) was added to the precipitate, the samples were vortexed for 20 s, and centrifuged at $7500\times g$ for 15 min at 4 °C. The supernatant was discarded and the samples were air-dried and dissolved in 20–50 μL (depending on pellet size) of DEPC-treated water.

The spectrophotometric analysis at $\lambda = 260$ nm (NanoDrop spectrophotometer; Thermo Fisher Scientific, Waltham, MA, USA) was performed in order to assess the concentration of the samples.

2.6. Microarray-Based Transcriptomic Profiling

Total RNA (100 ng) from each pooled sample was subjected to two rounds of sense cDNA amplification (Ambion® WT Expression Kit, Life Technologies, Carlsbad, CA, USA). The obtained cDNA was used for biotin labeling and fragmentation by Affymetrix GeneChip® WT Terminal Labeling and Hybridization (Affymetrix, Santa Clara, CA, USA). Biotin-labeled fragments of cDNA (5.5 μg) were hybridized to the Affymetrix® Porcine Gene 1.1 ST Array Strip (48 °C/20 h). Microarrays were then washed and stained according to the technical protocol using the Affymetrix GeneAtlas Fluidics Station. The array strips were scanned employing Imaging Station of the GeneAtlas System. Preliminary analysis of the scanned chips was performed using Affymetrix GeneAtlas™ Operating Software. The quality of gene expression data was confirmed according to the quality control criteria provided by the software. The obtained CEL files were imported into downstream data analysis software.

2.7. Reverse Transcription-Quantitative Polymerase Chain Reaction (RT-qPCR)

An amount of 1 µg of isolated RNA diluted in PCR-grade water (up to a final volume of 8 µL) from each sample was reverse transcribed by RT2 First Strand Kit (Qiagen®, Hilden, Germany) according to the protocol. With the exception of incubation, samples were kept on ice. First, to eliminate the genomic DNA, 2 µL of GE (5x ×gDNA Elimination Buffer) was added to 1 µg of isolated RNA. Samples were incubated at 42 °C for 5 min. Next, the reaction mix, which included 4 µL BC3 (5× RT Buffer 3), 1 µL P2 (Primer and External Control Mix), 2 µL RE3 (RT Enzyme Mix 3), and PCR-grade water up to the final volume of 10 µL, was prepared and 2 incubations were performed: at 42 °C for 15 min, and at 95 °C for 5 min. Next, samples were cooled on ice and 91 µL of H_2O was added to each reaction.

The validation of microarray data was/will be performed on LightCycler® 96 Instrument (Roche Diagnostics GmbH, Mannheim, Germany). cDNA synthesized in reverse transcription served as a template. Primers were designed in Primer3Plus software (version 0.4.0; Whitehead Institute for Biomedical Research, Massachusetts Institute of Technology, Cambridge, MA, USA) [37–39] using sequences of chosen transcript variants of the genes available in the Ensembl database [40] (Table 1). The components of the reaction mix were as follows: QUANTUM EvaGreen® PCR Kit (5×) (Syngen Biotech, Wroclaw, Poland) which was used as the master mix, 10 µM of oligodeoxynucleotides (SIGMA-ALDRICH, St Louis, MO, USA, Merck KGaA, Darmstadt, Germany), and PCR-grade water. Next, 9 µL of the reaction mix and 1 µL of the template were added to each respective well on a 96-well plate. The plate was sealed with a sealing foil, centrifuged at 1500 rpm 400 × *g* for 1 min, and placed in the thermocycler. The thermal profile of the reaction is presented in Table 2.

The $2^{-\Delta\Delta CT}$ method for relative gene expression analysis was applied [41,42]. cDNA synthesized from RNA isolated after 0 days of cell culture served as a control. The relative expression of the studied genes was normalized against the expression of 2 reference genes—*GAPDH* and *ACTB*.

Table 1. Oligodeoxynucleotides designed to detect the expression levels of the studied genes.

Gene Symbol	5′→3′ Sequence of the Forward Primer	5′→3′ Sequence of the Reverse Primer	Product Length (bp)
ASPM	AACAGATTACGTCGTGCTGC	CTGTCTCTAGGCCGATTCGT	205
BMPR1A	ATTTGGGAAATGGCTCGTCG	CCCAGCATTCCGACATTAGC	205
CTNNA1	CTTCTTGGCGGTCTCAGAGA	ACCCCTGGCTCATAGTTGTC	171
CTNNB1	CCTCTCATCAAGGCTACCGT	AGGGCTCCAGTACAACCTTC	218
DDX6	ACACTGCCTCAACACACTTT	GATTTCGATGTTCCTGCCTCA	220
DOCK7	TGCTTACTCCACCTGCATCA	ATAGTCCTTCCGCATCAGGG	198
EIF4E	GCAAACCTTCGGCTGATCTC	ATTAGCCATCGTCCTCCTCG	173
EPAS1	TGGGCTGGAGAGTTGAGAAG	AATAGTCTCCAGGGCTGCTG	151
FANCD2	CAGTGTATGCCGCTCCTAGA	GAGATTGCCCAGCCAGAAAG	246
FGF2	GCAGAAGAGAGAGGGGTTGT	CGTTCGTTTCAGTGCCACAT	195
FGFR2	ATCGAGATTTAGCCGCCAGA	TCCCACATTAACACCCCGAA	214
FOXO1	ATCACCAAGGCCATCGAGAG	AGTTCCTTCATTCTGCACGC	189

Table 1. *Cont.*

Gene Symbol	$5' \rightarrow 3'$ Sequence of the Forward Primer	$5' \rightarrow 3'$ Sequence of the Reverse Primer	Product Length (bp)
FRS2	TGCCTTTAAGTGTGCTCGTG	TGGGTAAGTTCTGAGCAGCA	182
HIF1A	CTCAGTCGTCACAGTCTGGA	CCACCTCTTTTGGCAAGCAT	212
HMGA2	ACCTCCCAATCTCCCGAAAG	GTTGTCCCTGGGCTGAAGT	161
ID4	ACTACATCCTGGACCTGCAG	CCTCCCTCTCTAGTGCTCCT	250
KDM1A	GTCCAGTTTGTGCCACCTCT	TGCCTACATGCCCGAACAAA	139
KIT	ATTGTGAATCTTCTCGGCGC	ACTCCGGGTTTCATGTCCAT	227
KITLG	TGGCCAGTTCTATCCATGCA	TGGTCAGGGGTAAAGGCAAA	175
KLF10	TGAGCTGCAGTTGGAAGTCT	TGTGAGGCTTGGCAGTATCT	246
LIF	GAACCTCTGAAAACTGCCGG	ACAGGAGTGATGGAAAGGGG	151
MCPH1	CAGCCGACCATGTTCATCAG	AGTTCTCAGAGGCACAGACC	217
MED14	CCACCATCCTCACTCACAGT	TCACTCCGGGTTCATTGGAA	198
MED21	GTCCTCCTGCCTCTTTCAGT	CCTCCAGACATGTAGCAGCT	228
MED27	ACTTGCATTCAGTCAACCGG	TTGTTCGACCACTTGTACGC	172
MED7	ACTAGTAGAAGGCACGCGAA	ACTGCCCTTCACGGTGATTA	150
MTF2	AAACTGCTGAGCCACCTTTG	TGCCTGGAAATGCTAGACGA	172
NF1	GAATCCCCACCACAGTACCA	AAGGAGATGTGGGTGTCAGG	226
NKAP	GAGTCCCAGGAAGAGTTGCT	CATAGCTGCACCTTCACCAG	162
NOTCH2	TTATGTCTCACCCCTGCCAG	ACTGTCCTGGAACGTCACAT	246
PAX6	TTGCCCGAGAAAGACTAGCA	GTGGGTTGTGGGATTGGTTG	196
PRRX1	GGACACACTACCCAGATGCT	TTTGAGGAGGGAAGCGTTCT	155
RAB10	ATGGCGAAGAAGACGTACGA	AGGAGGTTGTGATCGTGTGA	232
RACGAP1	ACCGCTGGAATACTGGAGTC	TGACAGGGAGCTGGATGAAG	184
RBPJ	ATGGGCAGTGGATGGAAGAA	TGTTTTGGCCGTGCAATAGT	156
RTF1	CCAGGCGACAGTGTAAACCT	TCGCTGGCTGACTTGGAATT	175
RUNX1	GTCCCAACTTCCTCTGCTCT	CTTCCACTCCGACCGACAAA	226
SFRP2	GCTCCAAAGGTATGTGAAGCC	GGTCTTGCTCTTGGTCTCCA	159
SMC3	TGGAGGACACTGAGGCAAAT	TCCTGTTGCCGCTCTAAGAA	248
SNAI2	GCCGAGAAGTTTCAGTGCAA	GGGTCCGAATGTGCATCTTC	169
SOX5	CAGCAGCAAGAACAGATCGC	AGCCAGTGTCCGTTGATCAG	147
SS18	CGGATATGACCAGGGACAGT	CTTGCTGCGTTTCACCTGAT	175
STAT3	AGCAGCAAAGAAGGAGGAGT	ACACGAGGATGTTGGTAGCA	166
TAF4B	GCCAGTCAGTTTCCTCAAGC	ACGAGTGTGCCAACCAATTC	227
TBX3	AGGGTGTTCGATGACAGACA	GACGTGGTGGTGGAGATCTT	233
TET1	TCTGGCAAGAAGAGAGCAGC	ATGGATGGGGTCGGTGAGTA	248
TGFB1	ACCATGCCAATTTCTGCCTG	GAACGCACGATCATGTTGGA	208
VPS72	TCCTTCGAGTACAAGAGCGG	GCACTTGCGCTTCTTATGGA	188
WNT2B	CAACGTGGGGACTTTGACTG	TGGCACTTACACTCCAGCTT	185
ACTB	CCCTGGAGAAGAGCTACGAG	CGTCGCACTTCATGATGGAG	156
GAPDH	CCAGAACATCATCCCTGCCT	CCTGCTTCACCACCTTCTTG	185

Table 2. Thermal profile of RT-qPCRs.

Step		Temperature (°C)	Time (s)	Number of Cycles
Preincubation		95	600	1
Amplification	Denaturation	95	15	
	Annealing	58	15	40
	Elongation	72	15	
Melting		95	60	
		40	60	
		70	1	1
		95	1	
Cooling		37	30	1

2.8. Bioinformatic Data Analysis and Visualization

All of the presented analyses were performed and prepared using the R programming language with the Bioconductor package [43,44]. Each CEL file was merged with a description file. In order to correct the background, normalize, and summarize results, the Robust Multiarray Averaging (RMA) algorithm was used. To determine the statistical significance of the analyzed genes, moderated t-statistics from the empirical Bayes method were computed. The obtained *p*-value was corrected for multiple comparisons using Benjamini and Hochberg's procedure. The selection of differentially expressed genes (DEGs) was based on adjusted *p*-value $\leq$ 0.05 and ratio > abs(1). The DEGs list was uploaded to DAVID (Database for Annotation, Visualization, and Integrated Discovery) [45–47]. DAVID was used to retrieve Gene Ontology Biological Process (GO BP) terms enriched in the list of identified genes [48,49]. Significantly changed GO BP terms were defined as those enriched in at least 5 genes and for which the adjusted *p*-value $\leq$ 0.05. Plots were generated using the GOplot package [50].

RT-qPCR analyses and graphs were performed and prepared using Microsoft Excel (Microsoft Corporation). The results are presented as mean values $\pm$ standard deviation (SD) of two technical replicates of each reaction.

3. Results

3.1. Isolation of Cells from Two Fragments of Porcine Cardiac Muscle and Long-Term In Vitro Primary Culture

Cell isolation, establishment, and maintenance of the cell culture from two fragments, namely, the right atrial appendage and right atrial wall, was successful. The cultures were observed every 3 days up to the 30th day (Figure 1). Cell morphology was similar in all cultures. Initially, cells had rather irregular shapes and gained a spindle-like appearance with the duration of the culture. Interestingly, we observed the formation of 3D clusters as captured for the culture of cells from the right atrial wall. The choice of those fragments was based on the fact that they can be easily collected during open heart surgery for coronary artery bypass grafting and, therefore, performed experiments could be later translated into humans. Two fragments—the right atrial appendage and right atrial wall—were selected for further characterization due to the ease of collection.

3.2. Identification of Possible Stemness Phenotype of the Cells Isolated from the Heart Tissue

In order to enable a comparison of specific regions of the heart as potential locations of concentration of cells with presumed stem cell characteristics, CD44, CD90, and CD105 were chosen and the cells were analyzed for their presence with flow cytometry in the 7th, 15th, and 30th days of primary cultures. Results from the 7th day of cell culture derived from the right atrium show some expression of cytoplasmic CD105 (MFI fold change of 1.82). A similar situation was observed for the 15th day of cell culture (MFI fold change of 1.84), while for the 30th day, expression of cytoplasmic CD105 and CD90 was detected (MFI fold change of 2.88 and 3.12, respectively) (Figure 2). As for the right atrial appendage, cytoplasmic CD105 and CD90 was expressed in cells derived from the 7th-day culture (MFI fold change 1.89 and 2.03, respectively), while cytoplasmic CD105 was also expressed in the 15th day of culture (MFI fold change 1.95) and CD90 on the surface and in the cytoplasm were detected in the cells from the 30th-day culture (MFI fold change 11.4 and 5.08, respectively) (Figure 3). It should be noted however that the number of events—i.e., the number of cells—is relatively low (frequently below 100), except for the sample derived from the right atrial appendage in the 30th day of culture.

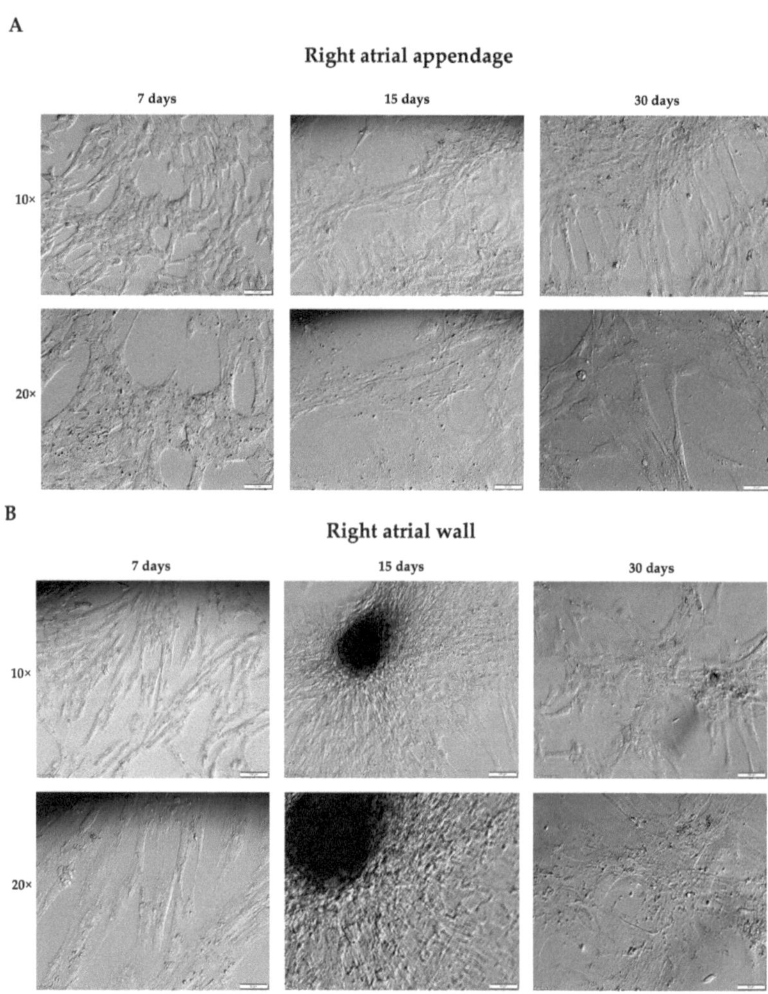

Figure 1. Representative images of long-term in vitro cell cultures established from 2 different heart muscle fragments—right atrial appendage (**A**) and right atrial wall (**B**). Nomarski phase/contrast images. Scale bars' length for magnification: 10×: 100 μm, 20×: 50 μm.

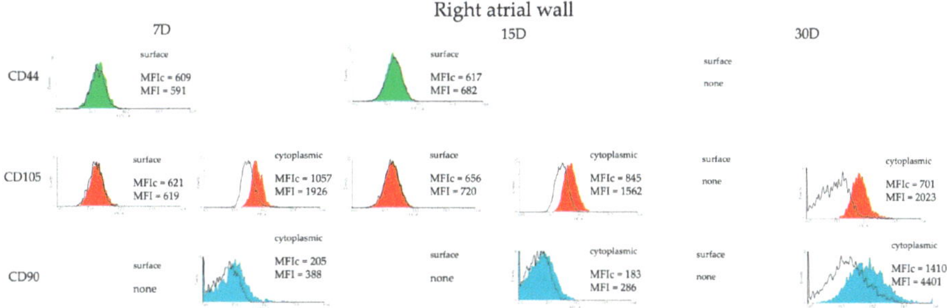

Figure 2. Phenotype of porcine CMs isolated from the right atrial wall in the 7th, 15th, and 30th day of culture analyzed by flow cytometry. CD44, CD105, and CD90 state for mesenchymal cell markers. MFIc—mean fluorescence intensity for control samples; MFI—mean fluorescence intensity for research samples.

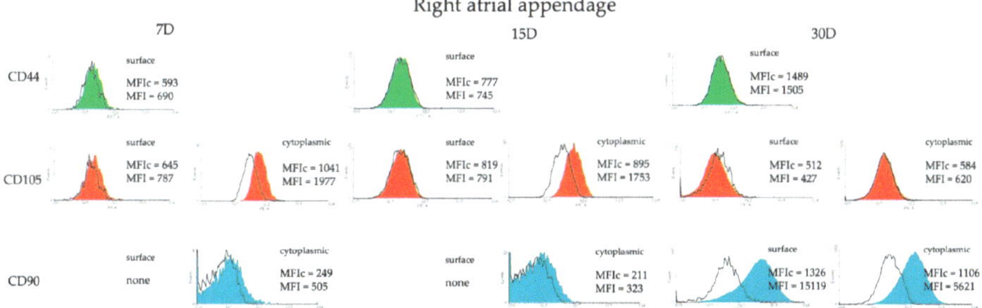

Figure 3. Phenotype of porcine CMs isolated from right atrial appendage in the 7th, 15th, and 30th day of culture analyzed by flow cytometry. CD44, CD105, and CD90 state for mesenchymal cell markers. MFIc—mean fluorescence intensity for control samples; MFI—mean fluorescence intensity for research samples.

3.3. Transcriptomic Characterization of the Cardiomyocytes during Long-Term In Vitro Culture

To further understand molecular changes occurring in CMs under the influence of in vitro culture conditions, whole transcriptome profiling by Affymetrix microarray was performed. Gene expression changes between 0 and 7, 15, and 30 days of cardiomyocytes culture were studied. A total of 12,257 transcripts were examined by Affymetrix® Porcine Gene 1.1 ST Array Strip. Genes were considered as differentially expressed (differentially expressed genes, DEGs) based on 2 parameters: ratio > abs |2| and adjusted *p*-value ≤ 0.05.

Followingly, biologically relevant processes in which the DEGs could be implicated were explored. For this purpose, DAVID (Database for Annotation, Visualization, and Integrated Discovery) software was utilized and enabled the extraction of gene ontology biological process terms (GO BP), which contain DEGs. DAVID searching was performed for up- and downregulated gene sets separately, and only gene sets in which the adjusted *p*-value ≤ 0.05 were selected. Two gene ontology biological process terms (GO BP) were chosen in this work: "stem cell population maintenance" and "stem cell proliferation". The chosen sets were then subjected to a hierarchical clusterization procedure and presented as heatmaps (Figure 4). The symbols, fold changes in expression, and adjusted *p*-values for the 10 most deregulated genes at each time point in culture are shown in Tables S1 and S2. The majority of genes which belong to the GO term "stem cell maintenance" in samples derived from right atrial wall tissue are upregulated in comparison to day 0, and this tendency remains until day 30. *FGF2*, *EIF4E*, *TBX3*, *LIF*, and *EPAS1* showed a gradual increase in expression levels with the duration of culture, with the highest expression in day 30 among all the genes in this GO term. There are, however, several genes in which expression in comparison to day 0 seems unchanged or was downregulated. Similar tendencies can be observed for genes in the group "stem cell proliferation". In this case, *TBX3* and *SOX5* were the most upregulated, while downregulation was noted for *PAX6*, *WNT2B*, *FGFR2*, and *SNAI2* (Figure 4). In the samples from the right atrial appendage, approximately half of the genes enriched in the "stem cell maintenance" GO term showed a slight increase in expression in relation to day 0, while the other half experienced initial upregulation, especially in day 7, slight in day 15, and subsequent downregulation in day 30. Similar observations can be made when analyzing the genes for the "stem cell proliferation" GO term in the same samples (Figure 4).

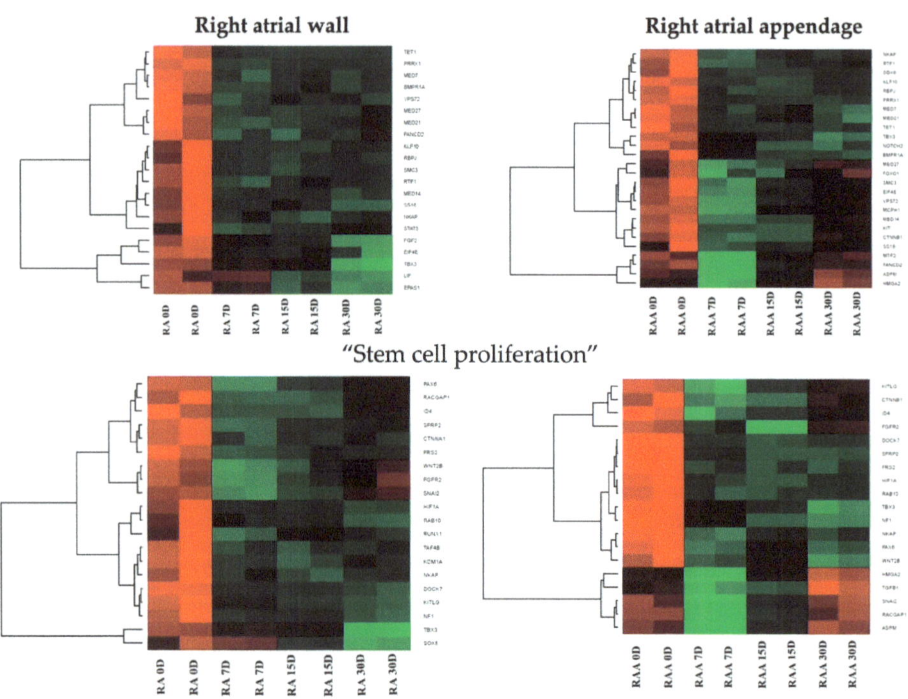

Figure 4. Heatmap representation of differentially expressed genes belonging to the chosen GO BP terms. The arbitrary signal intensity acquired from microarray analysis is represented by colors (green—higher expression; red—lower expression). D—day of the primary culture; RA—right atrial wall; RAA—right atrial appendage.

To check if the genes belong to more than one ontological group and to characterize these groups in a more detailed manner, the results of gene annotation enrichment analysis were visualized with the circle plot (GOCircle) (Figure 5), the relationships between genes and terms were plotted with the GOChord plotting function (Figure 6), and a heatmap of genes and terms was generated with the GOHeat function (Figure 7).

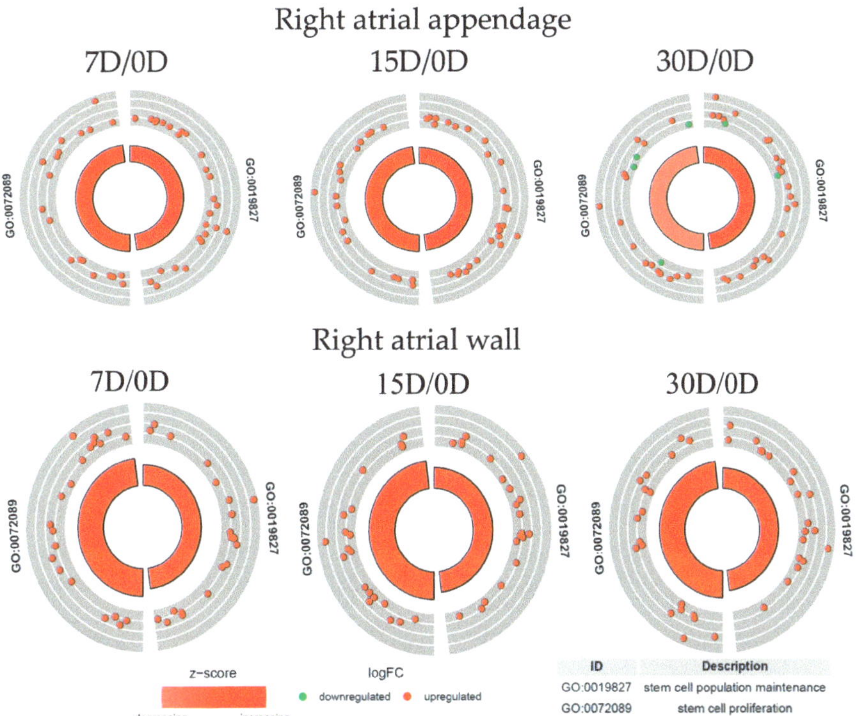

Figure 5. The circle plot shows the differentially expressed genes and z-score of the chosen GO BP terms "stem cell population maintenance" and "stem cell proliferation". The outer circle shows a scatter plot for each term of the logFC of the assigned genes. Red circles display upregulation and green circles downregulation. The inner circle shows the z-score of each GO BP term. The width of each bar corresponds to the number of genes within the GO BP term and the color corresponds to the z-score. The z-score refers to the difference between the number of upregulated genes and the number of downregulated genes divided by the square root of the count, where the count is the number of genes assigned to a term.

The circle plot reveals that the genes in the samples from the right atrial appendage show a general tendency of expression level upregulation. There are more genes enriched for the "stem cell proliferation" GO term than for "stem cell population maintenance". As for the samples from the right atrial appendage, although the tendency of gene expression upregulation can also be observed, some genes in the sample from the 30th day of cell culture show decreased expression levels. Furthermore, the "stem cell proliferation" GO BP term is slightly decreased (based on the z-score).

Moving to Figure 6, for samples from the right atrial wall, only 1 of the displayed genes was present in both GO terms, and all of the genes showed greater or lesser upregulation in day 7 compared to day 0. In the samples derived from the right atrial appendage, 5 of the presented genes were enriched in both GO terms, and only 2 of the displayed genes showed a tendency to decrease.

The heatmap presented in Figure 7 shows common differentially expressed genes for both GO terms in the 7th vs. 0 day of the culture. Genes *SFRP2* and *PRRX1* are the most highly expressed for both the right atrial wall and right atrial appendage.

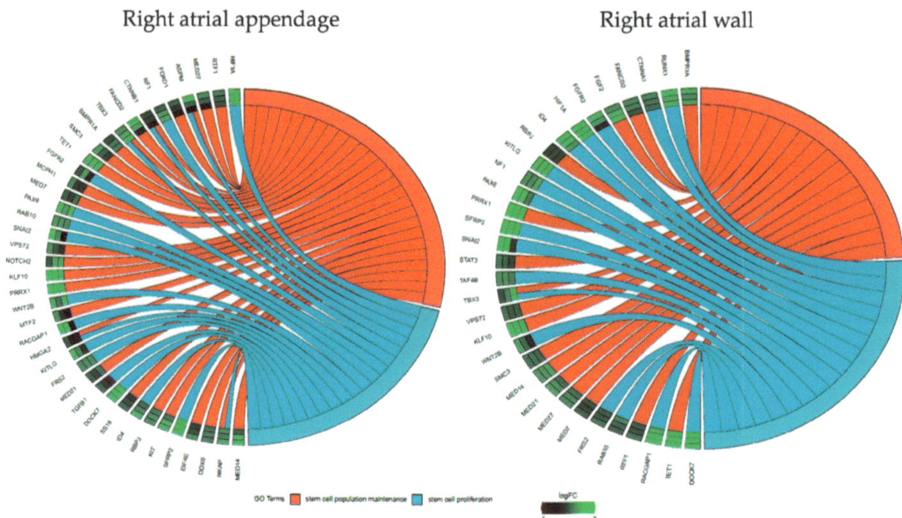

Figure 6. GOChord plot representing the relationship between differentially expressed genes and chosen GO BP terms: "stem cell population maintenance" and "stem cell proliferation". The ribbons indicate which gene belongs to which categories. The color of each block closest to the gene name corresponds to the logFC of each gene (7D/0D) (green—upregulated, red—downregulated).

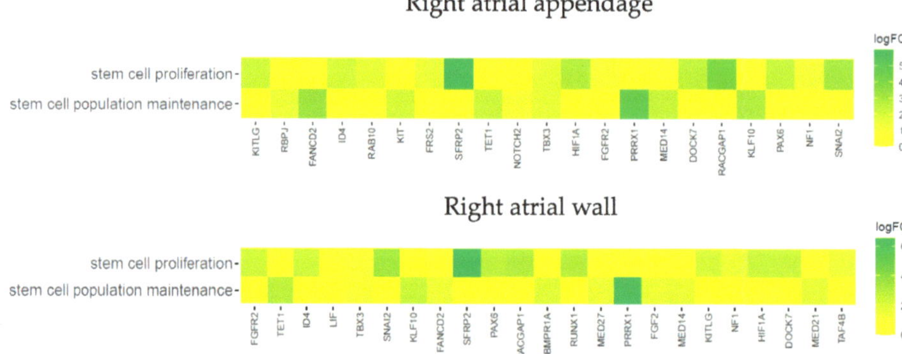

Figure 7. Heatmap presenting the relationship between differentially expressed genes and chosen GO BP terms: "stem cell population maintenance" and "stem cell proliferation". The color corresponds to the logFC of a particular gene (7D/0D).

The RT-qPCR (reverse transcription-quantitative polymerase chain reaction) technique was also utilized to validate the chosen results of high-throughput transcriptome profiling. The results are demonstrated as a bar chart (Figure 8).

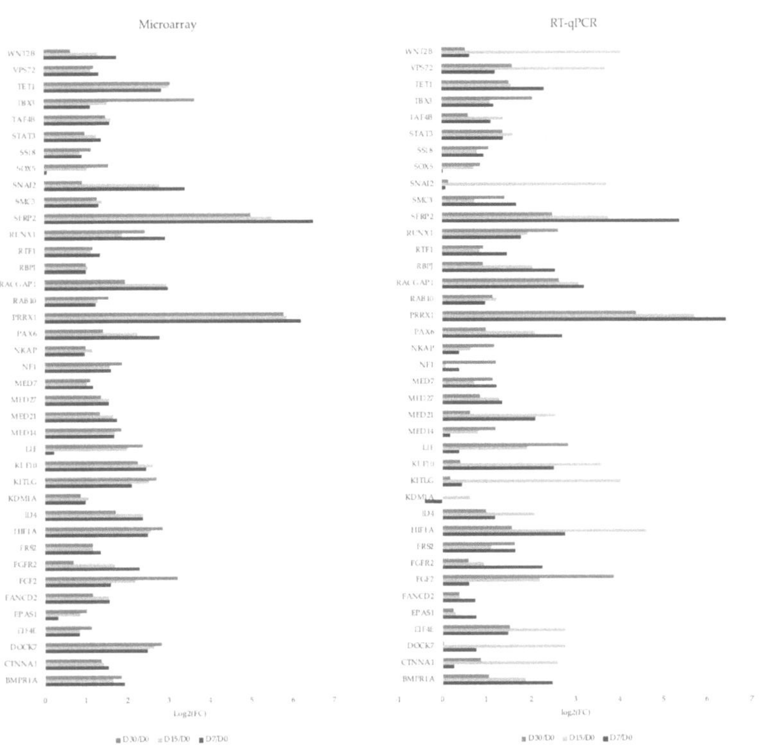

Figure 8. The results of the validation of the expression level of the analyzed genes by RT-qPCR presented as a bar chart. The values are presented as the log2FC of the tested culture duration vs. 0 days' culture duration.

4. Discussion

Despite recent progress in cardiology having caused a significant decrease in CVD (cardiovascular disease)-related deaths, the number of heart failure cases has increased [1]. There has been an increasing interest in regenerative medicine that would allow finding a cellular replacement for the cardiomyocytes lost during the disease [13].

In recent years, the dogma stating that the adult heart lacks any regenerative potential has been questioned by the evidence that cardiac muscle cells possess some small self-renewal capacity over the mammalian lifetime. Heart precursor cells have been described in adult hearts of several mammalian species. However, there is no unified methodology for the isolation of cardiac stem cells [14,28,51–53].

Primary cultures of cells from two of the porcine heart tissue fragments have been successfully established and, as evident by results from a previous study of the group, namely, the presence of α-MHC and α-actinin as well as GATA-4 in the cells analyzed by flow cytometry, CM from the right atrial wall and right atrial appendage have been isolated [14,54]. In the last two decades, there has been a significant increase in the number of types of cardiac progenitor cells reported. These cells have been identified and isolated based on the expression of a whole variety of markers on their surface which overlap to some extent and are highly mixed. Here, CD90, CD44, CD105, and GATA-4 have been proposed as putative stemness markers [54] and a small population of cardiac muscle-derived cells expressing these markers has been identified.

CD90 (THY1) has been described as a marker of a variety of stem cells, including mesenchymal stem cells [55–57]. As for other cells, CD90 was described as characteristic for endothelial cells, smooth muscle cells, or fibroblasts, among others [58–60]. According to some reports, CD90 can be present on cardiac progenitors [28,61–64], while others have shown that only CD90[neg] CSCs can have any therapeutic role and that they generally outperform unsorted cells [65]. The flow cytometry results have revealed the presence of this marker in the cytoplasm of cells derived from the right atrial wall in the 30th day of the IVC and derived from the right atrial appendage in the 7th day of the IVC. Cells derived from the right atrial appendage in the 30th day seem to point out the presence of this marker on their cell surface. As the populations of cardiac progenitor cells are usually small, this may suggest the presence of other cell types typical for the heart. It should be noted however that the number of events in case of these samples is relatively small and thus the interpretation cannot be held fully reliable—additional experiments involving more viable cells would be required.

CD44 is a surface glycoprotein which participates in the transduction of signals important for cell migration, cell adhesion, and cell–cell interaction, and it plays a role in sensing cues from the microenvironment [66]. It interacts with several components of the extracellular matrix, including hyaluronan [67]. Similarly, to CD90, it is a canonical marker of mesenchymal stem cells [55]. There were several reports of cardiac progenitor cells which are CD44[pos], including epicardium-derived progenitor cells and cardiac colony-forming unit fibroblasts [28,68]. According to FC, very few cells have shown CD44 expression. Conversely, it is widely known that cardiac stem cells constitute a small population of cells in the adult mammalian heart [13].

Endoglin, or CD105, is a type I transmembrane protein which serves as a BMP9/10 co-receptor and TGF-β signaling auxiliary receptor that is expressed primarily on vascular endothelial cells [69,70]. It is important for maintaining the proper structure of vasculature and is indispensable for heart development [71]. Furthermore, together with CD90 and CD44, CD105 is a marker of mesenchymal stem cells [55]. There are numerous reports of CD105 presence on cardiac progenitor cells, especially for cardiosphere-derived cells, Sca-1[pos] CPC, and EDPCs [25,28,61,63,64]. According to our FC analysis, there is a small population of cells expressing this marker.

GATA-4 is a transcription factor which regulates early cardiac development and specification [72–74]. Together with ISL1, MEF2C, and NKX2-5, it is often used for the identification of cardiac phenotype in heart precursor cells [11]. GATA-4 is expressed in CPCs including Sca1[pos] cells, Isl-1[pos] cells, and cardiac fibroblasts [31,51,52,75]. Apart from its role in cardiac morphogenesis at the embryonic stage, GATA-4 is crucial for cardiac hypertrophy by regulating transcriptional activation of cardiac-specific genes [74,76]. BRD4-GATA4 have also been recently revealed to regulate adult cardiac metabolism [77].

FC analysis has revealed that GATA-4 was expressed in all samples except the 30D right atrial appendage (MFI fold change in the range 2–4 compared to the control) [54]. It is therefore suggested that GATA-4 could regulate the functioning of adult cardiomyocytes in culture, including their growth.

Despite several phenotypic variations of cardiac progenitor cells having been identified, the precise role and molecular identity of CPCs in the adult heart remains enigmatic. A small pool of precursor cells is also a significant challenge and underlines the importance of the research that would allow molecular profiling of CPCs. There are studies which report characterization of the mammalian cardiac progenitor cells' transcriptome but the majority utilize different techniques than ours and aim at studying changes occurring at the embryonic stage [78–80]. Nevertheless, molecular changes at the transcriptomic and epigenomic level in putative adult murine heart-derived CPCs have been described [81,82]. Interestingly, in 2013, Dey et al. reported microarray-based transcriptional profiling of five cardiac (ckit[+], Sca1[+], and side population) and bone marrow-derived (ckit[+] and mesenchymal stem cell) progenitors from adult mice and found out that cardiac ckit[+] cells are the most distinct and seem to be the most primitive cell population [83]. However, the hypothesis that cells present in the bone marrow transdifferentiate to cardiomyocytes was disproven and the studies of Anversa et al., who first identified c-kit[+] cells, are now not held reliable [84,85]. Additionally, genetic lineage tracing studies have revealed that Sca1+ cells give rise rather to endothelial cells and not myocytes [86]. The porcine cardiac transcriptome has been recently reported as updated [87] and there are several studies in which transcriptomic profiling of porcine hearts was utilized to address different questions [88–90]. It appears that the current study is the first analysis that aims at transcriptome profiling of porcine cardiomyocytes during long-term in vitro culture.

SFRP2 was one of the most highly upregulated genes in cells derived from both heart fragments through the duration of the culture. For the right atrial wall, a certain decrease in expression levels with IVC time was observed; nevertheless, it remained as the 1st- or 2nd-most highly upregulated gene, while for the right atrial appendage, the expression increased in the 15th day compared to the 7th day and remained high until the end.

SFRP2, a secreted frizzled-related protein 2, is a member of the sFRP family which regulates both canonical and noncanonical Wnt signaling, influencing many fundamental processes such as proliferation, apoptosis, and differentiation [91]. β-catenin-dependent and β-catenin-independent Wnt signaling pathways are indispensable for heart development [92]. As for cardiomyocytes differentiation, research has shown that after the initial activation of β-catenin, required for mesoderm specification to cardiac progenitors, it is the inhibition of this molecule which allows for differentiation of these progenitors to cardiomyocytes [93]. It has also shown to be important in cardiac fibrosis as well as angiogenesis, thereby contributing to the restoration of the cardiac function after injury [94,95]. Studies on adult murine cardiac progenitor cells (Sca1$^+$) have shown that Sfrp2 may promote the differentiation of such cells into CMs after injury [96]. It has also been shown that myocardial injury may induce c-kit$^+$ cells to exhibit cardiomyocyte-specific gene expression and Sfrp2 enhances this effect [97], and, lastly, the same group has claimed that Sfrp2 induces cardiomyocyte differentiation in c-kit$^+$ cells in vivo in health and MI, as revealed by genetic lineage tracing and functional assessment of the developed CMs [98].

Although, at the early stages of cardiomyogenesis, *SFRP2* prevents the generation of cardiac progenitors from mesoderm cells and promotes the maintenance of cells in the undifferentiated state [99], as mentioned above, there is evidence in adult cardiac progenitors that it facilitates differentiation to cardiomyocytes and, furthermore, it promotes cardiac fibrosis. As the cells in this study were isolated from adult porcine hearts, it is suspected that upregulation of this molecule had rather a negative effect on maintaining the stemness phenotype/undifferentiated state.

PRRX1 (Paired Related Homeobox 1) was also one of the two most highly upregulated genes in samples from both heart fragments. For the right atrial wall, a situation was similar to *SFRP2*—compared to the 7th day, a certain decrease in *PRRX1* levels in the 15th and 30th days of the culture was observed, and the ratios were generally similar to *SFRP2*. For the right atrial appendage, the expression increased in the 15th day compared to the 7th day, but only slightly, and, overall, the expression levels were 2–3 times lower than for *SFRP2*.

PRRX1 belongs to the paired family of homeobox-containing transcription factor proteins and has nuclear localization.

There are several studies characterizing the expression of *PRRX1* in the developing hearts of vertebrates, mainly in mesenchymal tissues, including in the heart and arteries [100–102].

Prrx1 is involved in scar-free heart regeneration. A 2021 study by de Bakker et al. demonstrated that loss of *Prrx1b* in cryo-injured zebrafish hearts led to excessive fibrosis and a reduction of cardiomyocyte proliferation at the injury border zone. Single cell sequencing and lineage tracing have shown that *Prrx1* is expressed mostly in epicardial and epicardial-derived cells and that there is an excess of fibroblasts producing TGF-β ligand and ECM in injured prrx1b$^{-/-}$ hearts. Mechanistically, Prrx1 promotes Ngr1 and, consequently, cardiomyocytes' proliferation not only in in vivo zebrafish model but, noticeably, in human fetal epicardial-derived cells [103].

As a transcription factor, PRRX1 can physically interact and regulate expression of genes crucial for proper cardiac function. In a study investigating AF pathogenesis,

Guo et al. showed that *SHOX2* and *ISL1,2* were genes that played roles in cardiac pacing and conducting systems that undergo such regulation [104].

In another 2021 study, it was revealed that deletion of a noncoding variant genomic region at 1q24 associated with downregulation of *Prrx1* leads to a gain of more myogenic phenotype in atrial cardiomyocytes. This suggests that *Prrx1* is a negative regulator of cardiac muscle lineage-specific gene expression. They also described a negative feedback regulatory loop between Prrx1 and Mef2, an important regulator of cardiomyocytes' development and differentiation, which may indicate that this relationship is important in the maintenance of cardiomyocyte homeostasis [105].

It is difficult to unequivocally assess the potential influence of *PRRX1* on the long-term culture. Many of the previous studies have described the role of *PRRX1* in embryonic development, while here, the cells were collected from adult pigs. It seems that studies on AF seem to be the closest to the current system, as they were mostly conducted on mammalian–human or murine cells, although not porcine. *PRRX1* could potentially promote cardiomyocytes' proliferation and control expression of genes involved in the proper functioning of cells, as well as regulate the myogenic phenotype. The culture consists of cells which express markers characteristic for cardiomyocytes based upon flow cytometry data. At the same time, the authors do not have broader transcriptomic data which would possibly reveal a profile of expression of other genes important for the determination of such a phenotype. Furthermore, it is interesting that contradictory trends in the expression of this gene have been observed with the duration of culture between samples from the right atrial appendage and right atrial wall.

The other two most deregulated genes in the presented data are *RACGAP1* and *SNAI2*. *RACGAP1*, Rac GTPase Activating Protein 1, here enriched in the "stem cell proliferation" GO BP term, is crucial for cytokinesis, namely, for the myosin contractile ring formation as well as proper attachment of the midbody to the cell membrane [106,107]. Many reports describe its role in the development and progression of different cancers, where it may serve as a proliferation and poor prognosis marker [108–110]. It has also been shown that RACGAP1 promotes hematopoietic stem cells' differentiation and inhibits cell growth [111].

It has been shown that *brk1*, *nckap1*, and *wasf2*, which belong to WAVE2 complex and the regulators of small GTPase signaling, namely, *cul3a* and *racgap1*, are critical to cardiac development. A CRISPR KO of *racgap1* in zebrafish heart demonstrates atrial dilation and pericardial edema [112].

It therefore seems that *RACGAP1* could promote the proliferation of cultured cells. In samples from the right atrial wall, it can be clearly observed that the expression drops in the 30th day of culture, while for samples from the right atrial appendage, this tendency is even more significant; in the 15th day of culture, there is already 3-fold downregulation, which would suggest that, with time, possibly the cell proliferation decreases and further implies that further culture optimization is needed.

SNAI2, Snail Family Transcriptional Repressor 2, is an EMT-related transcription factor, promoting a migratory and invasive phenotype. Possibly the best-known target gene of this transcriptional repressor is CDH1, encoding E-cadherin, an epithelial cell adhesion molecule [113]. It is therefore not surprising that *SNAI2* plays a crucial role in developmental processes such as mesoderm formation, or neural crest migration, but also in cancer [114,115]. As for the heart development, Slug is a key for epicardial EndMT, in which a group of epicardial cells migrate to the heart and differentiate into smooth muscle cells and fibroblasts, as well as for endocardial EndMT, which is indispensable for valve development [116,117].

Apart from functions in embryonic development, *SNAI2* regulates adult stem/progenitor cell function—self-renewal, lineage commitment, and apoptosis in hematopoietic, mammary, epidermal, or mesenchymal tissues [118].

It is difficult to assess the role of *SNAI2* in the system presented here as it should consist mainly of adult porcine cardiomyocytes, not endothelial cells, or cells at the embryonic stage. In cultures derived from both heart fragments, the tendencies are relatively similar—gradual downregulation with the culture duration. If it therefore comes to any pro-angiogenic influence, it is downregulated with the culture time.

5. Conclusions

Primary in vitro long-term cultures of porcine cardiomyocytes isolated from two different fragments, the right atrial appendage and right atrial wall, were established and documented. The expression of potential stemness markers (GATA-4, CD44, CD90, and CD105) and CMs' markers (α-MHC and α-actinin) in cultures was characterized using flow cytometry [54]. GATA-4, α-MHC, and α-actinin were expressed in all cultures, at all time points, while CD90 was majorly expressed at later time points [54]. It can therefore be concluded that the isolation and culture of mature cardiomyocytes was successful. CD90 and GATA-4 expression may indicate not only putative stemness, because these proteins can serve as markers of some cardiac cell types or may play roles in the functioning of mature cardiomyocytes. The influence of cell culture conditions on potential stemness properties was also characterized by transcriptomic profiling. In total, 49 genes associated with such properties demonstrated differential expression, with the 4 most deregulated genes being *SFRP2*, *PRRX1*, *RCAGP1*, and *SNAI2*.

It would be interesting to perform all the experiments after fluorescence-activated or magnetic cell sorting with properly chosen markers characteristic for the cells which have been claimed to possess self-renewal and differentiation capacities and compare them to mature cardiomyocytes and previously characterized putative cardiac progenitor cells [119,120]. Conclusions based on the presence or absence of certain stem cell markers alone cannot be treated as a proof of potency [121]. It would be crucial to establish functional assays which would enable the assessment of such features [11–13,120].

Lastly, the low number of cells analyzed by flow cytometry as well as the results of transcriptomic profiling, which seem to point out the presence of sources of variation in the data other than the conditions itself, clearly show that more optimization of the culture conditions is needed.

Supplementary Materials: The following supporting information can be downloaded at https://www.mdpi.com/article/10.3390/genes14061223/s1, Table S1: Symbols, changes in expression, and adjusted *p*-values of the 10 most deregulated genes in the 7th, 15th, and 30th days of cell culture derived from the fragment of the right arial wall.; Table S2: Symbols, changes in expression, and adjusted *p*-values of the 10 most deregulated genes in the 7th, 15th, and 30th days of cell culture derived from the fragment of the right arial appendage.

Author Contributions: Conceptualization, R.B., M.J.N., D.B. and B.K.; methodology, M.J.N., M.K., M.Z. and B.K.; software, K.J. and M.K.; validation, M.J.N., P.M. and M.Z.; formal analysis, P.A. and P.D.; investigation, R.B., K.J. and M.K.; resources, D.B. and P.A.; data curation, D.B. and B.K.; writing—original draft preparation, R.B.; writing—review and editing, M.J.N., P.M. and B.K.; visualization, K.J. and M.K.; supervision, P.D. and B.K.; project administration, B.K.; funding acquisition, P.M. All authors have read and agreed to the published version of the manuscript.

Funding: This research was funded in part by USDA/NIFA multistate project NC 1184.

Institutional Review Board Statement: Not applicable.

Informed Consent Statement: Not applicable.

Data Availability Statement: Not applicable.

Conflicts of Interest: The authors declare no conflict of interest.

References

1. Braunwald, E. The War against Heart Failure: The Lancet Lecture. *Lancet* **2015**, *385*, 812–824. [CrossRef]
2. Yancy, C.W.; Jessup, M.; Bozkurt, B.; Butler, J.; Casey, D.E.; Colvin, M.M.; Drazner, M.H.; Filippatos, G.S.; Fonarow, G.C.; Givertz, M.M.; et al. 2017 ACC/AHA/HFSA Focused Update of the 2013 ACCF/AHA Guideline for the Management of Heart Failure: A Report of the American College of Cardiology/American Heart Association Task Force on Clinical Practice Guidelines and the Heart Failure Society of America. *J. Am. Coll. Cardiol.* **2017**, *70*, e137–e161.

3.   Ziaeian, B.; Fonarow, G.C. Epidemiology and Aetiology of Heart Failure. *Nat. Rev. Cardiol.* **2016**, *13*, 368–378. [CrossRef] [PubMed]
4.   Yancy, C.W.; Jessup, M.; Bozkurt, B.; Butler, J.; Casey, D.E.; Drazner, M.H.; Fonarow, G.C.; Geraci, S.A.; Horwich, T.; Januzzi, J.L.; et al. 2013 ACCF/AHA Guideline for the Management of Heart Failure: A Report of the American College of Cardiology Foundation/American Heart Association Task Force on Practice Guidelines. *Circulation* **2013**, *128*, e147–e239. [CrossRef] [PubMed]
5.   Bergmann, O.; Bhardwaj, R.D.; Bernard, S.; Zdunek, S.; Barnabé-Heide, F.; Walsh, S.; Zupicich, J.; Alkass, K.; Buchholz, B.A.; Druid, H.; et al. Evidence for Cardiomyocyte Renewal in Humans. *Science* **2009**, *324*, 98–102. [CrossRef]
6.   Bergmann, O.; Zdunek, S.; Felker, A.; Salehpour, M.; Alkass, K.; Bernard, S.; Sjostrom, S.L.; Szewczykowska, M.; Jackowska, T.; Dos Remedios, C.; et al. Dynamics of Cell Generation and Turnover in the Human Heart. *Cell* **2015**, *161*, 433–436. [CrossRef]
7.   Hsieh, P.C.H.; Segers, V.F.M.; Davis, M.E.; MacGillivray, C.; Gannon, J.; Molkentin, J.D.; Robbins, J.; Lee, R.T. Evidence from a Genetic Fate-Mapping Study That Stem Cells Refresh Adult Mammalian Cardiomyocytes after Injury. *Nat. Med.* **2007**, *13*, 970–974. [CrossRef]
8.   Hsueh, Y.C.; Wu, J.M.F.; Yu, C.K.; Wu, K.K.; Hsieh, P.C.H. Prostaglandin E2 Promotes Post-Infarction Cardiomyocyte Replenishment by Endogenous Stem Cells. *EMBO Mol. Med.* **2014**, *6*, 496–503. [CrossRef]
9.   Walsh, S.; Pontén, A.; Fleischmann, B.K.; Jovinge, S. Cardiomyocyte Cell Cycle Control and Growth Estimation in Vivo-An Analysis Based on Cardiomyocyte Nuclei. *Cardiovasc. Res.* **2010**, *86*, 365–373. [CrossRef]
10. Malliaras, K.; Zhang, Y.; Seinfeld, J.; Galang, G.; Tseliou, E.; Cheng, K.; Sun, B.; Aminzadeh, M.; Marbán, E. Cardiomyocyte Proliferation and Progenitor Cell Recruitment Underlie Therapeutic Regeneration after Myocardial Infarction in the Adult Mouse Heart. *EMBO Mol. Med.* **2013**, *5*, 191–209. [CrossRef]
11. Le, T.; Chong, J. Cardiac Progenitor Cells for Heart Repair. *Cell Death Discov.* **2016**, *2*, 16052. [CrossRef] [PubMed]
12. Ge, Z.; Lal, S.; Le, T.Y.L.; dos Remedios, C.; Chong, J.J.H. Cardiac Stem Cells: Translation to Human Studies. *Biophys. Rev.* **2015**, *7*, 127–139. [CrossRef] [PubMed]
13. Scalise, M.; Marino, F.; Cianflone, E.; Mancuso, T.; Marotta, P.; Aquila, I.; Torella, M.; Nadal-Ginard, B.; Torella, D. Heterogeneity of Adult Cardiac Stem Cells. *Adv. Exp. Med. Biol.* **2019**, *1169*, 141–178. [PubMed]
14. Ellison, G.M.; Vicinanza, C.; Smith, A.J.; Aquila, I.; Leone, A.; Waring, C.D.; Henning, B.J.; Stirparo, G.G.; Papait, R.; Scarfò, M.; et al. Adult C-Kitpos Cardiac Stem Cells Are Necessary and Sufficient for Functional Cardiac Regeneration and Repair. *Cell* **2013**, *154*, 827–842. [CrossRef] [PubMed]
15. Ellison, G.M.; Torella, D.; Dellegrottaglie, S.; Perez-Martinez, C.; Perez De Prado, A.; Vicinanza, C.; Purushothaman, S.; Galuppo, V.; Iaconetti, C.; Waring, C.D.; et al. Endogenous Cardiac Stem Cell Activation by Insulin-like Growth Factor-1/Hepatocyte Growth Factor Intracoronary Injection Fosters Survival and Regeneration of the Infarcted Pig Heart. *J. Am. Coll. Cardiol.* **2011**, *58*, 977–986. [CrossRef]
16. Fransioli, J.; Bailey, B.; Gude, N.A.; Cottage, C.T.; Muraski, J.A.; Emmanuel, G.; Wu, W.; Alvarez, R.; Rubio, M.; Ottolenghi, S.; et al. Evolution of the C-kit-Positive Cell Response to Pathological Challenge in the Myocardium. *Stem Cells* **2008**, *26*, 1315–1324. [CrossRef]
17. Hensley, M.T.; de Andrade, J.; Keene, B.; Meurs, K.; Tang, J.; Wang, Z.; Caranasos, T.G.; Piedrahita, J.; Li, T.S.; Cheng, K. Cardiac Regenerative Potential of Cardiosphere-Derived Cells from Adult Dog Hearts. *J. Cell Mol. Med.* **2015**, *19*, 1805–1813. [CrossRef] [PubMed]
18. Lee, S.T.; White, A.J.; Matsushita, S.; Malliaras, K.; Steenbergen, C.; Zhang, Y.; Li, T.S.; Terrovitis, J.; Yee, K.; Simsir, S.; et al. Intramyocardial Injection of Autologous Cardiospheres or Cardiosphere-Derived Cells Preserves Function and Minimizes Adverse Ventricular Remodeling in Pigs with Heart Failure Post-Myocardial Infarction. *J. Am. Coll. Cardiol.* **2011**, *57*, 455–465. [CrossRef]
19. Malliaras, K.; Li, T.S.; Luthringer, D.; Terrovitis, J.; Cheng, K.; Chakravarty, T.; Galang, G.; Zhang, Y.; Schoenhoff, F.; Van Eyk, J.; et al. Safety and Efficacy of Allogeneic Cell Therapy in Infarcted Rats Transplanted with Mismatched Cardiosphere-Derived Cells. *Circulation* **2012**, *125*, 100–112. [CrossRef]
20. Messina, E.; De Angelis, L.; Frati, G.; Morrone, S.; Chimenti, S.; Fiordaliso, F.; Salio, M.; Battaglia, M.; Latronico, M.V.G.; Coletta, M.; et al. Isolation and Expansion of Adult Cardiac Stem Cells from Human and Murine Heart. *Circ. Res.* **2004**, *95*, 911–921. [CrossRef]
21. Hierlihy, A.M.; Seale, P.; Lobe, C.G.; Rudnicki, M.A.; Megeney, L.A. The Post-Natal Heart Contains a Myocardial Stem Cell Population. *FEBS Lett.* **2002**, *530*, 239–243. [CrossRef] [PubMed]
22. Pfister, O.; Mouquet, F.; Jain, M.; Summer, R.; Helmes, M.; Fine, A.; Colucci, W.S.; Liao, R. CD31- but Not CD31+ Cardiac Side Population Cells Exhibit Functional Cardiomyogenic Differentiation. *Circ. Res.* **2005**, *97*, 52–61. [CrossRef] [PubMed]
23. Yamahara, K.; Fukushima, S.; Coppen, S.R.; Felkin, L.E.; Varela-Carver, A.; Barton, P.J.R.; Yacoub, M.H.; Suzuki, K. Heterogeneic Nature of Adult Cardiac Side Population Cells. *Biochem. Biophys. Res. Commun.* **2008**, *371*, 615–620. [CrossRef] [PubMed]
24. Matsuura, K.; Nagai, T.; Nishigaki, N.; Oyama, T.; Nishi, J.; Wada, H.; Sano, M.; Toko, H.; Akazawa, H.; Sato, T.; et al. Adult Cardiac Sca-1-Positive Cells Differentiate into Beating Cardiomyocytes. *J. Biol. Chem.* **2004**, *279*, 11384–11391. [CrossRef] [PubMed]

25. Van Vliet, P.; Roccio, M.; Smits, A.M.; Van Oorschot, A.A.M.; Metz, C.H.G.; Van Veen, T.A.B.; Sluijter, J.P.G.; Doevendans, P.A.; Goumans, M.J. Progenitor Cells Isolated from the Human Heart: A Potential Cell Source for Regenerative Therapy. *Neth. Heart J.* **2008**, *16*, 163–169. [CrossRef]
26. Winter, E.M.; Grauss, R.W.; Hogers, B.; Van Tuyn, J.; Van Der Geest, R.; Lie-Venema, H.; Steijn, R.V.; Maas, S.; Deruiter, M.C.; Devries, A.A.F.; et al. Preservation of Left Ventricular Function and Attenuation of Remodeling after Transplantation of Human Epicardium-Derived Cells into the Infarcted Mouse Heart. *Circulation* **2007**, *116*, 917–927. [CrossRef]
27. Zhou, Q.; Brown, J.; Kanarek, A.; Rajagopal, J.; Melton, D.A. In Vivo Reprogramming of Adult Pancreatic Exocrine Cells to β-Cells. *Nature* **2008**, *455*, 627–632. [CrossRef]
28. Chong, J.J.H.; Chandrakanthan, V.; Xaymardan, M.; Asli, N.S.; Li, J.; Ahmed, I.; Heffernan, C.; Menon, M.K.; Scarlett, C.J.; Rashidianfar, A.; et al. Adult Cardiac-Resident MSC-like Stem Cells with a Proepicardial Origin. *Cell Stem Cell* **2011**, *9*, 527–540. [CrossRef]
29. Chong, J.J.H.; Reinecke, H.; Iwata, M.; Torok-Storb, B.; Stempien-Otero, A.; Murry, C.E. Progenitor Cells Identified by PDGFR-α Expression in the Developing and Diseased Human Heart. *Stem Cells Dev.* **2013**, *22*, 1932–1943. [CrossRef]
30. Engleka, K.A.; Manderfield, L.J.; Brust, R.D.; Li, L.; Cohen, A.; Dymecki, S.M.; Epstein, J.A. Islet1 Derivatives in the Heart Are of Both Neural Crest and Second Heart Field Origin. *Circ. Res.* **2012**, *110*, 922–926. [CrossRef] [PubMed]
31. Laugwitz, K.L.; Moretti, A.; Lam, J.; Gruber, P.; Chen, Y.; Woodard, S.; Lin, L.Z.; Cai, C.L.; Lu, M.M.; Reth, M.; et al. Postnatal Isl1+ Cardioblasts Enter Fully Differentiated Cardiomyocyte Lineages. *Nature* **2005**, *433*, 647–653. [CrossRef]
32. Moretti, A.; Caron, L.; Nakano, A.; Lam, J.T.; Bernshausen, A.; Chen, Y.; Qyang, Y.; Bu, L.; Sasaki, M.; Martin-Puig, S.; et al. Multipotent Embryonic Isl1+ Progenitor Cells Lead to Cardiac, Smooth Muscle, and Endothelial Cell Diversification. *Cell* **2006**, *127*, 1151–1165. [CrossRef] [PubMed]
33. Bolli, R.; Chugh, A.R.; D'Amario, D.; Loughran, J.H.; Stoddard, M.F.; Ikram, S.; Beache, G.M.; Wagner, S.G.; Leri, A.; Hosoda, T.; et al. Cardiac Stem Cells in Patients with Ischaemic Cardiomyopathy (SCIPIO): Initial Results of a Randomised Phase 1 Trial. *Lancet* **2011**, *378*, 1847–1857. [CrossRef] [PubMed]
34. Makkar, R.R.; Smith, R.R.; Cheng, K.; Malliaras, K.; Thomson, L.E.J.; Berman, D.; Czer, L.S.C.; Marbán, L.; Mendizabal, A.; Johnston, P.V.; et al. Intracoronary Cardiosphere-Derived Cells for Heart Regeneration after Myocardial Infarction (CADUCEUS): A Prospective, Randomised Phase 1 Trial. *Lancet* **2012**, *379*, 895–904. [CrossRef] [PubMed]
35. Chomczynski, P.; Sacchi, N. Single-Step Method of RNA Isolation by Acid Guanidinium Thiocyanate-Phenol-Chloroform Extraction. *Anal. Biochem.* **1987**, *162*, 156–159. [CrossRef] [PubMed]
36. Chomczynski, P.; Sacchi, N. The Single-Step Method of RNA Isolation by Acid Guanidinium Thiocyanate-Phenol-Chloroform Extraction: Twenty-Something Years On. *Nat. Protoc.* **2006**, *1*, 581–585. [CrossRef]
37. Untergasser, A.; Cutcutache, I.; Koressaar, T.; Ye, J.; Faircloth, B.C.; Remm, M.; Rozen, S.G. Primer3-New Capabilities and Interfaces. *Nucleic Acids Res.* **2012**, *40*, e115. [CrossRef]
38. Koressaar, T.; Remm, M. Enhancements and Modifications of Primer Design Program Primer3. *Bioinformatics* **2007**, *23*, 1289–1291. [CrossRef]
39. Kõressaar, T.; Lepamets, M.; Kaplinski, L.; Raime, K.; Andreson, R.; Remm, M. Primer3-Masker: Integrating Masking of Template Sequence with Primer Design Software. *Bioinformatics* **2018**, *34*, 1937–1938. [CrossRef]
40. Yates, A.D.; Achuthan, P.; Akanni, W.; Allen, J.; Allen, J.; Alvarez-Jarreta, J.; Amode, M.R.; Armean, I.M.; Azov, A.G.; Bennett, R.; et al. Ensembl 2020. *Nucleic Acids Res.* **2020**, *48*, D682–D688. [CrossRef]
41. Livak, K.J.; Schmittgen, T.D. Analysis of Relative Gene Expression Data Using Real-Time Quantitative PCR and the 2-ΔΔCT Method. *Methods* **2001**, *25*, 402–408. [CrossRef] [PubMed]
42. Schmittgen, D.T.; Livak, K.J. Analyzing Real-Time PCR Data by the Comparative CT Method. *Nat. Protoc.* **2008**, *3*, 1101–1108. [CrossRef] [PubMed]
43. R Core Team. *R: A Language and Environment for Statistical Computing*; R Foundation for Statistical Computing: Vienna, Austria, 2020.
44. Huber, W.; Carey, V.J.; Gentleman, R.; Anders, S.; Carlson, M.; Carvalho, B.S.; Bravo, H.C.; Davis, S.; Gatto, L.; Girke, T.; et al. Orchestrating High-Throughput Genomic Analysis with Bioconductor. *Nat. Methods* **2015**, *12*, 115–121. [CrossRef]
45. Huang, D.W.; Sherman, B.T.; Lempicki, R.A. Systematic and Integrative Analysis of Large Gene Lists Using DAVID Bioinformatics Resources. *Nat. Protoc.* **2009**, *4*, 44–57. [CrossRef]
46. Huang, D.W.; Sherman, B.T.; Lempicki, R.A. Bioinformatics Enrichment Tools: Paths toward the Comprehensive Functional Analysis of Large Gene Lists. *Nucleic Acids Res.* **2009**, *37*, 1–13. [CrossRef] [PubMed]
47. Sherman, B.T.; Hao, M.; Qiu, J.; Jiao, X.; Baseler, M.W.; Lane, H.C.; Imamichi, T.; Chang, W. DAVID: A Web Server for Functional Enrichment Analysis and Functional Annotation of Gene Lists (2021 Update). *Nucleic Acids Res.* **2022**, *50*, W216–W221. [CrossRef]
48. Ashburner, M.; Ball, C.A.; Blake, J.A.; Botstein, D.; Butler, H.; Cherry, J.M.; Davis, A.P.; Dolinski, K.; Dwight, S.S.; Eppig, J.T.; et al. Gene Ontology: Tool for the Unification of Biology. *Nat. Genet.* **2000**, *25*, 25–29. [CrossRef]
49. The Gene Ontology Consortium; Acencio, M.; Lægreid, A.; Kuiper, M. The Gene Ontology Resource: 20 Years and Still GOing Strong. *Nucleic Acids Res.* **2019**, *47*, D330–D338.
50. Walter, W.; Sánchez-Cabo, F.; Ricote, M. GOplot: An R Package for Visually Combining Expression Data with Functional Analysis. *Bioinformatics* **2015**, *31*, 2912–2914. [CrossRef]

51. Beltrami, A.P.; Barlucchi, L.; Torella, D.; Baker, M.; Limana, F.; Chimenti, S.; Kasahara, H.; Rota, M.; Musso, E.; Urbanek, K.; et al. Adult Cardiac Stem Cells Are Multipotent and Support Myocardial Regeneration. *Cell* **2003**, *114*, 763–776. [CrossRef]

52. Oh, H.; Bradfute, S.B.; Gallardo, T.D.; Nakamura, T.; Gaussin, V.; Mishina, Y.; Pocius, J.; Michael, L.H.; Behringer, R.R.; Garry, D.J.; et al. Cardiac Progenitor Cells from Adult Myocardium: Homing, Differentiation, and Fusion after Infarction. *Proc. Natl. Acad. Sci. USA* **2003**, *100*, 12313–12318. [CrossRef] [PubMed]

53. Smith, R.R.; Barile, L.; Cho, H.C.; Leppo, M.K.; Hare, J.M.; Messina, E.; Giacomello, A.; Abraham, M.R.; Marbán, E. Regenerative Potential of Cardiosphere-Derived Cells Expanded from Percutaneous Endomyocardial Biopsy Specimens. *Circulation* **2007**, *115*, 896–908. [CrossRef]

54. Nawrocki, M.J.; Jopek, K.; Kaczmarek, M.; Zdun, M.; Mozdziak, P.; Jemielity, M.; Perek, B.; Bukowska, D.; Kempisty, B. Transcriptomic Profile of Genes Regulating the Structural Organization of Porcine Atrial Cardiomyocytes during Primary In Vitro Culture. *Genes* **2022**, *13*, 1205. [CrossRef] [PubMed]

55. Dominici, M.; Le Blanc, K.; Mueller, I.; Slaper-Cortenbach, I.; Marini, F.C.; Krause, D.S.; Deans, R.J.; Keating, A.; Prockop, D.J.; Horwitz, E.M. Minimal Criteria for Defining Multipotent Mesenchymal Stromal Cells. The International Society for Cellular Therapy Position Statement. *Cytotherapy* **2006**, *8*, 315–317. [CrossRef]

56. Weiss, T.S.; Dayoub, R. Thy-1 (CD90)-Positive Hepatic Progenitor Cells, Hepatoctyes, and Non-Parenchymal Liver Cells Isolated from Human Livers. *Methods Mol. Biol.* **2017**, *1506*, 75–89.

57. Nakamura, Y.; Muguruma, Y.; Yahata, T.; Miyatake, H.; Sakai, D.; Mochida, J.; Hotta, T.; Ando, K. Expression of CD90 on Keratinocyte Stem/Progenitor Cells. *Br. J. Dermatl.* **2006**, *154*, 1062–1070. [CrossRef] [PubMed]

58. Yang, J.; Zhan, X.Z.; Malola, J.; Li, Z.Y.; Pawar, J.S.; Zhang, H.T.; Zha, Z.G. The Multiple Roles of Thy-1 in Cell Differentiation and Regeneration. *Differentiation* **2020**, *113*, 38–48. [CrossRef]

59. Pinto, A.R.; Ilinykh, A.; Ivey, M.J.; Kuwabara, J.T.; D'antoni, M.L.; Debuque, R.; Chandran, A.; Wang, L.; Arora, K.; Rosenthal, N.A.; et al. Revisiting Cardiac Cellular Composition. *Circ. Res.* **2016**, *118*, 400–409. [CrossRef]

60. Hudon-David, F.; Bouzeghrane, F.; Couture, P.; Thibault, G. Thy-1 Expression by Cardiac Fibroblasts: Lack of Association with Myofibroblast Contractile Markers. *J. Mol. Cell Cardiol.* **2007**, *42*, 991–1000. [CrossRef]

61. Gambini, E.; Pompilio, G.; Biondi, A.; Alamanni, F.; Capogrossi, M.C.; Agrifoglio, M.; Pesce, M. C-Kit+ Cardiac Progenitors Exhibit Mesenchymal Markers and Preferential Cardiovascular Commitment. *Cardiovasc. Res.* **2011**, *89*, 362–373. [CrossRef] [PubMed]

62. Davis, D.R.; Zhang, Y.; Smith, R.R.; Cheng, K.; Terrovitis, J.; Malliaras, K.; Li, T.S.; White, A.; Makkar, R.; Marbán, E. Validation of the Cardiosphere Method to Culture Cardiac Progenitor Cells from Myocardial Tissue. *PLoS ONE* **2009**, *4*, e7195. [CrossRef] [PubMed]

63. Chan, H.H.L.; Meher Homji, Z.; Gomes, R.S.M.; Sweeney, D.; Thomas, G.N.; Tan, J.J.; Zhang, H.; Perbellini, F.; Stuckey, D.J.; Watt, S.M.; et al. Human Cardiosphere-Derived Cells from Patients with Chronic Ischaemic Heart Disease Can Be Routinely Expanded from Atrial but Not Epicardial Ventricular Biopsies. *J. Cardiovasc. Transl. Res.* **2012**, *5*, 678–687. [CrossRef] [PubMed]

64. van Tuyn, J.; Atsma, D.E.; Winter, E.M.; van der Velde-van Dijke, I.; Pijnappels, D.A.; Bax, N.A.M.; Knaän-Shanzer, S.; Gittenberger-de Groot, A.C.; Poelmann, R.E.; van der Laarse, A.; et al. Epicardial Cells of Human Adults Can Undergo an Epithelial-to-Mesenchymal Transition and Obtain Characteristics of Smooth Muscle Cells In Vitro. *Stem Cells* **2007**, *25*, 271–278. [CrossRef] [PubMed]

65. Cheng, K.; Ibrahim, A.; Hensley, M.T.; Shen, D.; Sun, B.; Middleton, R.; Liu, W.; Smith, R.R.; Marbán, E. Relative Roles of CD90 and C-Kit to the Regenerative Efficacy of Cardiosphere-Derived Cells in Humans and in a Mouse Model of Myocardial Infarction. *J. Am. Heart Assoc.* **2014**, *3*, e001260. [CrossRef]

66. Senbanjo, L.T.; Chellaiah, M.A. CD44: A Multifunctional Cell Surface Adhesion Receptor Is a Regulator of Progression and Metastasis of Cancer Cells. *Front. Cell Dev. Biol.* **2017**, *5*, 18. [CrossRef]

67. Thorne, R.F.; Wang, Y.; Zhang, Y.; Jing, X.; Zhang, X.D.; de Bock, C.E.; Oliveira, C.S. Evaluating Nuclear Translocation of Surface Receptors: Recommendations Arising from Analysis of CD44. *Histochem. Cell Biol.* **2020**, *153*, 77–87. [CrossRef]

68. Bollini, S.; Vieira, J.M.N.; Howard, S.; Dubè, K.N.; Balmer, G.M.; Smart, N.; Riley, P.R. Re-Activated Adult Epicardial Progenitor Cells Are a Heterogeneous Population Molecularly Distinct from Their Embryonic Counterparts. *Stem Cells Dev.* **2014**, *23*, 1719–1730. [CrossRef]

69. Lebrin, F.; Goumans, M.J.; Jonker, L.; Carvalho, R.L.C.; Valdimarsdottir, G.; Thorikay, M.; Mummery, C.; Arthur, H.M.; Ten Dijke, P. Endoglin Promotes Endothelial Cell Proliferation and TGF-β/ALK1 Signal Transduction. *EMBO J.* **2004**, *23*, 4018–4028. [CrossRef]

70. Singh, E.; Phillips, H.M.; Arthur, H.M. Dynamic Changes in Endoglin Expression in the Developing Mouse Heart. *Gene Expr. Patterns* **2021**, *39*, 119165. [CrossRef]

71. Arthur, H.M.; Ure, J.; Smith, A.J.H.; Renforth, G.; Wilson, D.I.; Torsney, E.; Charlton, R.; Parums, D.V.; Jowett, T.; Marchuk, D.A.; et al. Endoglin, an Ancillary TGFβ Receptor, Is Required for Extraembryonic Angiogenesis and Plays a Key Role in Heart Development. *Dev. Biol.* **2000**, *217*, 42–53. [CrossRef]

72. Sunagawa, Y.; Morimoto, T.; Takaya, T.; Kaichi, S.; Wada, H.; Kawamura, T.; Fujita, M.; Shimatsu, A.; Kita, T.; Hasegawa, K. Cyclin-Dependent Kinase-9 Is a Component of the P300/GATA4 Complex Required for Phenylephrine-Induced Hypertrophy in Cardiomyocytes. *J. Biol. Chem.* **2010**, *285*, 9556–9568. [CrossRef]

73. Bisping, E.; Ikeda, S.; Kong, S.W.; Tarnavski, O.; Bodyak, N.; McMullen, J.R.; Rajagopal, S.; Son, J.K.; Ma, Q.; Springer, Z.; et al. Gata4 Is Required for Maintenance of Postnatal Cardiac Function and Protection from Pressure Overload-Induced Heart Failure. *Proc. Natl. Acad. Sci. USA* **2006**, *103*, 14471–14476. [CrossRef] [PubMed]
74. Yamamura, S.; Izumiya, Y.; Araki, S.; Nakamura, T.; Kimura, Y.; Hanatani, S.; Yamada, T.; Ishida, T.; Yamamoto, M.; Onoue, Y.; et al. Cardiomyocyte Sirt (Sirtuin) 7 Ameliorates Stress-Induced Cardiac Hypertrophy by Interacting with and Deacetylating GATA4. *Hypertension* **2020**, *75*, 98–108. [CrossRef] [PubMed]
75. Furtado, M.B.; Costa, M.W.; Pranoto, E.A.; Salimova, E.; Pinto, A.R.; Lam, N.T.; Park, A.; Snider, P.; Chandran, A.; Harvey, R.P.; et al. Cardiogenic Genes Expressed in Cardiac Fibroblasts Contribute to Heart Development and Repair. *Circ. Res.* **2014**, *114*, 1422–1434. [CrossRef] [PubMed]
76. Akazawa, H.; Komuro, I. Roles of Cardiac Transcription Factors in Cardiac Hypertrophy. *Circ. Res.* **2003**, *92*, 1079–1088. [CrossRef] [PubMed]
77. Padmanabhan, A.; Alexanian, M.; Linares-Saldana, R.; González-Terán, B.; Andreoletti, G.; Huang, Y.; Connolly, A.J.; Kim, W.; Hsu, A.; Duan, Q.; et al. BRD4 (Bromodomain-Containing Protein 4) Interacts with GATA4 (GATA Binding Protein 4) to Govern Mitochondrial Homeostasis in Adult Cardiomyocytes. *Circulation* **2020**, *142*, 2338–2355. [CrossRef]
78. Jia, G.; Preussner, J.; Chen, X.; Guenther, S.; Yuan, X.; Yekelchyk, M.; Kuenne, C.; Looso, M.; Zhou, Y.; Teichmann, S.; et al. Single Cell RNA-Seq and ATAC-Seq Analysis of Cardiac Progenitor Cell Transition States and Lineage Settlement. *Nat. Commun.* **2018**, *9*, 4877. [CrossRef]
79. Masino, A.M.; Gallardo, T.D.; Wilcox, C.A.; Olson, E.N.; Williams, R.S.; Garry, D.J. Transcriptional Regulation of Cardiac Progenitor Cell Populations. *Circ. Res.* **2004**, *95*, 389–397. [CrossRef]
80. de Soysa, T.Y.; Ranade, S.S.; Okawa, S.; Ravichandran, S.; Huang, Y.; Salunga, H.T.; Schricker, A.; del Sol, A.; Gifford, C.A.; Srivastava, D. Single-Cell Analysis of Cardiogenesis Reveals Basis for Organ-Level Developmental Defects. *Nature* **2019**, *572*, 120–124. [CrossRef]
81. Zhang, Y.; Zhong, J.F.; Qiu, H.; MacLellan, W.R.; Marbán, E.; Wang, C. Epigenomic Reprogramming of Adult Cardiomyocyte-Derived Cardiac Progenitor Cells. *Sci. Rep.* **2015**, *5*, 17686. [CrossRef]
82. Zhang, Y.; Li, T.S.; Lee, S.T.; Wawrowsky, K.A.; Cheng, K.; Galang, G.; Malliaras, K.; Abraham, M.R.; Wang, C.; Marbán, E. Dedifferentiation and Proliferation of Mammalian Cardiomyocytes. *PLoS ONE* **2010**, *5*, e12559. [CrossRef] [PubMed]
83. Dey, D.; Han, L.; Bauer, M.; Sanada, F.; Oikonomopoulos, A.; Hosoda, T.; Unno, K.; De Almeida, P.; Leri, A.; Wu, J.C. Dissecting the Molecular Relationship among Various Cardiogenic Progenitor Cells. *Circ. Res.* **2013**, *112*, 1253–1262. [CrossRef] [PubMed]
84. Chien, K.R.; Frisén, J.; Fritsche-Danielson, R.; Melton, D.A.; Murry, C.E.; Weissman, I.L. Regenerating the Field of Cardiovascular Cell Therapy. *Nat. Biotechnol.* **2019**, *37*, 232–237. [CrossRef]
85. Orlic, D.; Kajstura, J.; Chimenti, S.; Bodine, D.M.; Leri, A.; Anversa, P. Bone Marrow Stem Cells Regenerate Infarcted Myocardium. *Proc. Pediatr. Transplant.* **2003**, *7*, 86–88. [CrossRef]
86. Vagnozzi, R.J.; Sargent, M.A.; Lin, S.C.J.; Palpant, N.J.; Murry, C.E.; Molkentin, J.D. Genetic Lineage Tracing of Sca-1+ Cells Reveals Endothelial but Not Myogenic Contribution to the Murine Heart. *Circulation* **2018**, *138*, 2931–2939. [CrossRef]
87. Müller, T.; Boileau, E.; Talyan, S.; Kehr, D.; Varadi, K.; Busch, M.; Most, P.; Krijgsveld, J.; Dieterich, C. Updated and Enhanced Pig Cardiac Transcriptome Based on Long-Read RNA Sequencing and Proteomics. *J. Mol. Cell Cardiol.* **2021**, *150*, 23–31. [CrossRef]
88. Ou, Q.; Jacobson, Z.; Abouleisa, R.R.E.; Tang, X.L.; Hindi, S.M.; Kumar, A.; Ivey, K.N.; Giridharan, G.; El-Baz, A.; Brittian, K.; et al. Physiological Biomimetic Culture System for Pig and Human Heart Slices. *Circ. Res.* **2019**, *125*, 628–642. [CrossRef] [PubMed]
89. Mathiyalagan, P.; Adamiak, M.; Mayourian, J.; Sassi, Y.; Liang, Y.; Agarwal, N.; Jha, D.; Zhang, S.; Kohlbrenner, E.; Chepurko, E.; et al. FTO-Dependent N6-Methyladenosine Regulates Cardiac Function during Remodeling and Repair. *Circulation* **2019**, *139*, 518–532. [CrossRef]
90. Tu, L.N.; Timms, A.E.; Kibiryeva, N.; Bittel, D.; Pastuszko, A.; Nigam, V.; Pastuszko, P. Transcriptome Profiling Reveals Activation of Inflammation and Apoptosis in the Neonatal Striatum after Deep Hypothermic Circulatory Arrest. *J. Thorac. Cardiovasc. Surg.* **2019**, *158*, 882–890.e4. [CrossRef]
91. Wu, Y.; Liu, X.; Zheng, H.; Zhu, H.; Mai, W.; Huang, X.; Huang, Y. Multiple Roles of SFRP2 in Cardiac Development and Cardiovascular Disease. *Int. J. Biol. Sci.* **2020**, *16*, 730–738. [CrossRef]
92. Tian, Y.; Cohen, E.D.; Morrisey, E.E. The Importance of Wnt Signaling in Cardiovascular Development. *Pediatr. Cardiol.* **2010**, *31*, 342–348. [CrossRef]
93. Hsueh, Y.C.; Hodgkinson, C.P.; Gomez, J.A. The Role of Sfrp and DKK Proteins in Cardiomyocyte Development. *Physiol. Rep.* **2021**, *9*, e14678. [CrossRef]
94. He, W.; Zhang, L.; Ni, A.; Zhang, Z.; Mirotsou, M.; Mao, L.; Pratt, R.E.; Dzau, V.J. Exogenously Administered Secreted Frizzled Related Protein 2 (Sfrp2) Reduces Fibrosis and Improves Cardiac Function in a Rat Model of Myocardial Infarction. *Proc. Natl. Acad. Sci. USA* **2010**, *107*, 21110–21115. [CrossRef] [PubMed]
95. Courtwright, A.; Siamakpour-Reihani, S.; Arbiser, J.L.; Banet, N.; Hilliard, E.; Fried, L.; Livasy, C.; Ketelsen, D.; Nepal, D.B.; Perou, C.M.; et al. Secreted Frizzle-Related Protein 2 Stimulates Angiogenesis via a Calcineurin/NFAT Signaling Pathway. *Cancer Res.* **2009**, *69*, 4621–4628. [CrossRef] [PubMed]
96. Schmeckpeper, J.; Verma, A.; Yin, L.; Beigi, F.; Zhang, L.; Payne, A.; Zhang, Z.; Pratt, R.E.; Dzau, V.J.; Mirotsou, M. Inhibition of Wnt6 by Sfrp2 Regulates Adult Cardiac Progenitor Cell Differentiation by Differential Modulation of Wnt Pathways. *J. Mol. Cell Cardiol.* **2015**, *85*, 215–225. [CrossRef] [PubMed]

97. Hodgkinson, C.P.; Gomez, J.A.; Baksh, S.S.; Payne, A.; Schmeckpeper, J.; Pratt, R.E.; Dzau, V.J. Insights from Molecular Signature of in Vivo Cardiac C-Kit(+) Cells Following Cardiac Injury and β-Catenin Inhibition. *J. Mol. Cell Cardiol.* **2018**, *123*, 64–74. [CrossRef]

98. Gomez, J.A.; Payne, A.; Pratt, R.E.; Hodgkinson, C.P.; Dzau, V.J. A Role for Sfrp2 in Cardiomyogenesis in Vivo. *Proc. Natl. Acad. Sci. USA* **2021**, *118*, e2103676118. [CrossRef]

99. Deb, A.; Davis, B.H.; Guo, J.; Ni, A.; Huang, J.; Zhang, Z.; Mu, H.; Dzau, V.J. SFRP2 Regulates Cardiomyogenic Differentiation by Inhibiting a Positive Transcriptional Autofeedback Loop of Wnt3a. *Stem Cells* **2008**, *26*, 35–44. [CrossRef]

100. Higuchi, M.; Yoshida, S.; Ueharu, H.; Chen, M.; Kato, T.; Kato, Y. PRRX1 and PRRX2 Distinctively Participate in Pituitary Organogenesis and a Cell-Supply System. *Cell Tissue Res.* **2014**, *357*, 323–335. [CrossRef]

101. Leussink, B.; Brouwer, A.; El Khattabi, M.; Poelmann, R.E.; Gittenberger-de Groot, A.C.; Meijlink, F. Expression Patterns of the Paired-Related Homeobox Genes MHox/Prx1 and S8/Prx2 Suggest Roles in Development of the Heart and the Forebrain. *Mech. Dev.* **1995**, *52*, 51–64. [CrossRef]

102. Bergwerff, M.; Gittenberger-De Groot, A.C.; Deruiter, M.C.; Van Iperen, L.; Meijlink, F.; Poelmann, R.E. Patterns of Paired-Related Homeobox Genes PRX1 and PRX2 Suggest Involvement in Matrix Modulation in the Developing Chick Vascular System. *Dev. Dyn.* **1998**, *213*, 59–70. [CrossRef]

103. de Bakker, D.E.M.; Bouwman, M.; Dronkers, E.; Simões, F.C.; Riley, P.R.; Goumans, M.J.; Smits, A.M.; Bakkers, J. Prrx1b Restricts Fibrosis and Promotes Nrg1-Dependent Cardiomyocyte Proliferation during Zebrafish Heart Regeneration. *Development* **2021**, *148*, dev198937. [CrossRef] [PubMed]

104. Guo, X.J.; Qiu, X.B.; Wang, J.; Guo, Y.H.; Yang, C.X.; Li, L.; Gao, R.F.; Ke, Z.P.; Di, R.M.; Sun, Y.M.; et al. Prrx1 Loss-of-Function Mutations Underlying Familial Atrial Fibrillation. *J. Am. Heart Assoc.* **2021**, *10*, e023517. [CrossRef] [PubMed]

105. Bosada, F.M.; Rivaud, M.R.; Uhm, J.S.; Verheule, S.; Van Duijvenboden, K.; Verkerk, A.O.; Christoffels, V.M.; Boukens, B.J. A Variant Noncoding Region Regulates Prrx1 and Predisposes to Atrial Arrhythmias. *Circ. Res.* **2021**, *129*, 420–434. [CrossRef]

106. Hirose, K.; Kawashima, T.; Iwamoto, I.; Nosaka, T.; Kitamura, T. MgcRacGAP Is Involved in Cytokinesis through Associating with Mitotic Spindle and Midbody. *J. Biol. Chem.* **2001**, *276*, 5821–5828. [CrossRef]

107. Zhao, W.M.; Fang, G. MgcRacGAP Controls the Assembly of the Contractile Ring and the Initiation of Cytokinesis. *Proc. Natl. Acad. Sci. USA* **2005**, *102*, 13158–13163. [CrossRef]

108. Yang, X.M.; Cao, X.Y.; He, P.; Li, J.; Feng, M.X.; Zhang, Y.L.; Zhang, X.L.; Wang, Y.H.; Yang, Q.; Zhu, L.; et al. Overexpression of Rac GTPase Activating Protein 1 Contributes to Proliferation of Cancer Cells by Reducing Hippo Signaling to Promote Cytokinesis. *Gastroenterology* **2018**, *155*, 1233–1249.e22. [CrossRef]

109. Milde-Langosch, K.; Karn, T.; Müller, V.; Witzel, I.; Rody, A.; Schmidt, M.; Wirtz, R.M. Validity of the Proliferation Markers Ki67, TOP2A, and RacGAP1 in Molecular Subgroups of Breast Cancer. *Breast Cancer Res. Treat* **2013**, *137*, 57–67. [CrossRef]

110. Imaoka, H.; Toiyama, Y.; Saigusa, S.; Kawamura, M.; Kawamoto, A.; Okugawa, Y.; Hiro, J.; Tanaka, K.; Inoue, Y.; Mohri, Y.; et al. RacGAP1 Expression, Increasing Tumor Malignant Potential, as a Predictive Biomarker for Lymph Node Metastasis and Poor Prognosis in Colorectal Cancer. *Carcinogenesis* **2015**, *36*, 346–354. [CrossRef]

111. Kawashima, T.; Hirose, K.; Satoh, T.; Kaneko, A.; Ikeda, Y.; Kaziro, Y.; Nosaka, T.; Kitamura, T. MgcRacGAP Is Involved in the Control of Growth and Differentiation of Hematopoietic Cells. *Blood* **2000**, *96*, 2116–2124. [CrossRef]

112. Edwards, J.J.; Rouillard, A.D.; Fernandez, N.F.; Wang, Z.; Lachmann, A.; Shankaran, S.S.; Bisgrove, B.W.; Demarest, B.; Turan, N.; Srivastava, D.; et al. Systems Analysis Implicates WAVE2 Complex in the Pathogenesis of Developmental Left-Sided Obstructive Heart Defects. *JACC Basic Transl. Sci.* **2020**, *5*, 376–386. [CrossRef]

113. Hajra, K.M.; Chen, D.Y.S.; Fearon, E.R. The SLUG Zinc-Finger Protein Represses E-Cadherin in Breast Cancer. *Cancer Res.* **2002**, *62*, 1613–1618.

114. Nieto, M.A.; Sargent, M.G.; Wilkinson, D.G.; Cooke, J. Control of Cell Behavior during Vertebrate Development by Slug, a Zinc Finger Gene. *Science* **1994**, *264*, 835–839. [CrossRef]

115. Nieto, M.A. The Snail Superfamily of Zinc-Finger Transcription Factors. *Nat. Rev Mol. Cell Biol.* **2002**, *3*, 155–166. [CrossRef] [PubMed]

116. Takeichi, M.; Nimura, K.; Mori, M.; Nakagami, H.; Kaneda, Y. The Transcription Factors Tbx18 and Wt1 Control the Epicardial Epithelial-Mesenchymal Transition through Bi-Directional Regulation of Slug in Murine Primary Epicardial Cells. *PLoS ONE* **2013**, *8*, e57829. [CrossRef] [PubMed]

117. Liang, X.; Wu, S.; Geng, Z.; Liu, L.; Zhang, S.; Wang, S.; Zhang, Y.; Huang, Y.; Zhang, B. LARP7 Suppresses Endothelial-to-Mesenchymal Transition by Coupling with TRIM28. *Circ. Res.* **2021**, *129*, 843–856. [CrossRef]

118. Zhou, W.; Gross, K.M.; Kuperwasser, C. Molecular Regulation of Snai2 in Development and Disease. *J. Cell Sci.* **2019**, *132*, jcs235127. [CrossRef]

119. Collantes, M.; Pelacho, B.; García-Velloso, M.J.; Gavira, J.J.; Abizanda, G.; Palacios, I.; Rodriguez-Borlado, L.; Álvarez, V.; Prieto, E.; Ecay, M.; et al. Non-Invasive in Vivo Imaging of Cardiac Stem/Progenitor Cell Biodistribution and Retention after Intracoronary and Intramyocardial Delivery in a Swine Model of Chronic Ischemia Reperfusion Injury. *J. Transl. Med.* **2017**, *15*, 56. [CrossRef] [PubMed]

120. Smith, A.J.; Lewis, F.C.; Aquila, I.; Waring, C.D.; Nocera, A.; Agosti, V.; Nadal-Ginard, B.; Torella, D.; Ellison, G.M. Isolation and Characterization of Resident Endogenous C-Kit+ Cardiac Stem Cells from the Adult Mouse and Rat Heart. *Nat. Protoc.* **2014**, *9*, 1662–1681. [CrossRef]
121. Dulak, J.; Szade, K.; Szade, A.; Nowak, W.; Józkowicz, A. Adult Stem Cells: Hopes and Hypes of Regenerative Medicine. *Acta Biochim. Pol.* **2015**, *62*, 329–337. [CrossRef]

Communication

Droplet Digital PCR Quantification of Selected Intracellular and Extracellular microRNAs Reveals Changes in Their Expression Pattern during Porcine In Vitro Adipogenesis

Adrianna Bilinska, Marcin Pszczola , Monika Stachowiak , Joanna Stachecka, Franciszek Garbacz, Mehmet Onur Aksoy and Izabela Szczerbal *

Department of Genetics and Animal Breeding, Poznan University of Life Sciences, 60-637 Poznan, Poland
* Correspondence: izabela.szczerbal@up.poznan.pl

Abstract: Extracellular miRNAs have attracted considerable interest because of their role in intercellular communication, as well as because of their potential use as diagnostic and prognostic biomarkers for many diseases. It has been shown that miRNAs secreted by adipose tissue can contribute to the pathophysiology of obesity. Detailed knowledge of the expression of intracellular and extracellular microRNAs in adipocytes is thus urgently required. The system of in vitro differentiation of mesenchymal stem cells (MSCs) into adipocytes offers a good model for such an analysis. The aim of this study was to quantify eight intracellular and extracellular miRNAs (miR-21a, miR-26b, miR-30a, miR-92a, miR-146a, miR-148a, miR-199, and miR-383a) during porcine in vitro adipogenesis using droplet digital PCR (ddPCR), a highly sensitive method. It was found that only some miRNAs associated with the inflammatory process (miR-21a, miR-92a) were highly expressed in differentiated adipocytes and were also secreted by cells. All miRNAs associated with adipocyte differentiation were highly abundant in both the studied cells and in the cell culture medium. Those miRNAs showed a characteristic expression profile with upregulation during differentiation.

Keywords: adipocytes; ddPCR; ECmiRNAs; mesenchymal stem cells; pig

Citation: Bilinska, A.; Pszczola, M.; Stachowiak, M.; Stachecka, J.; Garbacz, F.; Aksoy, M.O.; Szczerbal, I. Droplet Digital PCR Quantification of Selected Intracellular and Extracellular microRNAs Reveals Changes in Their Expression Pattern during Porcine In Vitro Adipogenesis. *Genes* **2023**, *14*, 683. https://doi.org/10.3390/genes14030683

Academic Editors: Katarzyna Piórkowska, Katarzyna Ropka-Molik and Alexander S. Graphodatsky

Received: 15 December 2022
Revised: 27 January 2023
Accepted: 8 March 2023
Published: 9 March 2023

1. Introduction

MicroRNAs (miRNAs) are a well-known class of small, noncoding RNAs that regulate post-transcriptional gene expression through mRNA destabilization or inhibition of translation [1]. To date, over 2500 miRNAs have been discovered in the human genome, and it is estimated that they regulate over 60% of protein-coding genes [2]. miRNAs thus play an essential role in all biological processes, including cell differentiation and development [3]. Changes in miRNA expression have been reported in altered physiological conditions and various diseases, so these molecules have been treated as promising therapeutic targets. miRNA-based therapies involve correcting altered miRNA expression levels using mimics or inhibitors [4]. Moreover, miRNAs can be used as biomarkers of pathophysiological conditions [5]. In particular, extracellular miRNAs (ECmiRNAs) can serve as good diagnostic markers due to their stability and ease of sample collection. ECmiRNAs have been detected in cell-free conditions, including cell culture media and biological fluids, such as serum, plasma, saliva, tears, urine, breast milk, etc. [6].

The role of miRNAs has been extensively studied in the context of the development of obesity. It has been shown that miRNAs are involved in the control of a range of processes, including adipogenesis, insulin resistance, and inflammation in adipose tissue [7].

Dysregulation of many miRNAs has been identified in the adipose tissue of obese individuals [8–10]. The presence of adipocyte-related miRNAs in adipocyte-derived microvesicles indicates their involvement in intercellular communication in both paracrine and endocrine manners [10–12]. Studies of miRNA in adipocyte tissue have also been conducted on the domestic pig (*Sus scrofa*), an important animal model for human obesity and also a major livestock species [13]. There are a number of reports on the functioning of individual miRNAs during the formation of fat tissue in the pig (summarized by Song et al. [14]). High-throughput miRNA profiling of porcine adipocyte tissue has also allowed the detection of a complex microRNA–mRNA regulatory network related to fat deposition in pigs [15–18]. A recent study of the identification of miRNAs in porcine adipose-derived and muscle-derived exosomes showed some miRNAs to be involved in skeletal muscle–adipose crosstalk [19].

Most of the research on porcine miRNAs has been carried out on adipose tissues, while studies on in vitro models of adipogenesis are scarce [20]. Due to the heterogeneous nature of adipose tissue—which is composed of several cell types, including adipocytes, preadipocytes, stem cells, endothelial cells, and various blood cells [21]—cultured adipocytes represent a good system for studying molecular events that occur during adipogenesis, including the secretion of miRNA by adipocytes [22]. The aim of this study was thus to quantify eight miRNAs (miR-21a, miR-26b, miR-30a, miR-92a, miR-146a, miR-148a, miR-199a, and miR-383) during porcine in vitro differentiation of mesenchymal stem cells (MSCs) into adipocytes. These miRNAs were selected on the basis of their role in differentiation and inflammation processes (Table 1). The expression of intracellular and extracellular microRNAs was evaluated using droplet digital PCR (ddPCR), a highly sensitive method.

Table 1. Characteristics of the analyzed miRNAs.

microRNA	Function	Reference
miR-21a	modulates inflammation, regulates adipogenic differentiation, and is upregulated in obesity	[23–25]
miR-26b	mediates the multiple differentiation of MSCs and promotes adipocyte differentiation	[26,27]
miR-30a	accelerates adipogenesis and promotes fatty acid and glucose metabolism in adipocytes	[28,29]
miR-92a	controls inflammatory response and inhibits adipose browning	[30,31]
miR-146a	plays a role in inflammatory process in various disorders and is upregulated in obesity	[32,33]
miR-148a	regulates MSC differentiation into adipocytes, a biomarker of obesity	[34,35]
miR-199a	regulates adipocyte differentiation and fatty acid composition during adipogenesis	[36,37]
miR-383	its expression correlates with various inflammatory diseases	[38]

2. Materials and Methods

2.1. Mesenchymal Stem Cell Culture

Mesenchymal stem cells were derived from the adipose tissue (AD-MSCs) of a three-month-old female Polish Large White pig. Tissue sample collection was approved by the Local Ethical Commission for Experiments on Animals at Poznan University of Life Sciences, Poznan, Poland (approval no. 57/2012). Following Stachecka et al. [39], the AD-MSCs were cultured in Advanced DMEM (Gibco, Life Technologies, Grand Island, NY, USA) supplemented with 10% FBS (*v/v*) (Sigma-Aldrich, St. Louis, MO, USA), 5 ng/mL FGF-2 (Promo-Cell GmbH, Heidelberg Germany), 2 mM L-glutamine (Gibco), 1 mM 2-mercaptoethanol (Sigma-Aldrich), 1 × antibiotic antimycotic solution (Sigma-Aldrich), and 1 × MEM NEAA (Gibco) at 37 °C in 5% CO_2. To avoid the possible influence of FBS-derived miRNAs on obtained results, the same part of filtered FBS was used during the whole cell culture experiment. The AD-MSCs were propagated by passaging using standard cell culture procedures, and their stemness was confirmed by staining for positive (CD44, CD90, CD105) and negative (CD45) markers (Abcam, Cambridge, UK).

2.2. Adipogenic Differentiation

Adipogenesis was induced by culturing early-passage MSCs in an adipogenic differentiation medium composed of Advanced DMEM (Gibco), 10% FBS (Sigma-Aldrich), 1 × antibiotic antimycotic solution (Sigma-Aldrich), 1 × MEM NEAA (Gibco), 5 ng/mL FGF-2 (PromoCell GmbH), 1 × linoleic acid albumin, 1 × ITS, 1 µm dexamethasone (Sigma-Aldrich), 100 µm indomethacin (Sigma-Aldrich), and 50 mM IBMX (Sigma-Aldrich). The cells were cultured for ten days. Adipogenic differentiation was monitored using visual examination of lipid droplet formation under a phase-contrast microscope (Nikon TS100 Eclipse, Melville, NY, USA) and BODIPY staining. Cells were fixed with 4% paraformaldehyde in PBS (*w/v*) for ten minutes at room temperature and washed thrice with PBS. The cells were then incubated with BODIPY 493/503 (Thermo Fisher, Waltham, MA, USA) in PBS (3 µg/mL) and washed thrice in PBS. The nuclei were counterstained with DAPI in Vectashield medium (Vector Laboratories, Newark, CA, USA) and examined under a fluorescence microscope (Nikon E600 Eclipse, Melville, NY, USA). Each measurement was performed in triplicate.

2.3. RNA Extraction from Cells and Culture Medium

Total RNA extraction from cells (approximately 2×10^6 in number) and the cell culture medium (200 µL) was performed on days 0, 2, 4, 6, 8, and 10 of adipogenesis using the miRNeasy Micro Kit (Qiagen, Hilden, Germany), following the manufacturer's protocol. The RNA samples isolated from cell culture medium were enriched in the fraction of miRNAs, both exosomal and non-exosomal ECmiRNAs. All samples were analyzed in duplicate. The RNA concentrations and quality were assessed using a NanoDrop 2000 spectrophotometer (Thermo Scientific, Wilmington, DE, USA) and Qubit RNA HS Assay Kit (Thermo Fisher Scientific) on a Qubit 2.0 Fluorometer (Thermo Fisher Scientific).

2.4. Real-Time PCR

One microgram of RNA was reversely transcribed using a Transcriptor High Fidelity cDNA Synthesis kit (Roche Diagnostic, Mannheim, Germany). Primer sets for quantitative real-time PCR for selected protein-coding marker and reference genes (Table S1) were designed using the PRIMER 3 software (http://simgene.com/Primer3 (accessed on 12 May 2022)). The relative transcript levels were assessed using a LightCycler 480 SYBR Green I Master kit (Roche Diagnostic) with a LightCycler 480 II (Roche Life Science). All samples were analyzed in triplicate. Standard curves were designed as tenfold dilutions of the PCR products. Relative transcript levels of the studied genes were calculated after normalization with the transcript level of a reference gene, ribosomal protein L27 (*RPL27*), which has shown stability during adipogenic differentiation [40,41].

2.5. miRNA-Specific Reverse Transcription

Reverse transcription was performed with 10 ng of total RNA using a TaqMan MicroRNA Reverse Transcription Kit (Applied Biosystems, Foster City, CA, USA). Reverse transcription reactions were conducted with the use of an RT primer specific to each tested miRNA. The following TaqMan MicroRNA Assays (Applied Biosystems) were employed: miR-21a-5p, (Assay ID: 000397), miR-26b-5p (Assay ID: 000406), miR-30a-5p (Assay ID: 000417), miR-92a-3p (Assay ID: 000431), miR-146a (Assay ID: 005896), miR-148a-3p (Assay ID: 000470), miR-199a-3p (Assay ID: 002304), and miR-383-5p (Assay ID: 000573). RNU6b (Assay ID: 001093) was used as the reference for normalizing the ddPCR results [42]. The reverse transcription reactions were performed following the manufacturers' recommendations.

2.6. Droplet Digital PCR (ddPCR)

miRNA quantification was performed using droplet digital PCR (ddPCR). All samples were analyzed in duplicate. Each PCR reaction consisted of 1 μL cDNA, 11 μL of 2 × ddPCR SuperMix for Probes (Bio-Rad, Hercules, CA, USA), 9 μL of H_2O, and 1 μL of TaqMan primers and probe from the corresponding TaqMan MicroRNA Assay (Applied Biosystems). The reaction mixtures were divided into approximately 20,000 droplets using a QX200 droplet generator (Bio-Rad) followed by PCR performed on a T100 Thermal Cycler (Bio-Rad) using the following conditions (ramp rate of 2 °C/s): initial denaturation at 95 °C for 10 min, 40 cycles at 94 °C for 30s, followed by 60 °C for 1 min and denaturation at 98 °C for 10 min. A QX200 droplet reader (Bio-Rad) was used to detect fluorescence, and the results were analyzed using QuantaSoft software (Bio-Rad). The fraction of positive droplets was quantified using the Poisson distribution.

Since cell culture media may carry miRNAs derived from supplements such as fetal bovine serum (FBS) [43], an experiment on the expression level of the investigated miRNAs in the pure cell culture medium, supplemented with 10% of FBS, was performed. Expression of miR-92a, miR-146a, and miR-26b was not observed, while expression of miR-21a, miR-383, miR-30a, miR-148a, and miR-199a was on very low level (Table S2), which was about 1% of the average expression level of extracellular miRNAs (Table S6). Thus, an additional normalization step was abandoned.

2.7. Statistical Analysis

Differences between expression levels were assessed separately for each miRNA and for each medium. To give the analyzed variables a normal distribution, the expression levels were transformed by taking the natural logarithm of the original values. The following model was then used to assess the differences between the expression levels on each day:

$$\log (Exp)_{ij} = \mu + DAY_j + sampleID_i + error_{ij},$$

where $\log (Exp)$ is the natural logarithm of the expression level recorded on the *j*th *DAY* for the *i*th *sampleID*. *DAY* was a categorical variable with six levels (0, 2, 4, 6, 8, 10). The *sampleID* and *error* were random terms. The *sampleID* was treated as random term to account for repeated observations of the sample on following days. The analyses were performed using the R environment [44]. The effects of the model were estimated using the lme4 package [45] and the significance of the differences between days was assessed using the lmerTest [46] and emmeans packages [47], making use of Satterthwaite's method [48] for approximating degrees of freedom. The *p*-values for comparing expression levels on particular days were adjusted for multiple comparisons using Tukey's method for comparing six estimates.

To assess whether there was a relationship between the expression level in the medium and cells, the previously used model was updated to include the log-transformed expression in the medium $\log (Exp_{medium})$. The following model was thus used:

$$\log (Exp)_{ij} = \mu + \log (Exp_{medium})_{ij} + DAY_j + sampleID_i + error_{ij}.$$

The regression line was obtained by applying the locally weighted scatterplot smoothing method available from the ggplot2 package [49].

3. Results

Eight miRNAs associated with inflammatory processes (miR-21a, miR-92a, miR-146a, miR-383) and adipocyte differentiation processes (miR-26b, miR-30a, miR-148a, miR-199a) were included in this study (Table 1). The abundances of these miRNAs were determined in cells and in cell culture medium over ten days of adipogenic differentiation (Figure 1).

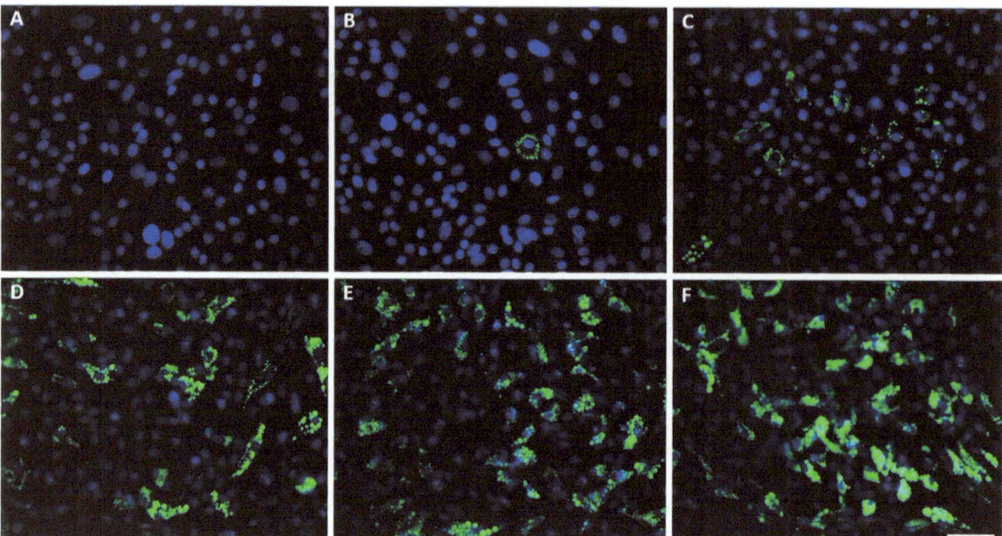

Figure 1. The adipocyte differentiation experiment. Mesenchymal stem cells (MSCs, day 0 of differentiation) were treated with adipogenic hormonal inducers and were cultured for ten days. Cells and media were collected for total RNA isolation on days 0 (**A**), 2 (**B**), 4 (**C**), 6 (**D**), 8 (**E**), and 10 (**F**). Representative images were taken after staining the lipid droplets with BODIPY 493/503 (green); the nuclei were counterstained with DAPI (blue). Scale bar: 50 μm.

The differentiation process was monitored by evaluating the accumulation of lipid droplets using BODIPY staining (Figures 1 and 2A, Table S3). On day 4, individual cells with lipid droplets were seen, while lipid accumulation was highly abundant from day 6. Adipocyte differentiation was also confirmed by the upregulation of expression of three marker genes: *CEBPA*, *FABP4*, and *PPARG* (Figure 2B, Table S3).

The expression of all the miRNAs was successfully detected with the ddPCR method (Figure 3).

It was found that, of the miRNAs associated with the inflammatory process, miR-21a showed the highest expression in differentiated adipocytes and was also highly secreted by these cells (Figure 4A,B; Tables S4 and S5). Both intracellular and extracellular miR-21a levels were upregulated during adipogenesis. miR-92a was also highly expressed by adipocytes, reaching its highest level on day 10 of differentiation. The abundance of extracellular miR-92a initially decreased on days 2–4, returned to its original level after day 6, and then decreased (Figure 4C,D; Tables S4 and S5). The expression of miR-146a in the studied differentiation system was quite low (Figure 4E,F; Tables S4 and S5). Cellular miR-146a was upregulated during adipogenesis, but was not secreted by the differentiated cells. The lowest expression level was found for miR-383, and this was comparable in the cells and in the cell culture medium (Figure 4G,H; Tables S4 and S5).

In terms of miRNAs associated with adipogenesis, all the molecules we examined here were highly expressed during differentiation (Figure 5; Tables S4 and S5). The highest expression in cells was found for miR-26b, next to miR-199a, miR-148a, and miR-30a. Of these, miR-26b, miR-148a, and miR-30a had the highest expression levels at the end of differentiation (day 10), while for miR-199a this occurred on day 4 of adipogenesis. All extracellular miRNAs had similar expression profiles, reaching the highest level on day 6 of differentiation. miR-199a, miR-30a, and miR-148a were secreted at comparable levels, while the amount of miR-26b was lower.

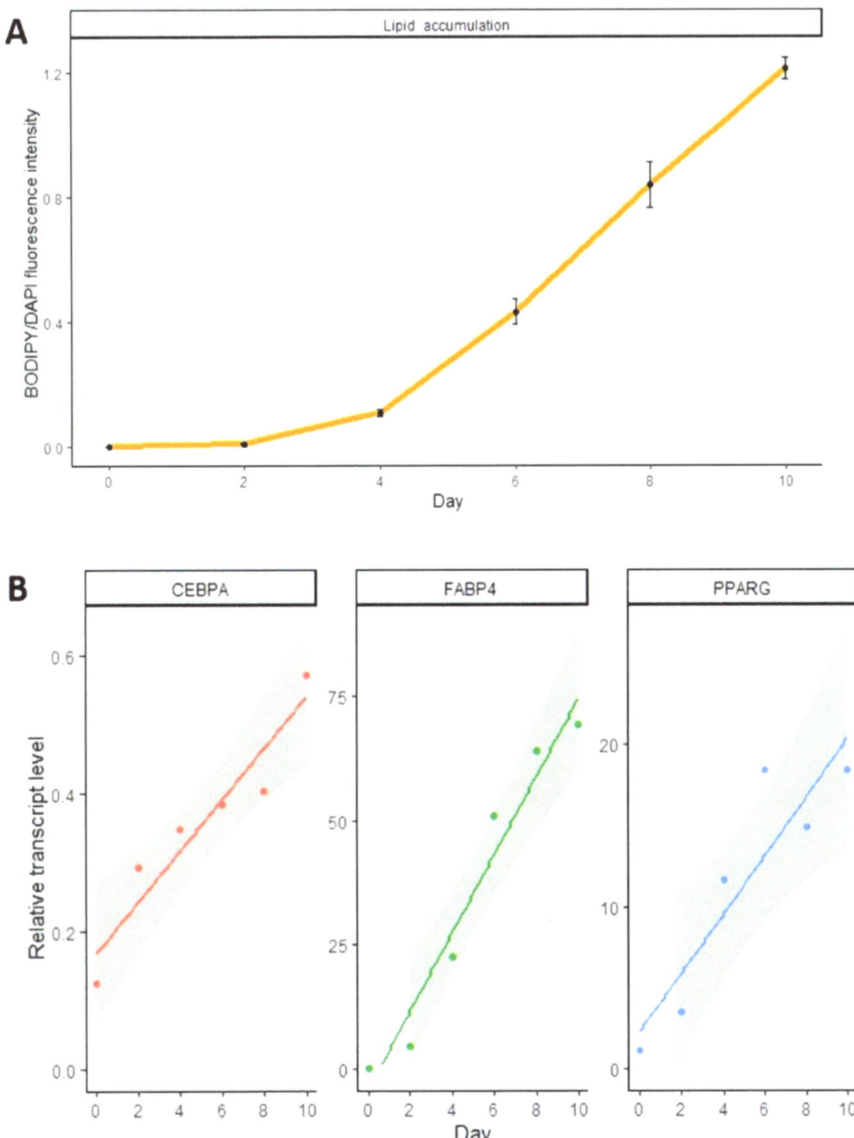

Figure 2. Monitoring of adipocyte differentiation. (**A**) Lipid droplet accumulation was quantified by measuring BODIPY 493/503/DAPI fluorescent intensity. Error bars show SDs. The error bars represent standard deviations. (**B**) Measurements of the relative transcript level of three genes: *CEBPA*, *FABP4*, and *PPARG*. Each dot represents relative expression (mean of 3 repeats). The line is simple linear regression based on the collected data points and the gray area represents the 0.95 confidence interval for the regression line.

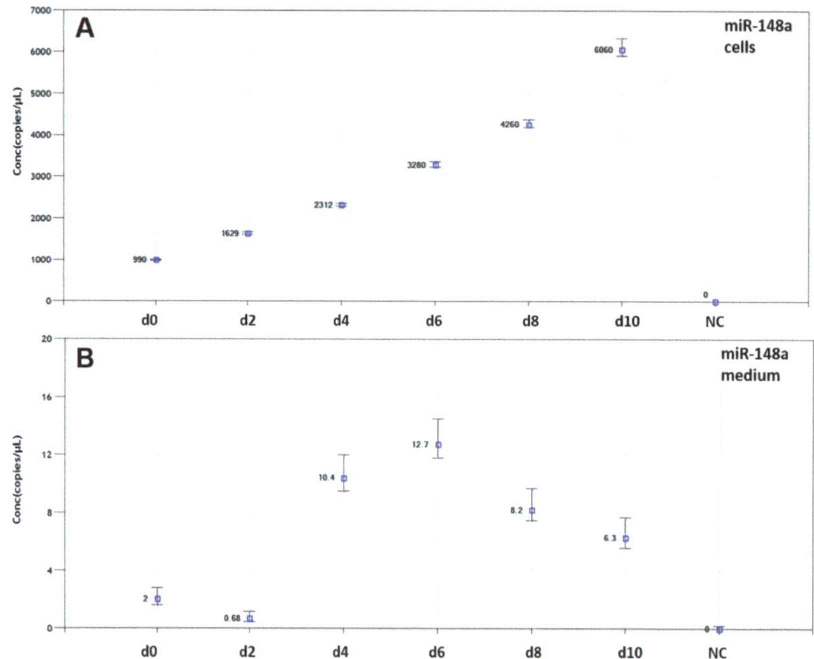

Figure 3. Examples of the detection of miRNAs using the ddPCR method. Absolute quantification (concentration in copies/μL) of miR-148a in cells (**A**) and cell culture medium (**B**) at days 0 (d0), 2 (d2), 4 (d4), 6 (d6), 8 (d8), and 10 (d10). NC: negative control (sample without cDNA template). Error bars indicate the Poisson 95% confidence intervals.

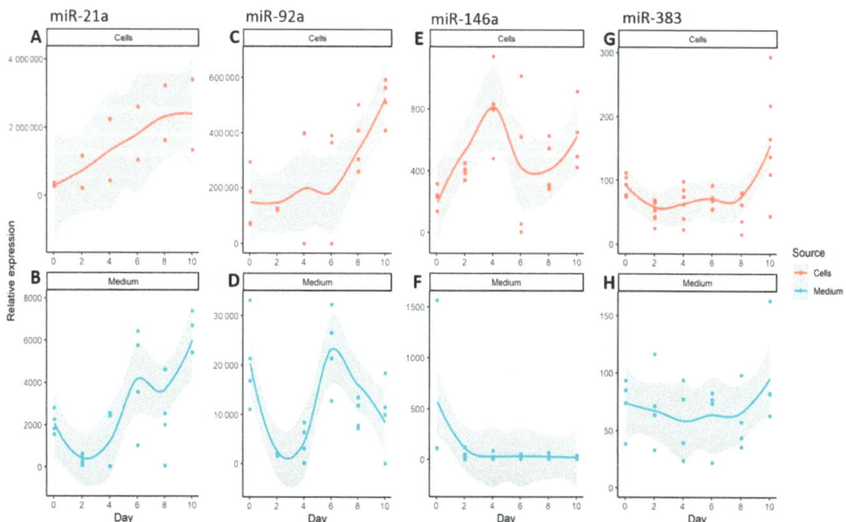

Figure 4. Relative expression levels of extracellular and intracellular miRNAs related to inflammation during the ten days of adipocyte differentiation. The expression of miR-21a (**A,B**), miR-92a (**C,D**), miR-146a (**E,F**), and miR-383 (**G,H**) was normalized using RNU6B. Dots represent the actual measurements, the line is a local regression based on the collected data points, and the gray area represents the 0.95 confidence interval for the regression line.

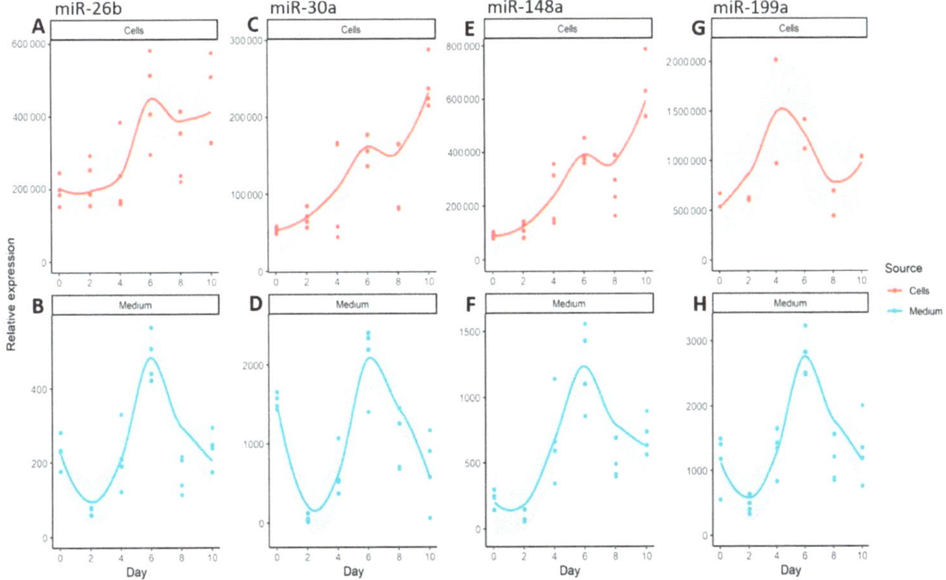

Figure 5. Relative expression levels of extracellular and intracellular miRNAs related to adipogenesis during the ten days of adipocyte differentiation. The expression of miR-26b (**A,B**), miR-30a (**C,D**), miR-199a (**E,F**), miR-148a (**G,H**) was normalized using RNU6B. Dots represent the actual measurements, the line is a local regression based on the collected data points, and the gray area represents the 0.95 confidence interval for the regression line.

Comparing the expression levels of intracellular and extracellular miRNAs showed higher expression in cells than in the medium for all the studied miRNAs except miR-383 (Table S6). No relationship was found between the expression level of intracellular and extracellular miRNAs (Table S7).

4. Discussion

To better understand the role of miRNA in adipocyte formation, we examined the expression of the selected intracellular and extracellular miRNAs during adipogenesis, using the domestic pig as a model organism. We employed ddPCR, as a robust method for absolute quantification of miRNAs [50]. This method has proven to be especially useful for quantifying extracellular miRNAs [51,52]. Here, we showed the usefulness of ddPCR for detecting less abundant ECmiRNAs in adipogenic spent medium.

We found that, of the studied miRNAs, miR-21a showed the highest expression in differentiated cells, and its expression was very high in the cell culture medium. It has been reported that miR-21 is frequently upregulated in many chronic diseases, including obesity [25]. It plays a pivotal role in the functioning of adipose tissue through its regulation of many biological processes, such as thermogenesis, browning of adipose tissue, angiogenesis, apoptosis, and adipogenesis [53]. A previous study of MSCs isolated from human adipose tissue showed that miR-21 expression increased in the early stages of adipogenic differentiation and gradually decreased after day 3 [24]; in our differentiation system, miR-21a was upregulated throughout the entire differentiation period. Studies of miR-21 mimics and inhibitors as therapeutic agents in obesity treatment have also provided varying results [53,54]. miR-21 has also been depicted as secreted by macrophages of adipose tissue [10], while in this study we confirmed its secretion by adipocytes. Further studies, including of both the intracellular and extracellular form of miR-21a, are thus needed.

To date, little has been learned about the role of miR-92a in adipogenesis. There are reports of its involvement in brown adipocyte differentiation [55]. Exosomal miR-92a abundance has also been observed in human serum after cold-induced brown adipose tissue activity [56]. In 3T3L1 cells, the miR-17–92 cluster accelerated adipocyte differentiation by negatively regulating the tumor suppressor Rb2/p130 during the early stages of adipogenesis [57]. Here, we provide evidence that miR-92a alone is upregulated during porcine adipogenesis and is secreted by adipocytes.

Recently, miR-146a has been recognized as a potential regulator of porcine intramuscular preadipocytes [58]. The authors reporting this observed that miR-146a-5p mimics inhibited preadipocyte proliferation and differentiation, while the miR-146a-5p inhibitors promoted cell proliferation and adipogenic differentiation. Both that study and our present one found a similar expression pattern for these miRNAs, with the expression peaking in the early stages of adipogenesis (on days 2 or 4 of differentiation, respectively). Interestingly, miR-146a was one of the studied miRNAs that was not secreted by differentiated cells (as was miR-383).

Of the studied miRNAs, miR-26b, miR-30a, miR-148a, and miR-199a have been reported as involved in adipocyte formation through their promotion or acceleration of adipogenesis, which they achieve by regulating numerous target genes [27,28,34,59]. Their expression was found to gradually increase after the induction of adipocyte differentiation, as in our study. Only miR-199a reached its highest expression at day 4 of adipogenesis, and its expression then decreased, as confirmed by its function in the proliferation and differentiation of porcine preadipocytes [60]. This miRNA was highly expressed in cells and also secreted more. It seems that this molecule has a comprehensive set of functions and plays a role in a range of different processes, such as angiogenesis, aging, apoptosis, proliferation, and myogenic differentiation [61,62].

All the extracellular miR-26b, miR-30a, miR-148a, miR-199a, and miR-92a showed the same expression profiles, with their expression peaking on day 6 of differentiation. It was previously reported that exosomal miRNAs are secreted from hypertrophic adipocytes and transferred to small adipocytes to promote lipogenesis and hypertrophy of emerging adipocytes [63,64]. This may be one reason why high expression of these extracellular miRNAs is observed in the intermediate days of differentiation, when new adipocytes arise at an intense rate. As we found no strong correlation between the expression of intracellular and extracellular miRNAs, it can be anticipated that the secretion of miRNA is an independent process regulated by other mechanisms, such as the formation of extracellular vesicles such as exosomes or transportation via protein–miRNA complexes [65,66].

Our study revealed relatively high intercellular variation of miRNA expression (Tables S4–S6, which may be related to the heterogeneity of cell populations in terms of differentiation timing. It has been shown in previous studies that a high standard deviation is found for low-expressed miRNA [67]. Application of new methods, such as single-cell microRNA–mRNA co-sequencing, revealed that microRNA expression variability might be responsible alone for non-genetic cell-to-cell heterogeneity [68]. The authors found that miRNAs with low expression levels showed inherently large standard deviations, while the variation of high-abundance miRNAs gradually decreased as the expression level increased.

It has been shown that the miRNA expression profile can serve as a signature of cell identity, through the expression of a unique miRNA. However, this would seem to be difficult to apply to adipose tissue, as it expresses a wide range of miRNAs [65,69]. Our study of cultured adipocytes allowed us to obtain more detailed knowledge of the relationship between the intracellular and extracellular microRNAs that are expressed during the formation of adipocyte cells. Taking into account the fact that miRNAs from adipose tissue participate in intercellular and interorgan communications, and that their aberrant expression may lead to pathological conditions, further comprehensive studies of extracellular and intracellular miRNAs are needed.

5. Conclusions

We showed that miRNAs associated with adipogenesis and inflammation processes are expressed by differentiated adipocytes. Both intracellular and extracellular miRNAs have characteristic expression profiles during porcine adipogenesis. We found that there is no relationship between the expression level of intracellular miRNAs and the levels of extracellular miRNAs. ddPCR proved a useful method of quantifying miRNAs during in vitro adipogenesis, especially for less abundant extracellular miRNAs.

Supplementary Materials: The following supporting information can be downloaded from https://www.mdpi.com/article/10.3390/genes14030683/s1. Table S1: PCR primers used to study the expression level of adipocyte differentiation marker genes. Table S2: Expression levels of miRNAs in cell culture medium supplemented with 10% of fetal bovine serum (FBS). SE—standard error; Table S3: Lipid droplet content and relative gene expression of *CEBPA*, *FABP4*, and *PPARG* genes across the consecutive days of experiment (supplementary to Figure 2). The regression slope and *p*-values were obtained using the locally weighted scatterplot smoothing method of the ggplot2 package. Table S4: Expression levels of intracellular miRNAs over the course of differentiation. The comparisons were made between days of the experiment for one miRNA at a time, and so the table should be read column by column. The same letters indicate an absence of significant difference (at the 0.05 level) between the expression levels; Table S5: Expression levels of extracellular miRNAs over the course of differentiation. The comparisons were made between days of the experiment for one miRNA at the time, so the table should be read column by column. The same letters indicate an absence of significant difference (at the 0.05 level) between the expression levels; Table S6: Expression levels of intracellular and extracellular miRNAs. The comparisons were made between miRNAs, so the table should be read row by row. The same letters indicate an absence of significant difference (at the 0.05 level) between the expression levels; Table S7: Relationship between the expression in the medium and cells based on regression slopes for the log-transformed expression in the medium on the log-transformed expression in cells for different miRNA.

Author Contributions: Conceptualization: I.S., M.P. and M.S.; methodology: A.B., M.P., M.S., J.S. and I.S.; formal analysis: A.B., M.P., M.S., J.S. and I.S.; investigation: A.B., M.P., M.S., J.S., F.G., M.O.A. and I.S.; writing—original draft preparation: I.S., M.P., M.S. and A.B.; writing—review and editing: I.S.; visualization: A.B., M.P., M.S., J.S., F.G., M.O.A. and I.S.; supervision: I.S.; project administration: I.S.; funding acquisition: I.S. All authors have read and agreed to the published version of the manuscript.

Funding: This research was funded by the statutory fund of the Department of Genetics and Animal Breeding, Faculty of Veterinary Medicine and Animal Science, Poznan University of Life Sciences, Poland (no. 506.534.09.00).

Institutional Review Board Statement: Not applicable.

Informed Consent Statement: Not applicable.

Data Availability Statement: The data supporting the findings of the study are available from the corresponding author (I.S.), upon reasonable request.

Acknowledgments: We wish to thank Magdalena Noak for her technical assistance.

Conflicts of Interest: The authors declare no conflict of interest.

References

1. Bartel, D.P. MicroRNAs: Target recognition and regulatory functions. *Cell* **2009**, *136*, 215–233. [CrossRef] [PubMed]
2. Shu, J.; Silva, B.V.R.E.; Gao, T.; Xu, Z.; Cui, J. Dynamic and Modularized MicroRNA Regulation and Its Implication in Human Cancers. *Sci. Rep.* **2017**, *7*, 13356. [CrossRef] [PubMed]
3. Yao, S. MicroRNA biogenesis and their functions in regulating stem cell potency and differentiation. *Biol. Proced. Online* **2016**, *18*, 8. [CrossRef] [PubMed]
4. Walayat, A. Therapeutic implication of miRNA in human disease. In *Antisense Therapy*; Yang, M., Ed.; IntechOpen: London, UK, 2019. [CrossRef]
5. Condrat, C.E.; Thompson, D.C.; Barbu, M.G.; Bugnar, O.L.; Boboc, A.; Cretoiu, D.; Suciu, N.; Cretoiu, S.M.; Voinea, S.C. miRNAs as Biomarkers in Disease: Latest Findings Regarding Their Role in Diagnosis and Prognosis. *Cells* **2020**, *9*, 276. [CrossRef] [PubMed]

6.   Sohel, M.H. Extracellular/Circulating MicroRNAs: Release Mechanisms, Functions and Challenges. *Achiev. Life Sci.* **2016**, *10*, 175–186. [CrossRef]

7.   Hilton, C.; Neville, M.J.; Karpe, F. MicroRNAs in adipose tissue: Their role in adipogenesis and obesity. *Int. J. Obes.* **2013**, *37*, 325–332. [CrossRef]

8.   Arner, P.; Kulyté, A. MicroRNA regulatory networks in human adipose tissue and obesity. *Nat. Rev. Endocrinol.* **2015**, *11*, 276–288. [CrossRef]

9.   Oger, F.; Gheeraert, C.; Mogilenko, D.; Benomar, Y.; Molendi-Coste, O.; Bouchaert, E.; Caron, S.; Dombrowicz, D.; Pattou, F.; Duez, H.; et al. Cell-specific dysregulation of microRNA expression in obese white adipose tissue. *J. Clin. Endocrinol. Metab.* **2014**, *99*, 2821–2833. [CrossRef]

10.   Heyn, G.S.; Corrêa, L.H.; Magalhães, K.G. The Impact of Adipose Tissue-Derived miRNAs in Metabolic Syndrome, Obesity, and Cancer. *Front. Endocrinol.* **2020**, *11*, 563816. [CrossRef]

11.   Ogawa, R.; Tanaka, C.; Sato, M.; Nagasaki, H.; Sugimura, K.; Okumura, K.; Nakagawa, Y.; Aoki, N. Adipocyte-derived microvesicles contain RNA that is transported into macrophages and might be secreted into blood circulation. *Biochem. Biophys. Res. Commun.* **2010**, *398*, 723–729. [CrossRef]

12.   Thomou, T.; Mori, M.A.; Dreyfuss, J.M.; Konishi, M.; Sakaguchi, M.; Wolfrum, C.; Rao, T.N.; Winnay, J.N.; Garcia-Martin, R.; Grinspoon, S.K.; et al. Adipose-derived circulating miRNAs regulate gene expression in other tissues. *Nature* **2017**, *542*, 450–455. [CrossRef]

13.   Stachowiak, M.; Szczerbal, I.; Switonski, M. Genetics of Adiposity in Large Animal Models for Human Obesity-Studies on Pigs and Dogs. *Prog. Mol. Biol. Transl. Sci.* **2016**, *140*, 233–270. [CrossRef]

14.   Song, Z.; Cooper, D.K.C.; Cai, Z.; Mou, L. Expression and Regulation Profile of Mature MicroRNA in the Pig: Relevance to Xenotransplantation. *BioMed Res. Int.* **2018**, *2018*, 2983908. [CrossRef]

15.   Ma, J.; Jiang, Z.; He, S.; Liu, Y.; Chen, L.; Long, K.; Jin, L.; Jiang, A.; Zhu, L.; Wang, J.; et al. Intrinsic features in microRNA transcriptomes link porcine visceral rather than subcutaneous adipose tissues to metabolic risk. *PLoS ONE* **2013**, *8*, e80041. [CrossRef]

16.   Liu, X.; Gong, J.; Wang, L.; Hou, X.; Gao, H.; Yan, H.; Zhao, F.; Zhang, L.; Wang, L. Genome-Wide Profiling of the Microrna Transcriptome Regulatory Network to Identify Putative Candidate Genes Associated with Backfat Deposition in Pigs. *Animals* **2019**, *9*, 313. [CrossRef]

17.   Xing, K.; Zhao, X.; Liu, Y.; Zhang, F.; Tan, Z.; Qi, X.; Wang, X.; Ni, H.; Guo, Y.; Sheng, X.; et al. Identification of Differentially Expressed MicroRNAs and Their Potential Target Genes in Adipose Tissue from Pigs with Highly Divergent Backfat Thickness. *Animals* **2020**, *10*, 624. [CrossRef]

18.   Ropka-Molik, K.; Pawlina-Tyszko, K.; Żukowski, K.; Tyra, M.; Derebecka, N.; Wesoły, J.; Szmatoła, T.; Piórkowska, K. Identification of Molecular Mechanisms Related to Pig Fatness at the Transcriptome and miRNAome Levels. *Genes* **2020**, *11*, 600. [CrossRef]

19.   Li, W.; Wen, S.; Wu, J.; Zeng, B.; Chen, T.; Luo, J.; Shu, G.; Wang, S.-B.; Zhang, Y.; Xi, Q. Comparative Analysis of MicroRNA Expression Profiles Between Skeletal Muscle- and Adipose-Derived Exosomes in Pig. *Front. Genet.* **2021**, *12*, 631230. [CrossRef]

20.   Eirin, A.; Riester, S.M.; Zhu, X.-Y.; Tang, H.; Evans, J.M.; O'Brien, D.; van Wijnen, A.J.; Lerman, L.O. MicroRNA and mRNA cargo of extracellular vesicles from porcine adipose tissue-derived mesenchymal stem cells. *Gene* **2014**, *551*, 55–64. [CrossRef]

21.   Lynes, M.D.; Tseng, Y.-H. Deciphering adipose tissue heterogeneity. *Ann. N. Y. Acad. Sci.* **2018**, *1411*, 5–20. [CrossRef]

22.   Ruiz-Ojeda, F.J.; Rupérez, A.I.; Gomez-Llorente, C.; Gil, A.; Aguilera, C.M. Cell Models and Their Application for Studying Adipogenic Differentiation in Relation to Obesity: A Review. *Int. J. Mol. Sci.* **2016**, *17*, 1040. [CrossRef] [PubMed]

23.   Sheedy, F.J. Turning 21: Induction of miR-21 as a Key Switch in the Inflammatory Response. *Front. Immunol.* **2015**, *6*, 19. [CrossRef] [PubMed]

24.   Kim, Y.J.; Hwang, S.J.; Bae, Y.C.; Jung, J.S. MiR-21 regulates adipogenic differentiation through the modulation of TGF-beta signaling in mesenchymal stem cells derived from human adipose tissue. *Stem Cells* **2009**, *27*, 3093–3102. [CrossRef]

25.   Keller, P.; Gburcik, V.; Petrovic, N.; Gallagher, I.J.; Nedergaard, J.; Cannon, B.; Timmons, J.A. Gene-chip studies of adipogenesis-regulated microRNAs in mouse primary adipocytes and human obesity. *BMC Endocr. Disord.* **2011**, *11*, 7. [CrossRef] [PubMed]

26.   Yang, C.; Luo, M.; Chen, Y.; You, M.; Chen, Q. MicroRNAs as Important Regulators Mediate the Multiple Differentiation of Mesenchymal Stromal Cells. *Front. Cell Dev. Biol.* **2021**, *9*, 619842. [CrossRef]

27.   Li, G.; Ning, C.; Ma, Y.; Jin, L.; Tang, Q.; Li, X.; Li, M.; Liu, H. miR-26b Promotes 3T3-L1 Adipocyte Differentiation Through Targeting PTEN. *DNA Cell Biol.* **2017**, *36*, 672–681. [CrossRef]

28.   Cui, S.; Soni, C.B.; Xie, J.; Li, Y.; Zhu, H.; Wu, F.; Zhi, X. MiR-30a-5p accelerates adipogenesis by negatively regulating Sirtuin. *Int. J. Clin. Exp. Pathol.* **2018**, *11*, 5203–5212.

29.   Saha, P.K.; Hamilton, M.P.; Rajapakshe, K.; Putluri, V.; Felix, J.B.; Masschelin, P.; Cox, A.R.; Bajaj, M.; Putluri, N.; Coarfa, C.; et al. miR-30a targets gene networks that promote browning of human and mouse adipocytes. *Am. J. Physiol. Endocrinol. Metab.* **2020**, *319*, E667. [CrossRef]

30.   Lai, L.; Song, Y.; Liu, Y.; Chen, Q.; Han, Q.; Chen, W.; Pan, T.; Zhang, Y.; Cao, X.; Wang, Q. MicroRNA-92a negatively regulates Toll-like receptor (TLR)-triggered inflammatory response in macrophages by targeting MKK4 kinase. *J. Biol. Chem.* **2013**, *288*, 7956–7967. [CrossRef]

31.   Zhang, Y.; Song, K.; Qi, G.; Yan, R.; Yang, Y.; Li, Y.; Wang, S.; Bai, Z.; Ge, R.-L. Adipose-derived exosomal miR-210/92a cluster inhibits adipose browning via the FGFR-1 signaling pathway in high-altitude hypoxia. *Sci. Rep.* **2020**, *10*, 14390. [CrossRef]

32. Xie, Y.; Chu, A.; Feng, Y.; Chen, L.; Shao, Y.; Luo, Q.; Deng, X.; Wu, M.; Shi, X.; Chen, Y. MicroRNA-146a: A Comprehensive Indicator of Inflammation and Oxidative Stress Status Induced in the Brain of Chronic T2DM Rats. *Front. Pharmacol.* **2018**, *9*, 478. [CrossRef]

33. Roos, J.; Enlund, E.; Funcke, J.-B.; Tews, D.; Holzmann, K.; Debatin, K.-M.; Wabitsch, M.; Fischer-Posovszky, P. miR-146a-mediated suppression of the inflammatory response in human adipocytes. *Sci. Rep.* **2016**, *6*, 38339. [CrossRef]

34. Shi, C.; Zhang, M.; Tong, M.; Yang, L.; Pang, L.; Chen, L.; Xu, G.; Chi, X.; Hong, Q.; Ni, Y.; et al. miR-148a is Associated with Obesity and Modulates Adipocyte Differentiation of Mesenchymal Stem Cells through Wnt Signaling. *Sci. Rep.* **2015**, *5*, 9930. [CrossRef]

35. Shi, C.; Pang, L.; Ji, C.; Wang, J.; Lin, N.; Chen, J.; Chen, L.; Yang, L.; Huang, F.; Zhou, Y.; et al. Obesity-associated miR-148a is regulated by cytokines and adipokines via a transcriptional mechanism. *Mol. Med. Rep.* **2016**, *14*, 5707–5712. [CrossRef]

36. Tan, Z.; Du, J.; Shen, L.; Liu, C.; Ma, J.; Bai, L.; Jiang, Y.; Tang, G.; Li, M.; Li, X.; et al. miR-199a-3p affects adipocytes differentiation and fatty acid composition through targeting SCD. *Biochem. Biophys. Res. Commun.* **2017**, *492*, 82–88. [CrossRef]

37. Gu, N.; You, L.; Shi, C.; Yang, L.; Pang, L.; Cui, X.; Ji, C.; Zheng, W.; Guo, X. Expression of miR-199a-3p in human adipocytes is regulated by free fatty acids and adipokines. *Mol. Med. Rep.* **2016**, *14*, 1180–1186. [CrossRef]

38. Yi, Q.; Xie, W.; Sun, W.; Sun, W.; Liao, Y. A Concise Review of MicroRNA-383: Exploring the Insights of Its Function in Tumorigenesis. *J. Cancer* **2022**, *13*, 313–324. [CrossRef]

39. Stachecka, J.; Walczak, A.; Kociucka, B.; Ruszczycki, B.; Wilczyński, G.; Szczerbal, I. Nuclear organization during in vitro differentiation of porcine mesenchymal stem cells (MSCs) into adipocytes. *Histochem. Cell Biol.* **2018**, *149*, 113–126. [CrossRef]

40. Piórkowska, K.; Oczkowicz, M.; Różycki, M.; Ropka-Molik, K.; Piestrzyńska-Kajtoch, A. Novel porcine housekeeping genes for real-time RT-PCR experiments normalization in adipose tissue: Assessment of leptin mRNA quantity in different pig breeds. *Meat Sci.* **2011**, *87*, 191–195. [CrossRef]

41. Stachecka, J.; Nowacka-Woszuk, J.; Kolodziejski, P.A.; Szczerbal, I. The importance of the nuclear positioning of the PPARG gene for its expression during porcine in vitro adipogenesis. *Chromosome Res.* **2019**, *27*, 271–284. [CrossRef]

42. Peltier, H.J.; Latham, G.J. Normalization of microRNA expression levels in quantitative RT-PCR assays: Identification of suitable reference RNA targets in normal and cancerous human solid tissues. *RNA* **2008**, *14*, 844–852. [CrossRef] [PubMed]

43. Auber, M.; Fröhlich, D.; Drechsel, O.; Karaulanov, E.; Krämer-Albers, E.-M. Serum-free media supplements carry miRNAs that co-purify with extracellular vesicles. *J. Extracell. Vesicles* **2019**, *8*, 1656042. [CrossRef] [PubMed]

44. R Core Team. *R: A Language and Environment for Statistical Computing 2021*; R Core Team: Vienna, Austria, 2021. Available online: https://www.r-project.org/ (accessed on 22 February 2022).

45. Bates, D.; Mächler, M.; Bolker, B.; Walker, S. Fitting Linear Mixed-Effects Models Using lme4. *J. Stat. Softw.* **2015**, *67*, 1–48. [CrossRef]

46. Kuznetsova, A.; Brockhoff, P.B.; Christensen, R.H.B. lmerTest Package: Tests in Linear Mixed Effects Models. *J. Stat. Softw.* **2017**, *82*, 1–26. [CrossRef]

47. Lenth, R. V emmeans: Estimated Marginal Means, aka Least-Squares Means 2022. Available online: https://github.com/rvlenth/emmeans (accessed on 22 February 2022).

48. Hrong-Tai Fai, A.; Cornelius, P.L. Approximate F-tests of multiple degree of freedom hypotheses in generalized least squares analyses of unbalanced split-plot experiments. *J. Stat. Comput. Simul.* **1996**, *54*, 363–378. [CrossRef]

49. Wickham, H. *ggplot2: Elegant Graphics for Data Analysis*; Springer: New York, NY, USA, 2016; ISBN 978-0-387-98141-3.

50. Precazzini, F.; Detassis, S.; Imperatori, A.S.; Denti, M.A.; Campomenosi, P. Measurements Methods for the Development of MicroRNA-Based Tests for Cancer Diagnosis. *Int. J. Mol. Sci.* **2021**, *22*, 1176. [CrossRef]

51. Wang, C.; Ding, Q.; Plant, P.; Basheer, M.; Yang, C.; Tawedrous, E.; Krizova, A.; Boulos, C.; Farag, M.; Cheng, Y.; et al. Droplet digital PCR improves urinary exosomal miRNA detection compared to real-time PCR. *Clin. Biochem.* **2019**, *67*, 54–59. [CrossRef]

52. Zhao, G.; Jiang, T.; Liu, Y.; Huai, G.; Lan, C.; Li, G.; Jia, G.; Wang, K.; Yang, M. Droplet digital PCR-based circulating microRNA detection serve as a promising diagnostic method for gastric cancer. *BMC Cancer* **2018**, *18*, 676. [CrossRef]

53. Lhamyani, S.; Gentile, A.-M.; Giráldez-Pérez, R.M.; Feijóo-Cuaresma, M.; Romero-Zerbo, S.Y.; Clemente-Postigo, M.; Zayed, H.; Olivera, W.O.; Bermúdez-Silva, F.J.; Salas, J.; et al. miR-21 mimic blocks obesity in mice: A novel therapeutic option. *Mol. Ther. Nucleic Acids* **2021**, *26*, 401–416. [CrossRef]

54. Seeger, T.; Fischer, A.; Muhly-Reinholz, M.; Zeiher, A.M.; Dimmeler, S. Long-term inhibition of miR-21 leads to reduction of obesity in db/db mice. *Obesity* **2014**, *22*, 2352–2360. [CrossRef]

55. Zhang, Z.; Jiang, H.; Li, X.; Chen, X.; Huang, Y. MiR-92a regulates brown adipocytes differentiation, mitochondrial oxidative respiration, and heat generation by targeting SMAD7. *J. Cell. Biochem.* **2019**, *121*, 3825–3836. [CrossRef]

56. Chen, Y.; Buyel, J.J.; Hanssen, M.J.W.; Siegel, F.; Pan, R.; Naumann, J.; Schell, M.; van der Lans, A.; Schlein, C.; Froehlich, H.; et al. Exosomal microRNA miR-92a concentration in serum reflects human brown fat activity. *Nat. Commun.* **2016**, *7*, 11420. [CrossRef]

57. Wang, Q.; Li, Y.C.; Wang, J.; Kong, J.; Qi, Y.; Quigg, R.J.; Li, X. miR-17-92 cluster accelerates adipocyte differentiation by negatively regulating tumor-suppressor Rb2/p130. *Proc. Natl. Acad. Sci. USA* **2008**, *105*, 2889–2894. [CrossRef]

58. Zhang, Q.; Cai, R.; Tang, G.; Zhang, W.; Pang, W. MiR-146a-5p targeting SMAD4 and TRAF6 inhibits adipogenensis through TGF-β and AKT/mTORC1 signal pathways in porcine intramuscular preadipocytes. *J. Anim. Sci. Biotechnol.* **2021**, *12*, 12. [CrossRef]

59. Shuai, Y.; Yang, R.; Mu, R.; Yu, Y.; Rong, L.; Jin, L. MiR-199a-3p mediates the adipogenic differentiation of bone marrow-derived mesenchymal stem cells by regulating KDM6A/WNT signaling. *Life Sci.* **2019**, *220*, 84–91. [CrossRef]
60. Shi, X.-E.; Li, Y.-F.; Jia, L.; Ji, H.-L.; Song, Z.-Y.; Cheng, J.; Wu, G.-F.; Song, C.-C.; Zhang, Q.-L.; Zhu, J.-Y.; et al. MicroRNA-199a-5p affects porcine preadipocyte proliferation and differentiation. *Int. J. Mol. Sci.* **2014**, *15*, 8526–8538. [CrossRef]
61. Wang, Q.; Ye, B.; Wang, P.; Yao, F.; Zhang, C.; Yu, G. Overview of microRNA-199a Regulation in Cancer. *Cancer Manag. Res.* **2019**, *11*, 10327–10335. [CrossRef]
62. Fukuoka, M.; Fujita, H.; Numao, K.; Nakamura, Y.; Shimizu, H.; Sekiguchi, M.; Hohjoh, H. MiR-199-3p enhances muscle regeneration and ameliorates aged muscle and muscular dystrophy. *Commun. Biol.* **2021**, *4*, 427. [CrossRef]
63. Müller, G.; Schneider, M.; Biemer-Daub, G.; Wied, S. Microvesicles released from rat adipocytes and harboring glycosylphosphati-dylinositol-anchored proteins transfer RNA stimulating lipid synthesis. *Cell Signal.* **2011**, *23*, 1207–1223. [CrossRef]
64. Engin, A.B.; Engin, A. Adipogenesis-related microRNAs in obesity. *exRNA* **2022**, *4*, 16. [CrossRef]
65. Mori, M.A.; Ludwig, R.G.; Garcia-Martin, R.; Brandão, B.B.; Kahn, C.R. Extracellular miRNAs: From Biomarkers to Mediators of Physiology and Disease. *Cell Metab.* **2019**, *30*, 656–673. [CrossRef] [PubMed]
66. Hong, P.; Yu, M.; Tian, W. Diverse RNAs in adipose-derived extracellular vesicles and their therapeutic potential. *Mol. Ther. Nucleic Acids* **2021**, *26*, 665–677. [CrossRef] [PubMed]
67. Gentien, D.; Piqueret-Stephan, L.; Henry, E.; Albaud, B.; Rapinat, A.; Koscielny, S.; Scoazec, J.-Y.; Vielh, P. Digital Multiplexed Gene Expression Analysis of mRNA and miRNA from Routinely Processed and Stained Cytological Smears: A Proof-of-Principle Study. *Acta Cytol.* **2021**, *65*, 88–98. [CrossRef] [PubMed]
68. Wang, N.; Zheng, J.; Chen, Z.; Liu, Y.; Dura, B.; Kwak, M.; Xavier-Ferrucio, J.; Lu, Y.-C.; Zhang, M.; Roden, C.; et al. Single-cell microRNA-mRNA co-sequencing reveals non-genetic heterogeneity and mechanisms of microRNA regulation. *Nat. Commun.* **2019**, *10*, 95. [CrossRef]
69. Huang, Z.; Xu, A. Adipose Extracellular Vesicles in Intercellular and Inter-Organ Crosstalk in Metabolic Health and Diseases. *Front. Immunol.* **2021**, *12*, 608680. [CrossRef]

Communication

Genetic Diversity and Population Structure of the Native Pulawska and Three Commercial Pig Breeds Based on Microsatellite Markers

Anna Radko *, Anna Koseniuk and Grzegorz Smołucha

National Research Institute of Animal Production, Department of Animal Molecular Biology, Krakowska Street 1, 32-083 Balice, Poland
* Correspondence: anna.radko@iz.edu.pl

Abstract: Swine DNA profiling is highly important for animal identification and parentage verification and also increasingly important for meat traceability. This work aimed to analyze the genetic structure and genetic diversity in selected Polish pig breeds. The study used a set of 14 microsatellite (STR) markers recommended by ISAG for parentage verification in the native Puławska pig (PUL, n = 85) and three commercial pig breeds: Polish Large White (PLW, n = 74), Polish Landrace (PL, n = 85) and foreign breed Duroc (DUR, n = 84). Genetic differentiation among breeds accounted for 18% of the total genetic variability (AMOVA). Bayesian structure analysis (STRUCTURE) indicated that the four distinct genetic clusters obtained corresponded to the four breeds studied. The genetic Reynolds distances (θw) showed a close relationship between PL and PLW breeds and the most distant for DUR and PUL pigs. The genetic differentiation values (F_{ST}) were lower between PL and PLW and higher between PUL and DUR. The principal coordinate analysis (PCoA) supported the classification of the populations into four clusters.

Keywords: *Sus scrofa*; STR; genetic differentiation

check for **updates**

Citation: Radko, A.; Koseniuk, A.; Smołucha, G. Genetic Diversity and Population Structure of the Native Pulawska and Three Commercial Pig Breeds Based on Microsatellite Markers. *Genes* **2023**, *14*, 276. https://doi.org/10.3390/genes14020276

Academic Editors: Shaochen Sun and J. Peter W. Young

Received: 30 November 2022
Revised: 5 January 2023
Accepted: 18 January 2023
Published: 20 January 2023

1. Introduction

Maintaining high-quality meat production and the preservation of food safety are directly related to the biodiversity and individual and breed identification of animals. Swine DNA profiling is highly important for animal identification, parentage verification and, more recently, meat traceability [1–3]. Pork is the most frequently chosen meat by consumers; therefore, maintaining high-quality standards in pork production is very important—not only for commercial pigs but especially for the population of native pig breeds. Native breed pigs can provide meat that is both high in quality and functional. Local, primitive breeds give rise to native pig breeds. Breed purity was maintained through the careful selection of individuals for mating, while breeding work was based on selective breeding. Many of the traits inherited from primitive ancestors are present in pigs from native breeds, such as adaptability to local environmental conditions, feed availability and extensive farming. In addition to high fertility, the animals have good maternal care and display good breeding performance. The animals live for a long time and are resistant to stress and pathogens. Products obtained from their meat possess unique sensory and nutritional quality. The native breeds of pigs in Poland are White Złotnicka, Złotnicka Spotted and Puławska. Polish native breeds are valued not only for the fact that they are bred in Poland, but also because their meat is used to produce traditional Polish cured meats with specific technological properties and taste qualities. Regarding the chosen Polish native pigs breed to study, the Puławska breed originates from the Lublin region. These pigs have numerous favorable characteristics, such as good health, resistance to disease, low fodder requirements and fattening tendency. Moreover, the Puławska breed is recognized for its early maturation, rapid growth, high feed utilization, high fattening and slaughter values and adaptability to local environmental conditions. Through cross-breeding this pig with the Large White and Berkshire breeds, the Pulawska population was transformed into a dual-purpose fat–meat type. Now, the Puławska breed is used for commercial crossing [4]. Puławska, as a rare breed, was included in the Farm Animals Genetic Resources Conservation Program. Among the Polish local breeds, the distinguishing quality traits of the Puławska pig are emphasized. Meat obtained from these animals is characterized by higher culinary value than that from mass production. Recently, products from this breed have been very popular on the Polish market, e.g., "traditional cold cuts from Puławiak". In order for authentic Pulawska breed meat and meat products to be verified, it should be included in DNA testing for the individual identification of animals and provision for the retrieval of this information as and when required, which is very important for meat traceability. This means that the meat is produced from an identified animal and has information about its origin. According to ISAG recommendations published in the 1990s, microsatellite markers (short tandem repeats, STR) can be used to prove the parentage of farm animals. Moreover, STRs are applied to study genetic structure and diversity, as well as to track the ancestry of diverse farm animal species [5–9], including pigs [10–17]. The ISAG conference in 2012 outlined the first microsatellite panel of 24 markers for parentage verification: IGF1, S0002, S0005, S0026, S0068, S0090, S0101, S0155, S0178, S0215, S0225, S0226, S0227, S0228, S0355, S0386, SW024, SW072, SW240, SW632, SW857, SW911, SW936, SW951 [18]. An updated list of recommended markers was released in 2014. The STRs were divided into core and additional panels. Fifteen microsatellite loci made up the main panel: S0005, S0090, S0101, S0155, S0227, S0228, S0355, S0386, SW24, SW240, SW72, SW857, SW911, SW936 and SW951. The additional panel includes seven microsatellites: IGF1, S0002, S0026, S0215, S0225, S0226 and SW632 [19]. In the study, we analyzed DNA microsatellite marker polymorphisms of the core STR panel [18,19] in Polish native Puławska pigs and three commercial pig breeds: Polish Large White, Polish Landrace and a foreign breed, Duroc.

The National Breeding Program includes the following breeds: Puławska pig (PUL), Polish Large White (PLW), Polish Landrace (PL) and foreign breeds Duroc (DU), Hampshire and Pietrain. In Poland, these breeds are used for crossbreeding and fattening in pig production and are some of the most economically important (https://www.polsus.pl/index.php/en/pig-breeding, accessed on 29 November 2022). The aims of this study were to assess the level of genetic diversity and determine the population structure of the native Puławska pig and three commercial pig breeds, PL, PLW and DUR, by using a set of 14 STRs. The 14 STRs are recommended for individual pig identification and parentage verification. No studies present the assessment of the polymorphisms of STR markers recommended for the identification of pigs in the Polish population. The study by Szmatola et al. [20] was based only on five STRs not used in routine testing. With the values adjusted for sample sizes, they discovered four breeds to have high levels of genetic diversity: 0.740 for Polish Landrace, 0.697 for Pietrain, 0.692 for Polish Large White and 0.688 for Puławska. However, the Duroc breed has the smallest amount of effective alleles, allelic richness and genetic diversity (0.589). Their findings indicate there has been some gene flow between breeds, particularly between Polish Landrace and Polish Large White. The Duroc breed was shown to have the lowest admixture, confirming its great purity. The authors conclude that further research should probably be performed using more microsatellites and by analyzing mitochondrial DNA. Here, we test the possibility of using 14 STR markers to predict the pig breed, which may be useful in the future for meat traceability.

2. Materials and Methods

2.1. Material

Blood samples were collected from pigs undergoing routine parentage testing at NRIAP. A total of 338 pigs were studied, including Puławska pigs (PUL, $n = 85$) and three selected commercial breeds: Polish Large White (PLW, $n = 74$), Polish Landrace (PL, $n = 85$) and Duroc (DUR, $n = 84$).

DNA was extracted from blood samples using the Sherlock AX Kit (A&A Biotechnology, Gdynia, Poland), following the manufacturer's protocol. Extracted DNA was quantified using a NanoDrop 2000 spectrophotometer (Thermo Scientific, Wilmington, DE, USA).

In the analysis, we selected 14 loci from the recommended ISAG main panel of 15 markers for the identification of individuals and parentage testing in the pig: S0090, S0101, S0155, S0227, S0228, S0355, S0386, SW24, SW240, SW72, SW857, SW911, SW936 and SW951. The markers and used primer sequences are presented by Radko et al. [20,21].

2.2. Methods

The one-multiplex reaction containing the 14 STR loci was amplified using the Type-It Microsatellite PCR Kit (Qiagen Inc, Hilden, Germany) reagents and fluorescently labeled primers. The reaction mixtures, with a final reagent volume of 12.5 µL, contained 50 ng DNA. The Veriti® Thermal Cycler amplifier was used for the PCR reaction (Applied Biosystems, Foster City, CA, USA) with the following thermal profile: 5 min initial denaturation at 95 °C, followed by 28 cycles of denaturation at 95 °C for 30 s, annealing at 57 °C for 90 s, elongation of primers at 72 °C for 30 s and final elongation of primers at 60 °C for 30 min. The PCR products were analyzed using an ABI 3500xl capillary sequencer (Applied Bio-Systems, Foster City, CA, USA). The amplified DNA fragments were subjected to electrophoresis in 7% denaturing POP-7 polyacrylamide gel in the presence of a size standard of 500 LIZ (Thermo Fisher Scientific) and a reference sample with a known DNA profile for allele standardization. The results of the electrophoretic separation were analyzed using the GeneMapper® Software 4.0 (Applied Biosystems, Foster City, CA, USA).

2.3. Data Analysis

Analysis of molecular variance and genetic differentiation.

Analysis of molecular variance (AMOVA) and pairwise estimates of genetic differentiation (F_{ST}) across populations were performed using the GenAlEx ver. 6.51 software [21].

Population Structure and Genetic Distance

Population structure was analyzed using a Bayesian clustering algorithm implemented in STRUCTURE software version 2.3.4 [22–24], considering an admixture model with correlated allele frequencies between breeds. The lengths of the burn-in and Monte Carlo Markov Chain (MCMC) simulations were 100,000 and 500,000, respectively, in 5 runs for each number of clusters (K) ranging between 2 and 5. The results were exported to STRUCTURE HARVESTER [25] to plot the likelihood membership coefficient (ΔK) values.

Genetic distance was analyzed using pairwise estimates of genetic differentiation—FST and Reynolds distance—θw [26]. The individual-animal-based neighbor-joining dendrogram was generated from the estimated pairwise genetic distances between shared alleles using the DARwin ver. 6 software (http://darwin.cirad.fr/, accessed on 29 November 2022).

The population relationships based on principal coordinate analysis (PCoA) were obtained using the GenAlEx ver. 6.51 software [21].

3. Results and Interpretation

The development of reliable molecular tools for genetically distinguishing between two breeds of species and identifying breed components in food products has become increasingly important due to the increasing demand for improved quality control. For the purpose of the identification of animals and products, microsatellites (STRs) are widely used as molecular markers. STR markers used in this context have to present high diversity. The genetic diversity of microsatellite loci is determined based on genetic parameters such as the PIC index and heterozygosity. These parameter values show the usefulness of markers for further research, including individual identification and genetic population diversity. Previous studies have shown that all 14 STR markers, recommended by ISAG for pig identification, were polymorphic in the sampled groups [27]. Based on the STRs, we calculated the medium genetic differentiation for the breeds studied. Interestingly, the average value of heterozygosity (HE) and the polymorphic information content (PIC) were above 0.5 for all breeds except DUR (PIC = 0.477) [27]. These polymorphism results demonstrate the potential of the analyzed STRs for the individual identification of pigs.

The F-statistic is commonly chosen for studying population structures. It is frequently applied to decompose the genetic variance into within-individual, within-population and among-population components [28]. The analysis of molecular variance (AMOVA) is commonly implemented for estimating the F-statistic [29,30].

3.1. Analysis of Molecular Variance (AMOVA)

AMOVA is an important element of molecular analysis that allows the statistical inference of genetic variation among and within populations. In our study, the variance analysis (AMOVA) was performed using all 14 polymorphic STRs and revealed that variation among individuals was greater than the variation in the inter-population.

The average genetic differentiation between the breeds was 18% ($p < 0.001$), while the total variability was 82%. Details of AMOVA results are presented in Table 1 and Figure 1.

Table 1. Analysis of molecular variance (AMOVA) based on 14 STR markers. The AMOVA result revealed that individual variation was greater (82%) than the variation in the inter-population (18%).

Source of Variation	df	Sum of Squares	Variance Components	% Variation
Among populations	3	481.869	0.957	18%
Within populations	328	1396.500	4.258	82%
Total	655	3112.256	5.512	100%

AMOVA on model base population of four pig breeds (PLW—Polish Large White; PL—Polish Landrace; PUL—Puławska Pig; DUR—Duroc); df—degree of freedom.

Percentages of Molecular Variance

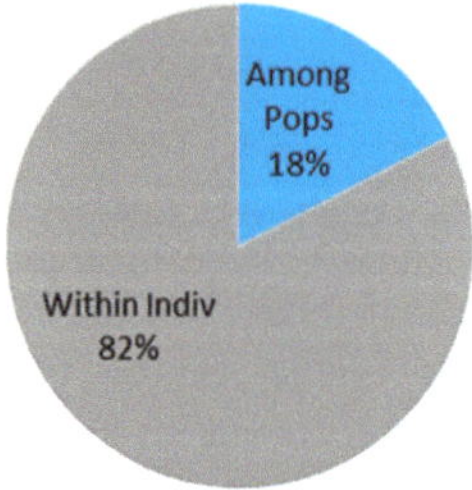

Figure 1. Percentages of AMOVA based on 14 STR markers.

In the pig population studied, the AMOVA revealed that most of the variance was attributed to differences within populations, among individuals, and 18% could be attributed to differences among the four pig groups. A similar genetic differentiation was observed with other breeds in other studies using STR markers [13,29].

3.2. Structure Analysis

STRUCTURE is the first approach giving an insight into the population structure resulting from the sample set and providing a prelude to other genetic analyses. Assigning individuals to breeds is often useful in population genetics studies, in which obtaining a population classification can provide an inference of individual ancestry that may not have been adequately defined beforehand [22,31,32]. The population structure and degree of admixture of the four pig breeds were evaluated using Bayesian model-based clustering in the STRUCTURE software. The structure of the breeds studied was determined based on the degree of admixture for each individual using the correlated allele frequencies model implemented in the software STRUCTURE. The obtained ΔK value was highest at K = 4.

On the basis of the 14 STRs, the results of STRUCTURE revealed the subdivision of the pig breeds into four genetic clusters (Figure 2). The average proportion of assignment to the cluster of above 95% was found for all pig breeds. The highest assignment value was found in the DUR (Q = 0.9716). Such a high probability may allow the assignment of unknown individuals to particular breeds. The between-individual tree of Figure 3 shows the same results—four clusters grouping the individuals that belong to the same breed.

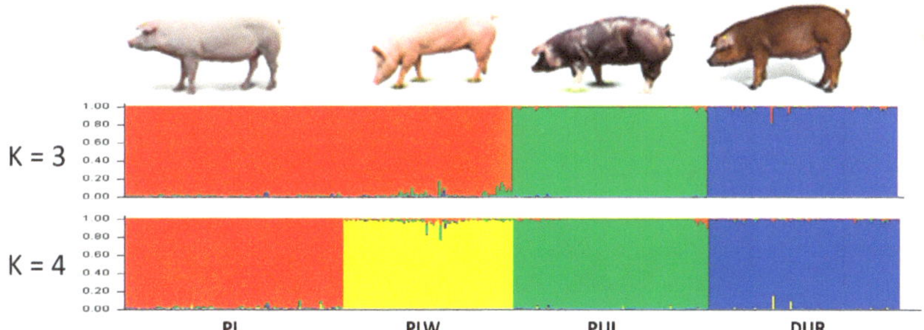

Figure 2. STRUCTURE analysis of 14 STR genotypes from pigs studied. The samples were grouped by the 4 breeds (K = 4). PLW—Polish Large White; PL—Polish Landrace; PUL—Puławska Pig; DUR—Duroc.

3.3. Genetic Differentiation

The study of the genetic differentiation of the breeds derived by the population structure analysis considered measures of two common estimates of differentiation—F_{ST} and the Reynolds genetic distance (θw). The pairwise F_{ST} values between breeds varied from 0.106 (between PL and PLW) to 0.283 (between PUL and DUR). Similarly, genetic distance was the greatest between PUL and DUR (θw = 0.430) and the smallest between PL and PLW (θw = 0.109) (Table 2).

Table 2. Reynolds genetic distance (θw) and pairwise estimates of genetic differentiation (F_{ST}) among 4 studied pig breeds (PUL—Puławska; PLW—Polish Large White; PL—Polish Landrace; DUR—Duroc). The θw values are above the diagonal, and F_{ST} estimates are below the diagonal.

	PL	PLW	PUL	DUR
PL		0.109	0.275	0.422
PLW	0.106		0.288	0.429
PUL	0.146	0.147		0.430
DUR	0.201	0.222	0.283	

The pairwise F_{ST} and the Reynolds distances among the breeds showed that the national PL and PLW breeds formed one cluster, while Duroc was relatively distant from the other breeds. This indicated that Poland's pig breeds are separated by the largest genetic distance from the foreign Duroc breed. This reflects the fact that Duroc originated in the USA by the crossing of Red Guinea pigs and Iberian pigs with the Berkshire breed and was introduced to Poland relatively recently—in the 1970s.

The close genetic relationship between the PL and PLW breeds was proven by STRUC-TURE analysis with K = 3 (Figure 2), the pairwise F_{ST} and θw values. This was further supported by the results of the principal coordinate analysis (PCoA). The obtained results of PCoA analysis show four pig clusters—Duroc (DUR), Puławska (PUL) and the Polish Landrace (PL) and Polish Large White (PLW) together (Figure 4). The PL and PLW breeds were included in one cluster, which confirmed the genetic relationship between these breeds and confirmed the clear distinction of the DUR breed from the Polish population.

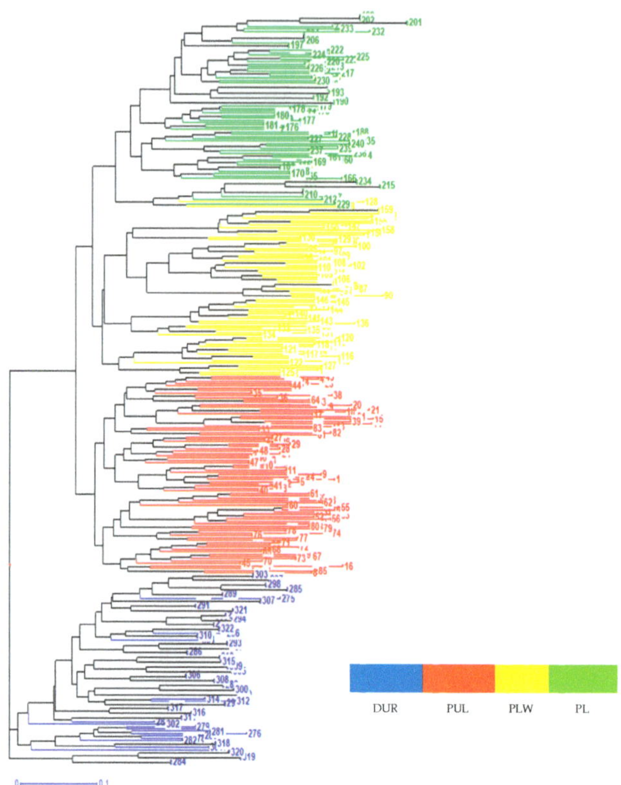

Figure 3. Genetic relationships among pigs with the use of a neighbor-joining tree obtained from a method based on genetic distance. PLW—Polish Large White; PL—Polish Landrace; PUL—Puławska Pig; DUR—Duroc.

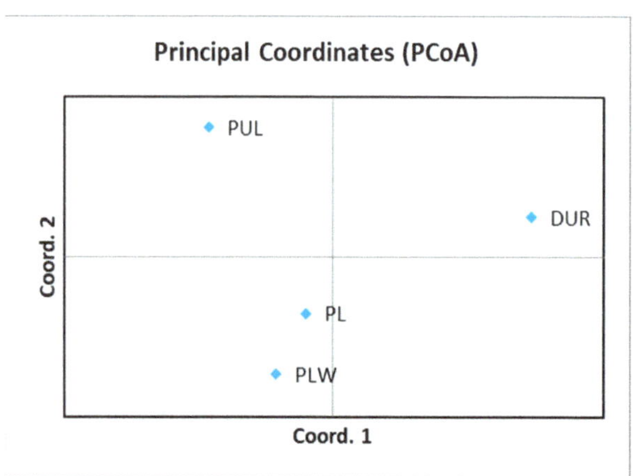

Figure 4. Principal coordinate analysis (PCoA) of the four pig breeds. DUR—Duroc; PUL—Puławska Pig; PL—Polish Landrace; PLW—Polish Large White. A two dimensional plot of the PCoA analysis show the clustering of four breeds. The first and second coordinates account for 56.9% and 24.5%, respectively, of the total variation.

Figure 4 illustrates the population relationships based on the PCoA using 14 STR markers. The first principal coordinate distinguished clearly the DUR breed from the PUL, PL and PLW breeds. The second principal coordinate separated the PL and PLW breeds from the other two pig breeds. The first, second and third principal coordinates (PCoA) represented 56.9%, 24.5% and 18.6%, respectively, of the total variation.

Nowadays, single-nucleotide polymorphism markers are increasingly used for biodiversity studies and the identification and parentage control of livestock, including pigs [33–36]. SNP genotyping has been used already to develop an SNP panel for discriminating breeds, meat and other pig products [37,38]. However, STR markers are still widely used in routine studies due to their reliability, sensitivity and cheaper analysis methods. Therefore, STR analysis in the pig population should continue.

4. Conclusions

The present study analyzed the genetic differentiation of selected pig breeds, Polish Large White, Polish Landrace, Puławska pigs and Duroc, and the possibility of using DNA tests for pig breed prediction.

The analysis of the genetic structure of the pig populations based on 14 STRs showed a clear division of the population into four groups, representing the four selected breeds for study. Our study demonstrates that a panel of microsatellite markers recommended by ISAG for the individual identification of pigs also may be useful for pig breed prediction and, in the future, for meat traceability. It is especially important for the population of native or local pigs, such as the Puławska breed, which can provide meat that is both high in quality and functional.

The presented work can be the first step to developing a system to determine whether meat comes from a pig of the declared breed and whether the meat was produced consistently with the declaration on the packaging.

Author Contributions: Conceptualization, A.R.; methodology, A.K. and G.S.; statistical analysis, A.R.; writing—original draft preparation, A.R.; writing—review and editing, A.R., A.K. and G.S. All authors have read and agreed to the published version of the manuscript.

Funding: This research received no external funding.

Institutional Review Board Statement: Not applicable.

Informed Consent Statement: Not applicable.

Data Availability Statement: The data presented in this study are available within the article.

Conflicts of Interest: The authors declare no conflict of interest.

References

1. Lim, H.T.; Seo, B.Y.; Jung, E.J.; Yoo, C.K.; Zhong, T.; Cho, I.C.; Yoon, D.H.; Lee, J.G.; Jeon, J.T. Establishment of a Microsatellite Marker Set for Individual, Pork Brand and Product Origin Identification in Pigs. *J. Anim. Sci. Technol.* **2009**, *51*, 201–206. [CrossRef]
2. Oh, J.-D.; Song, K.-D.; Seo, J.-H.; Kim, D.-K.; Kim, S.-H.; Seo, K.-S.; Lim, H.-T.; Lee, J.-B.; Park, H.-C.; Ryu, Y.-C.; et al. Genetic Traceability of Black Pig Meats Using Microsatellite Markers. *AJAS* **2014**, *27*, 926–931. [CrossRef] [PubMed]
3. Zhao, J.; Li, T.; Zhu, C.; Jiang, X.; Zhao, Y.; Xu, Z.; Yang, S.; Chen, A. Selection and Use of Microsatellite Markers for Individual Identification and Meat Traceability of Six Swine Breeds in the Chinese Market. *Food Sci. Technol. Int.* **2018**, *24*, 292–300. [CrossRef] [PubMed]
4. Babicz, M.; Hałabis, M.; Skałecki, P.; Domaradzki, P.; Litwińczuk, A.; Kropiwiec-Domańska, K.; Łukasik, M. Breeding and Performance Potential of Puławska Pigs—A Review. *Ann. Anim. Sci.* **2020**, *20*, 343–354. [CrossRef]
5. Rychlik, T.; Krawczyk, A. Kontrola wiarygodności rodowodów owiec w oparciu o markery genetyczne klasy I. *Wiad. Zootech.* **2008**, *R.XLVI*, 3–8.
6. Radko, A.; Rychlik, T.; Slota, E. Genetic Characterization of the Wrzosówka Sheep Breed on the Basis of 14 Microsatellite DNA Markers. *Med. Weter.* **2006**, *62*, 1073–1075.
7. Radko, A. Microsatellite DNA Polymorphism and Its Usefulness for Pedigree Verification of Cattle Raised in Poland. *Ann. Anim. Sci.* **2008**, *4*, 311–332.
8. Ząbek, T.; Fornal, A. Evaluation of the 17-Plex STR Kit for Parentage Testing of Polish Coldblood and Hucul Horses. 2009. Available online: https://depot.ceon.pl/handle/123456789/2227 (accessed on 29 November 2022).

9. Salamon, D.; Gutierrez-Gil, B.; Arranz, J.J.; Barreta, J.; Batinic, V.; Dzidic, A. Genetic Diversity and Differentiation of 12 Eastern Adriatic and Western Dinaric Native Sheep Breeds Using Microsatellites. *Animal* **2014**, *8*, 200–207. [CrossRef]
10. Laval, G.; Iannuccelli, N.; Legault, C.; Milan, D.; Groenen, M.A.; Giuffra, E.; Andersson, L.; Nissen, P.H.; Jørgensen, C.B.; Beeckmann, P.; et al. Genetic Diversity of Eleven European Pig Breeds. *Genet. Sel. Evol.* **2000**, *32*, 187–203. [CrossRef]
11. Fabuel, E.; Barragán, C.; Silió, L.; Rodríguez, M.C.; Toro, M.A. Analysis of Genetic Diversity and Conservation Priorities in Iberian Pigs Based on Microsatellite Markers. *Heredity* **2004**, *93*, 104–113. [CrossRef]
12. Vicente, A.A.; Carolino, M.I.; Sousa, M.C.O.; Ginja, C.; Silva, F.S.; Martinez, A.M.; Vega-Pla, J.L.; Carolino, N.; Gama, L.T. Genetic Diversity in Native and Commercial Breeds of Pigs in Portugal Assessed by Microsatellites. *J. Anim. Sci.* **2008**, *86*, 2496–2507. [CrossRef] [PubMed]
13. Sollero, B.P.; Paiva, S.R.; Faria, D.A.; Guimarães, S.E.F.; Castro, S.T.R.; Egito, A.A.; Albuquerque, M.S.M.; Piovezan, U.; Bertani, G.R.; Mariante, A.d.S. Genetic Diversity of Brazilian Pig Breeds Evidenced by Microsatellite Markers. *Livest. Sci.* **2009**, *123*, 8–15. [CrossRef]
14. Lin, Y.-C.; Hsieh, H.-M.; Lee, J.C.-I.; Hsiao, C.-T.; Lin, D.-Y.; Linacre, A.; Tsai, L.-C. Establishing a DNA Identification System for Pigs (Sus Scrofa) Using a Multiplex STR Amplification. *Forensic. Sci. Int. Genet.* **2014**, *9*, 12–19. [CrossRef] [PubMed]
15. Kramarenko, S.S.; Lugovoy, S.I.; Kharzinova, V.R.; Lykhach, V.Y.; Kramarenko, A.S.; Lykhach, A.V. Genetic Diversity of Ukrainian Local Pig Breeds Based on Microsatellite Markers. *Regul. Mech. Biosyst.* **2018**, *9*, 177–182. [CrossRef]
16. Margeta, P.; Skorput, D.; Salamon, D.; Mencik, S.; Gvozdanović, K.; Karolyi, D.; Lukovic, Z.; Salajpal, K. 12-Plex Highly Polymorphic Microsatellite Marker Set for Parentage Analysis in Banija Spotted Pigs. *J. Cent. Eur. Agric.* **2019**, *20*, 50–54. [CrossRef]
17. Charoensook, R.; Gatphayak, K.; Brenig, B.; Knorr, C. Genetic Diversity Analysis of Thai Indigenous Pig Population Using Microsatellite Markers. *Asian-Australas J. Anim. Sci.* **2019**, *32*, 1491–1500. [CrossRef]
18. ISAG Conference 2012, Cairns, Australia, Pig Molecular Comparison Test Report. Available online: https://www.isag.us/docs/piggengenomics2012.pdf (accessed on 29 November 2022).
19. ISAG Conference 2014, Xi'an, China, Pig Genetics and Genomics. Available online: https://www.isag.us/docs/pig%20genetics%20genomics%20and%20ct_2014.pdf (accessed on 22 August 2014).
20. Szmatoła, T.; Ropka-Molik, K.; Tyra, M.; Piórkowska, K.; Żukowski, K.; Oczkowicz, M.; Blicharski, T. The Genetic Structure of Five Pig Breeds Maintained in Poland. *Ann. Anim. Sci.* **2016**, *16*, 1019–1027. [CrossRef]
21. Smouse, P.E.; Banks, S.C.; Peakall, R. Converting Quadratic Entropy to Diversity: Both Animals and Alleles Are Diverse, but Some Are More Diverse than Others. *PLoS ONE* **2017**, *12*, e0185499. [CrossRef] [PubMed]
22. Pritchard, J.K.; Stephens, M.; Donnelly, P. Inference of Population Structure Using Multilocus Genotype Data. *Genetics* **2000**, *155*, 945–959. [CrossRef]
23. Hubisz, M.J.; Falush, D.; Stephens, M.; Pritchard, J.K. Inferring weak population structure with the assistance of sample group information. *Mol. Ecol. Resour.* **2009**, *9*, 1322–1332. [CrossRef]
24. Falush, D.; Stephens, M.; Pritchard, J.K. Inference of population structure using multilocus genotype data: Linked loci and correlated allele frequencies. *Genetics* **2003**, *164*, 1567–1587. [CrossRef] [PubMed]
25. Earl, D.A.; vonHoldt, B.M. STRUCTURE HARVESTER: A Website and Program for Visualizing STRUCTURE Output and Implementing the Evanno Method. *Conserv. Genet. Resour.* **2012**, *4*, 359–361. [CrossRef]
26. Reynolds, J.; Weir, B.S.; Cockerham, C.C. Estimation of the Coancestry Coefficient: Basis for a Short-Term Genetic Distance. *Genetics* **1983**, *105*, 767–779. [CrossRef] [PubMed]
27. Radko, A.; Smołucha, G.; Koseniuk, A. Microsatellite DNA Analysis for Diversity Study, Individual Identification and Parentage Control in Pig Breeds in Poland. *Genes* **2021**, *12*, 595. [CrossRef] [PubMed]
28. Wright, S. Variability within and among Natural Populations. In *Evolution and the Genetics of Populations*; University of Chicago Press: Chicago, IL, USA, 1978; Volume 4.
29. Meirmans, P.G. Using the Amova Framework to Estimate a Standardized Genetic Differentiation Measure. *Evolution* **2006**, *60*, 2399–2402. [CrossRef]
30. SanCristobal, M.; Chevalet, C.; Haley, C.S.; Joosten, R.; Rattink, A.P.; Harlizius, B.; Groenen, M.A.; Amigues, Y.; Boscher, M.-Y.; Russell, G.; et al. Genetic Diversity within and between European Pig Breeds Using Microsatellite Markers. *Anim. Genet.* **2006**, *37*, 189–198. [CrossRef]
31. Royal, C.D.; Novembre, J.; Fullerton, S.M.; Goldstein, D.B.; Long, J.C.; Bamshad, M.J.; Clark, A.G. Inferring Genetic Ancestry: Opportunities, Challenges, and Implications. *Am. J. Hum. Genet.* **2010**, *86*, 661–673. [CrossRef]
32. Porras-Hurtado, L.; Ruiz, Y.; Santos, C.; Phillips, C.; Carracedo, Á.; Lareu, M. An Overview of STRUCTURE: Applications, Parameter Settings, and Supporting Software. *Front. Genet.* **2013**, *4*, 98. [CrossRef]
33. Dadousis, C.; Muñoz, M.; Óvilo, C.; Fabbri, M.C.; Araújo, J.P.; Bovo, S.; Potokar, M.Č.; Charneca, R.; Crovetti, A.; Gallo, M.; et al. Admixture and breed traceability in European indigenous pig breeds and wild boar using genome-wide SNP data. *Sci. Rep.* **2022**, *12*, 7346. [CrossRef]
34. Muñoz, M.; Bozzi, R.; García-Casco, J.; Núñez, Y.; Ribani, A.; Franci, O.; García, F.; Škrlep, M.; Schiavo, G.; Bovo, S.; et al. Genomic diversity, linkage disequilibrium and selection signatures in European local pig breeds assessed with a high density SNP chip. *Sci. Rep.* **2019**, *9*, 13546. [CrossRef]

35. Silió, L.; Barragán, C.; Fernández, A.I.; García-Casco, J.; Rodríguez, M.C. Assessing effective population size, coancestry and inbreeding effects on litter size using the pedigree and SNP data in closed lines of the Iberian pig breed. *J. Anim. Breed. Genet.* **2016**, *133*, 145–154. [CrossRef] [PubMed]

36. Zanella, R.; Peixoto, J.O.; Cardoso, F.F.; Cardoso, L.L.; Biegelmeyer, P.; Cantão, M.E.; Otaviano, A.; Freitas, M.S.; Caetano, A.R.; Ledur, M.C. Genetic diversity analysis of two commercial breeds of pigs using genomic and pedigree data. *Genet. Sel. Evol.* **2016**, *48*, 24. [CrossRef] [PubMed]

37. Rogberg-Muñoz, A.; Wei, S.; Ripoli, M.; Guo, B.; Carino, M.; Lirón, J.; Giovambattista, G. Effectiveness of a 95 SNP panel for the screening of breed label fraud in the Chinese meat market. *Meat Sci.* **2016**, *111*, 47–52. [CrossRef]

38. Muñoz, M.; García-Casco, J.; Alves, E.; Benítez, R.; Barragán, C.; Caraballo, C.; Fernández, A.; García, F.; Núñez, Y.; Óvilo, C.; et al. Development of a 64 SNV panel for breed authentication in Iberian pigs and their derived meat products. *Meat Sci.* **2020**, *167*, 108152. [CrossRef] [PubMed]

Article

Lnc*PLAAT3*-AS Regulates *PLAAT3*-Mediated Adipocyte Differentiation and Lipogenesis in Pigs through miR-503-5p

Zhiming Wang [1,†], Jin Chai [1,†], Yuhao Wang [1], Yiren Gu [2], Keren Long [1], Mingzhou Li [1,*] and Long Jin [3,*]

[1] Key Laboratory of Livestock and Poultry Multiomics, Ministry of Agriculture and Rural Affairs, College of Animal Science and Technology, Sichuan Agricultural University, Chengdu 611130, China
[2] Sichuan Key Laboratory of Animal Breeding and Genetics, Sichuan Institute of Animal Science, Chengdu 610066, China
[3] Sichuan Provincial Key Laboratory of Animal Breeding and Genetics, Institute of Animal Genetics and Breeding, Sichuan Agricultural University, Chengdu 611130, China
* Correspondence: mingzhou.li@sicau.edu.cn (M.L.); longjin@sicau.edu.cn (L.J.)
† These authors contributed equally to this work.

Abstract: Animal fat deposition has a significant impact on meat flavor and texture. However, the molecular mechanisms of fat deposition are not well understood. Lnc*PLAAT3*-AS is a naturally occurring transcript that is abundant in porcine adipose tissue. Here, we focus on the regulatory role of lnc*PLAAT3*-AS in promoting preadipocyte proliferation and adipocyte differentiation. By overexpressing or repressing lnc*PLAAT3* expression, we found that lnc*PLAAT3*-AS promoted the transcription of its host gene *PLAAT3*, a regulator of adipocyte differentiation. In addition, we predicted the region of lnc*PLAAT3*-AS that binds to miR-503-5p and showed by dual luciferase assay that lnc*PLAAT3*-AS acts as a sponge to absorb miR-503-5p. Interestingly, miR-503-5p also targets and represses *PLAAT3* expression and helps regulate porcine preadipocyte proliferation and differentiation. Taken together, these results show that lnc*PLAAT3*-AS upregulates *PLAAT3* expression by absorbing miR-503-5p, suggesting a potential regulatory mechanism based on competing endogenous RNAs. Finally, we explored lnc*PLAAT3*-AS and *PLAAT3* expression in adipose tissue and found that both molecules were expressed at significantly higher levels in fatty pig breeds compared to lean pig breeds. In summary, we identified the mechanism by which lnc*PLAAT3*-AS regulates porcine preadipocyte proliferation and differentiation, contributing to our understanding of the molecular mechanisms of lipid deposition in pigs.

Keywords: primary adipocyte cells; adipose tissue; lncRNA-AS; RNA-seq

Citation: Wang, Z.; Chai, J.; Wang, Y.; Gu, Y.; Long, K.; Li, M.; Jin, L. Lnc*PLAAT3*-AS Regulates *PLAAT3*-Mediated Adipocyte Differentiation and Lipogenesis in Pigs through miR-503-5p. *Genes* **2023**, *14*, 161. https://doi.org/10.3390/genes14010161

Academic Editors: Katarzyna Piórkowska and Katarzyna Ropka-Molik

Received: 24 November 2022
Revised: 3 January 2023
Accepted: 3 January 2023
Published: 6 January 2023

1. Introduction

In recent years, genome annotation has identified many lncRNAs, although the functions of most of these lncRNAs are still unknown [1]. Some antisense lncRNAs have been demonstrated to have functions. For example, the *ZFPM2* antisense RNA 1 (*ZFPM2*-AS1) lncRNA has been reported to regulate the migration and invasion of hepatocellular carcinoma cells by mediating the miR-139/GDF10 axis [2]. *GAS6* antisense RNA 1 (*GAS6*-AS1) regulates cell proliferation and invasion in clear cell renal cell carcinoma (*ccRCC*) by mediating the AMPK/mTOR signaling pathway, suggesting that *GAS6*-AS1 may be a potential therapeutic target in *ccRCC* [3]. *Nqo1*-AS1 upregulates Nqo1 expression by binding to the *Nqo1* 3′UTR and increasing *Nqo1* mRNA stability, thereby attenuating cigarette smoke-induced oxidative stress [4]. Although many studies have been conducted on antisense lncRNAs [5], the role of lncRNAs in porcine adipogenesis remains largely unknown.

Pigs are an essential animal in the agricultural economy and an important source of meat worldwide [6]. In addition, domestic pigs are essential model animals that are frequently used in medical research due to having a similar genome size, gene structure and function, anatomical structure of the digestive organs, metabolic patterns, visceral organ metabolism, and dietary habits (omnivorous diet) to humans [7,8]. In nutritional, metabolic, and cardiovascular studies, as well as in several other areas of biomedical research, pigs have proven to be valuable animal models [9].

It has been previously reported that the knockdown of *PLAAT3* significantly inhibits lipid deposition in mice [10]. Previous studies have shown that lnc*PLAAT3*-AS enhances *PLAAT3* mRNA stability by forming an RNA–RNA dimer with the *PLAAT3* transcript [11]. However, there is no direct evidence of a specific regulatory role for lnc*PLAAT3*-AS in adipogenesis or for the associated molecular mechanisms. To analyze the role of lnc*PLAAT3*-AS in lipogenic differentiation, we overexpressed or knocked down lnc*PLAAT3*-AS in porcine primary preadipocytes. The results from these experiments showed that lnc*PLAAT3*-AS regulates the expression of cell cycle-related genes and promotes the differentiation of preadipocytes. lnc*PLAAT3*-AS promoted adipocyte differentiation by acting as a sponge for miR-503-5p, thereby repressing the activity of this miRNA and promoting the increased expression of *PLAAT3*. In conclusion, our findings demonstrate the molecular mechanism by which a lncRNA regulates adipogenesis and provide a potential molecular approach to improving lean muscle mass in livestock production by inhibiting fat deposition.

2. Materials and Methods

2.1. Ethics Statement

All procedures involving animals in this study were managed and operated in strict accordance with the Regulations for the Management of Laboratory Animal Affairs (Ministry of Science and Technology, Beijing, China, June 2004) and were also approved by the Animal Care and Use Group Institution, College of Animal Science and Technology, Sichuan Agricultural University, China, under license number NO.20210162.

2.2. Cell Culture and Differentiation

The piglet was humanely killed and the fat tissue from the backs of the female Rongchang piglets was collected to obtain preadipocytes. After removing the soft tissue and small tubes, the remaining fat tissue was washed three times with phosphate-buffered salt water (PBS, Gibco, Carlsbad, CA, USA). Next, about 60 g of the fat tissue was quickly minced and digested at 37 °C for 1 h with 0.25% type I collagenase (Invitrogen, Carlsbad, CA, USA). The digested tissue solution was then centrifuged at 1000 rpm for 8 min. The cells were passed through 100 um strainers, and the pellet (containing cleaned preadipocytes) was resuspended in Dulbecco's changed Eagle medium (Gibco, Carlsbad, CA, USA) containing 20% fetal calf serum (Gibco, Carlsbad, CA, USA). The preadipocytes were cultured containing 5% CO^2 and maintained at 37 °C. When preadipocytes reached 80% confluence, they were digested with trypsin and passaged in new neo-culture plates. After the preadipocytes had grown and spread to an appropriate density in the cell culture plates, they were treated with insulin (Sigma, Saint Louis, MI, USA), 3-isobutyl 1-methylxanthine (Sigma, Saint Louis, MI, USA), and dexamethasone (Sigma, Saint Louis, MI, USA) to induce differentiation of the preadipocytes into mature adipocytes [12]. All cell experiments were repeated three times independently, and the cells used on all three occasions were from the same pig.

2.3. Cell Transfection

Cells were transfected using lipofectamine 3000 (Lipo3000, Invitrogen) according to the manufacturer's instructions, with opti-MEM (Gibco, Carlsbad, CA, USA) as an auxiliary transfection reagent. Taking 6-well plates containing 2 mL of medium per well as an example, 100 µL of opti-MEM was first incubated with 5 µL of Lipo3000, separately, for 5 min. Subsequently, the two solutions were combined and incubated for 10 min, after which the combined solution was added to the culture medium to transfect the cells.

2.4. Western Blot Assay

Cells were collected from six-well plates using a cell spatula. The extracted protein was isolated using a radio immunoprecipitation assay (RIPA) kit (Bio-Rad, Hercules, CA, USA). After the proteins were extracted, the BCA kit was used to determine their concentration. Absin-prepared SDS-PAGE gels (Absin Bioscience Inc., Shanghai, China) were used to separate the proteins. Using a BIO-RAD protein transfer machine, the products were then transferred onto polyvinylidene difluoride membranes and incubated with primary antibodies. Protein bands were detected using ECL (Service) and visualized using the ECL Western Blot Detection Kit (Thermo Scientific, Rockford, IL, USA) and the Image Quant LAS 4000 kit (GE HealthCare, Chicago, IL, USA). All western blotting experiments were repeated at least three times.

2.5. Luciferase Reporter Assay

Wild type (WT) and MUT lncPLAAT3-AS plasmids were constructed using the pcDNA3.1 vector. They were transfected with miR-503-5p mimics, miR-503-5p inhibitor, miR-503-5p mimics negative control (mimics NC), and miR-503-5p inhibitor negative control (inhibitor NC) using the lipofectamine 3000 reagents into PK15 cells as described in Section 2.4. After 48 h, the samples were collected, and the fluorescence strength was detected using the Dual-Glo Luciferase Assay System (Promega, Madison, WI, USA) following the manufacturer's instructions.

2.6. Assessment of Cell Proliferation via CCK-8 and EdU Assays

The growth and spread of cells were detected using the Cell Counting Kit 8 (CCK-8, Biosharp, Hefei, China) and 5-ethynyl-20-deoxyuridine (EdU, Ribobio, Guangzhou, China). After transfection, 10 μL of CCK-8 reagent was added to each well of the cell culture plate, and the absorbance at 450 nm was measured after 24 h. Preadipocytes were seeded in 12-well plates for the EdU assay. Following transfection, 100 mL of 50 mM EdU reagent was added to each well and grown for at least 24 h before images were taken with a Nikon TE2000 microscope (Nikon, Tokyo, Japan).

2.7. RNA Isolation and Reverse Transcription

Trizol (Invitrogen) was used to extract the total RNA from cells. The total RNA was reverse transcribed to cDNA using the HiScript III RT Super Mix (Vazyme, Nanjing, China), oligo-dTs according to the manufacturer's instructions. Denaturing gel electrophoresis and spectrophotometry (Thermo, Waltham, MA, USA) was used to measure RNA mass and concentration, and the final product was diluted to the appropriate volume with water.

2.8. Real-Time Quantitative PCR

Taq Pro Universal SYBR qPCR Master Mix reagents (Vazyme, Nanjing, China) were used for real-time quantitative PCR (RT-qPCR). Primer5 software was used to design RT-qPCR primers, and their sequences are shown in Table 1. RT-qPCR was performed on a Bio-Rad CFX96 Real-Time PCR detection system. Each sample was analyzed in triplicate. The relative expression levels of each gene tested within the samples were calculated using the $2^{-\Delta\Delta CT}$ method.

Table 1. Primer information for quantitative real-time PCR (qRT-PCR).

Gene	Primer Sequence (5′ to 3′)	Product Size (bp)
PLA2G16	F: ATATGTGGTCCACCTGGCTCCCC R: ATTGCTTTTGCCGCTTGTTTCTG	395
PLA2G16-AS	F: GGACTCTGCGGCCATTTAAC R: GCTTTGGGACAATGAGTCGC	213

Table 1. *Cont.*

Gene	Primer Sequence (5′ to 3′)	Product Size (bp)
GAPDH	F: CCCCTTCATTGACCTCCACT R: CCATTTGATGTTGGCGGGAT	192
CCND1	F: GCATGTTCGTGGCCTCTAAG R: CGTGTTTGCGGATGATCTGT	228
CDK4	F: TCAGCACAGTTCGTGAGGTG R: GTCCATCAGCCGGACAACAT	77
P53	F: GGGACGGAACAGCTTTGA R: TTTGCACTGGCGAGGAG	161
PPARγ	F: CTCCAAGAATACCAAAGTGCGA R: GCCTGATGCTTTATCCCCACA	150
C/EBPα	F: CAAGAACAGCAACGAGTACCG R: GTCACTGGTCAACTCCAGCAC	124
FABP4	F: GAAGTGGGAGTGGGCTTT R: TTATGGTGCTCTTGACTTTCCT	190
ACS	F: GCAGGCAGGCTCAGTTT R: CTCTGTTCAGGGGAGGGT	129
ACADL	F: TGTCTCCAGCTGCATGAAACGA R: AGCTGCACACAGTCATAAGCCA	107
DGAT	F: CCTACCGCGATCTCTACTACTT R: GGGTGAAGAACAGCATCTCAA	126
FAS	F: CCAACCAGCAACACCAA R: CAGGTACGGGAATGAGGA	100
ssc-miR-503	UAGCAGCGGGAACAGUACUGCAG	
U6	F: CGCTTCGGCAGCACATATAC R: TTCACGAATTTGCGTGTCAT	87

2.9. Oil Red O Staining

After induction of adipogenesis, cells were washed two to three times with PBS (Gibco, Carlsbad, CA, USA) and then fixed for 30 min in 4% paraformaldehyde. The samples were rinsed twice with 60% isopropanol and dried for 30 min before being treated with 1 mL of the oil red O dye working solution. A microscope was used to observe the oil red O staining after adding 1 mL PBS (Gibco, Carlsbad, CA, USA) to the culture plate.

2.10. RNA-Seq and Collection of Sequencing Data

Ampure XP beads (Beckman Coulter, Brea, CA, USA) were used to purify the PCR-amplified cDNA fragments with adapters. The switching mechanism at the 5′ end of the RNA transcript primer with oligo(dG) at the 3′ end was added in advance to the cDNA synthesis reaction. The switching mechanism at the 5′ end of the RNA transcript primer oligo(dG) pairs with the protruding C's at the end of the synthetic cDNA to form an extension template for the cDNA, and the reverse transcriptase automatically switches the template to use the SMART primer as an extension template to continue extending the cDNA strand to the end of the primer. All resulting cDNA strands have an oligo(dT)-containing primer sequence at one end and a known SMART primer sequence at the other end, which can be amplified using universal primers after the second strand has been synthesized. A Bioanalyzer Agilent Technologies 2100 (Agilent Technologies, Santa Clara, CA, USA) was used to validate the cDNA library. Following heat-denatured PCR products, splint oligos were used to circularize them. The final library was considered to be the single-stranded circular DNA. After the final library was amplified with phi29 (Thermo Fisher Scientific, Waltham, MA, USA), over 300 copies of each molecule were produced as DNA nanoballs (DNBs) [13,14]. A Bioanalyzer Agilent Technologies 2100 was used to validate the cDNA library. Following heat-denatured PCR products, splint oligos were used to circularize them. The final library was considered to be the single-stranded circular DNA. After the final library was amplified with phi29 (Thermo Fisher Scientific, Waltham, MA, USA), over 300 copies of each molecule were produced as DNBs. The DNBs loaded into the patterned nanoarray were read using the BGISEQ500 platform (BGI, Shenzhen, China).

FAStQC software (Version 0.11.9) was used to quality control (QC) the raw data obtained from sequencing, and the Q30 values and gas chromatograph (GC) content of the sequencing data were calculated. Hisat software (Version 2 2.2.1) was used to perform genome alignment of the clean reads, and 90.55–91.20% could be aligned to the reference genome, indicating valid reads and good sequencing results. The differentially expressed genes were characterized, and Gene Ontology (GO) and Kyoto Encyclopedia of Genes and Genomes (KEGG) functional enrichment analyses were performed

A total of seven Chinese pig breeds were downloaded, including Chenghua, Neijiang, Tibetan, Qingyu, Wujin, Yacha, and Yanan, and one Western breed, Yorkshire. RNA-seq data were downloaded from the National Library of Medicine database (https://www.ncbi.nlm.nih.gov/geo/, accessed on 20 July 2022, accession numbers SRP090525) [15,16].

2.11. Data Analysis and Statistics

At least three independent replicates were performed for each experiment. MRNA and miRNA expression levels were calculated using the $2^{-\Delta\Delta Ct}$ method, and data are presented as the mean ± SEM of each group. ANOVA and Student's *t*-test were applied to assess the differences in expression levels between groups using GraphPad Prism 8.0 (GraphPad Software), with $p < 0.05$ considered significant and $p < 0.01$ highly significant. A * indicates significant differences, and ** indicates highly significant differences.

3. Results

3.1. LncPLAAT3-AS Regulates Porcine Primary Preadipocyte Proliferation

To assess the role of lnc*PLAAT3*-AS in porcine primary preadipocyte proliferation, we transfected porcine primary preadipocytes with a lnc*PLAAT3*-AS overexpression plasmid or siRNA-lnc*PLAAT3*-AS. As shown in Figure 1A, lnc*PLAAT3*-AS expression was significantly decreased after transfection with siRNA-lnc*PLAAT3*-AS and significantly increased after transfection with the lnc*PLAAT3*-AS overexpression plasmid compared with the negative control ($p < 0.01$). Next, we performed an EdU and proliferation assay and found that the percentage of EdU-positive cells increased after overexpression of lnc*PLAAT3*-AS, while transfection with siRNA-lnc*PLAAT3*-AS siRNA had the opposite effect. The results from the CCK-8 assay results were similar: the cell fluorescence value at 450 nm increased by 50% after overexpression of lnc*PLAAT3*-AS compared with the negative control ($p < 0.01$) (Figure 1B,D). We then examined the expression of genes involved in cell proliferation and apoptosis by RT-qPCR results and found that *CCND1*, *CCNE1*, and *CDK4* were significantly upregulated after transfection with the lnc*PLAAT3*-AS overexpression plasmid compared with the negative control, while transfection with siRNA-lnc*PLAAT3*-AS had the opposite effect ($p < 0.01$). In addition, lnc*PLAAT3*-AS overexpression resulted in a significant decrease in the expression of cell cycle protein-dependent kinase inhibitor (*p53*), a marker of inhibited cell proliferation ($p < 0.01$) (Figure 1C). Taken together, these results suggest that lnc*PLAAT3*-AS plays a role in promoting the proliferation of porcine primary preadipocytes.

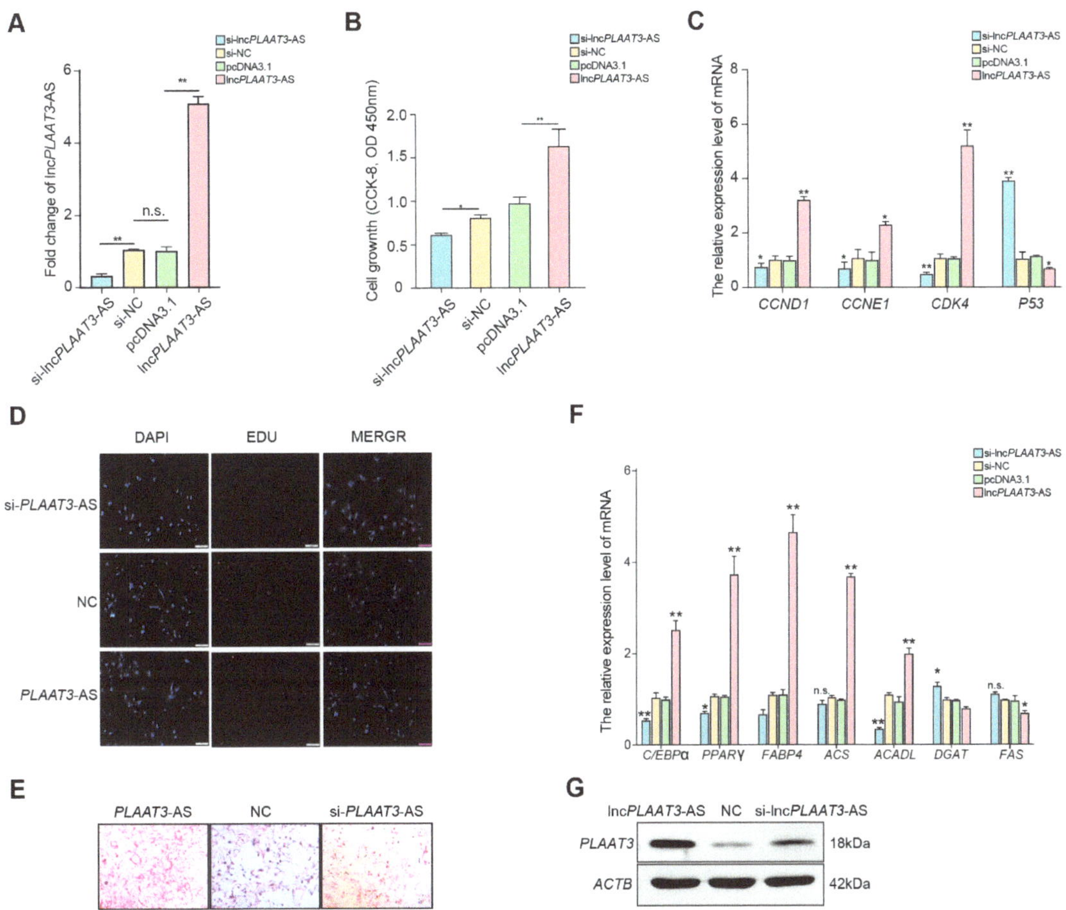

Figure 1. Lnc*PLAAT3*-AS promotes preadipocyte proliferation and differentiation. Primary adipocytes were transfected with a lnc*PLAAT3*-AS overexpression plasmid, an siRNA against lnc*PLAAT3*-AS (si-lnc*PLAAT3*-AS), or a negative control (si-NC) construct; (**A**) transfection efficiency was measured by qRT-PCR; (**B**) cell proliferation was evaluated by CCK-8 assay; (**C**) expression of genes associated with cell proliferation was measured by qRT-PCR; (**D**) representative images of an EdU assay of PK15 cells transfected with lnc*PLAAT3*-AS mimics or lnc*PLAAT3*-AS inhibitor; (**E**) oil red O staining; (**F**) expression of marker genes related to adipogenesis, fatty acid oxidation, and fatty acid transportation synthesis (scale bars 100 μm); (**G**) *PLAAT3* expression after transfection with each of the three constructs described above. All results are presented as means ± SEM; n = 3; * $p < 0.05$; ** $p < 0.01$; n.s., no significant difference.

3.2. LncPLAAT3 Promotes Porcine Primary Preadipocyte Differentiation

To assess the role of lnc*PLAAT3*-AS in porcine preadipocyte differentiation, we transfected porcine primary preadipocytes with one of the three constructs described in Section 3.1 and then induced lipogenic differentiation for 8 days. As shown in Figure 1E, oil red O staining demonstrated that inhibition of lnc*PLAAT3* expression significantly reduced lipid droplet formation compared with the negative control group, while overexpression of lnc*PLAAT3*-AS had the opposite effect. To confirm these results, we examined the expression levels of *CEBPα*, *PPARγ*, and *FABP4*, which are markers of adipocyte differentiation [17]. We found that transfection with the lnc*PLAAT3*-AS overexpression vector significantly enhanced the expression of all three of these genes, while transfection with the siRNA inhibited their expression (Figure 1F). These suggest that lnc*PLAAT3*-AS promotes adipocyte differentiation. In addition, we examined the expression of genes related to fatty acid synthesis (*ACS* and *ACADL*) [18] and fatty acid oxidation (*DGAT* and *FAS*). As expected, *ACS* and *ACADL* expression were significantly increased and *DGAT* and *FAS* expression were significantly decreased in the lnc*PLAAT3* overexpression group compared to the control group, while the opposite effect was observed in the siRNA-treated group (Figure 1F). These experimental results were further validated at the protein level within the samples (Figure 1G). Based on the above results, it is reasonable to conclude that lnc*PLAAT3*-AS promotes preadipocyte differentiation.

3.3. Differences in Gene Expression in Adipocytes after lncPLAAT3 Overexpression or Knockdown

To characterize the regulatory role of lnc*PLAAT3*-AS in primary preadipocytes, we performed RNA-seq analysis of cells in which lnc*PLAAT3*-AS was overexpressed or knocked down, as well as the negative control, with each sample yielding an average of 6.72 G of data (Figure S1A,B and Table S1). The correlation coefficient and principal component analyses showed a clear separation among the three groups (Figure 2A,B). Compared with the negative control group, there were 824 differentially expressed genes (DEGs) in the lnc*PLAAT3*-AS overexpression group and 2219 DEGs in the lnc*PLAAT3*-AS siRNA group (Figures 2C,D and S2A, and Table S2). The potential functions and signaling pathways of all DEGs were determined by GO and KEGG enrichment analyses. Upregulated genes in the lnc*PLAAT3*-AS overexpression group compared with the control group (such as *ADORA1*, *CCR5*, and *CD36*) were associated with an inflammatory response, cellular response to lipid, and regulation of lipid storage; downregulated genes in the lnc*PLAAT3*-AS overexpression group compared with the control group (such as *CSF2*, *TFPI*, and *NOS3*) were associated with a cellular response to lipopolysaccharide, lipid homeostasis, and positive regulation of lipid localization. Upregulated genes in the lnc*PLAAT3*-AS siRNA group compared with the control group (such as *ABL2*, *ADCY1*, and *ARG1*) were associated with a response to lipopolysaccharide and Semaphorin-plexin signaling, while downregulated genes in the lnc*PLAAT3*-AS siRNA group compared with the control group (such as *ACTC1*, *ACTN2*, and *ADGRB1*) were associated with cellular component morphogenesis (Figures 2E–H and S2B,C). Interestingly, 20 genes closely related to fat metabolism, such as *CD36*, *CD68*, *CCR5*, and *TREM2*, exhibited the same expression patterns as *PLAAT3*, indicating that *PLAAT3* plays an important role in adipogenesis (Figure 2I,J) [19–22]. For example, the scavenger receptor *CD36* participates in the high-affinity tissue uptake of long-chain fatty acids (FAs) and contributes to lipid accumulation and metabolic dysfunction in excess [18].

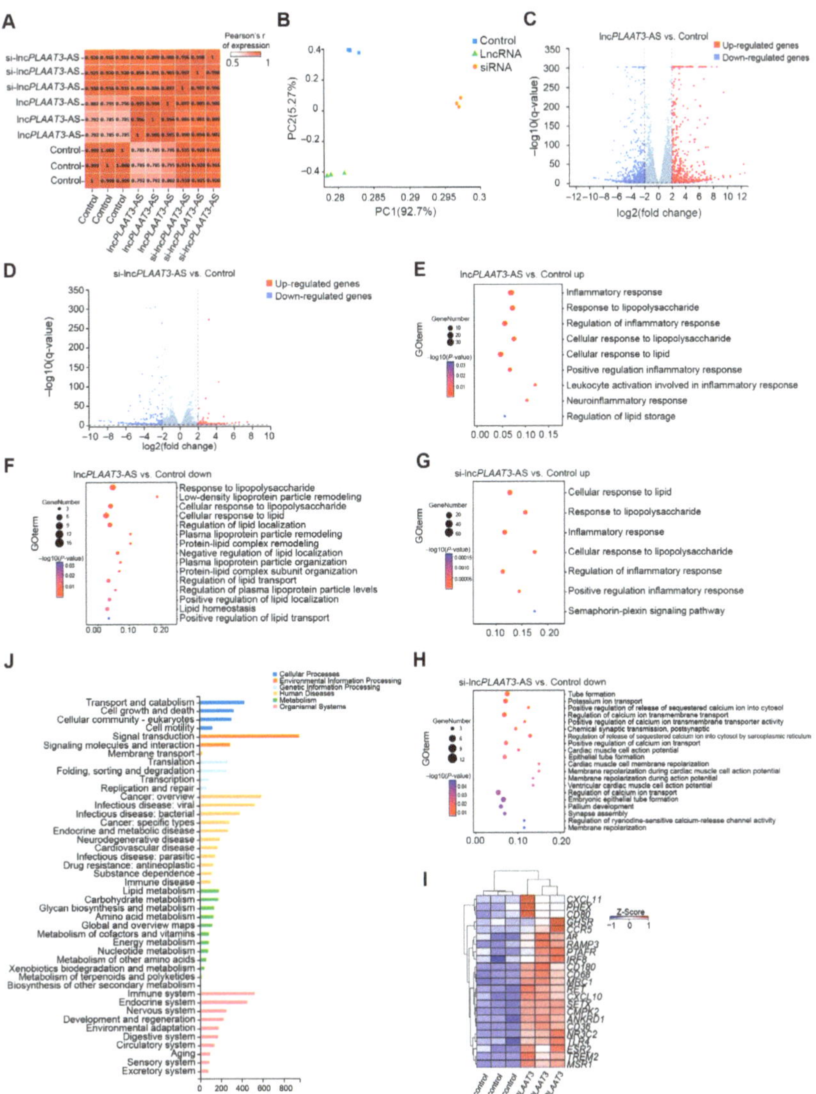

Figure 2. (**A**) Sample correlation heatmap; (**B**) principal component analysis; (**C,D**) volcano plot of differentially expressed genes. Positive values indicate upregulation, while negative values indicate downregulation. The x-axis represents the log2 scale of fold change. A significant difference in expression is indicated by the y-axis, which is the −log10 scale of the adjusted *p* values. The red dots in the figure indicate genes with significantly upregulated expression (at least two-fold change), blue dots indicate genes with significantly downregulated expression (at least two-fold change), and gray dots indicate genes with no significant difference in expression level. (**E–H**) The top 20 GO pathways for the differentially expressed genes according to the following comparisons: upregulated in lnc*PLAAT3*-AS vs. control, downregulated in lnc*PLAAT3*-AS vs. control, upregulated in si-lnc*PLAAT3*-AS vs. control, downregulated in si-lnc*PLAAT3*-AS vs. control; (**I**) differentially expressed gene pathway enrichment analysis; (**J**) All enriched GO and KEGG pathways.

3.4. LncPLAAT3-AS Targets miR-503-5p during Adipocyte Differentiation

Numerous studies have shown that lncRNAs can regulate various cellular responses by sponging on miRNAs [3,4,6,23]. Therefore, we used a lncTar online prediction software (http://www.cuilab.cn/lnctar (accessed on 23 November 2022)) and a dual-luciferase reporter system to screen for candidate miRNA binding partners and identified miR-503-5p as a potential lnc*PLAAT3*-AS target gene (Figure 3A). To determine whether this was an authentic interaction, we assayed the chemiluminescence values of PK15 cells cotransfected with the wild-type and mutant lnc*PLAAT3*-AS vectors described above and an miR-503-5p mimic or an NC mimic, using an enzyme marker. As expected, the relative fluorescence values of the cells cotransfected with the miR-503-5p mimic and the lnc*PLAAT3*-AS-WT vector were significantly lower than those of the other three groups (Figure 3B,C). These results validated the target gene prediction for lnc*PLAAT3*-AS. Moreover, we verified the interaction between miR-503-5p and *PLAAT3* using a dual-luciferase reporter system and found that miR-503-5p targets *PLAAT3* (Figure 3D). Next, we asked whether miR-503-5p is involved in porcine preadipocyte proliferation and differentiation. Then, we performed the mock transfection of porcine preadipocytes or transfected them with miR-503-5p and mock NC, inhibitor, or inhibitor NC (Figure 3E) to examine whether miR-503-5p affects preadipocyte proliferation. Transfection with the miR-503-5p mimics significantly inhibited preadipocyte proliferation, while the inhibitor-treated group had the opposite response (Figure 3F). CCK-8 and EdU assays confirmed this observation (Figure S3A,B), as did fluorescence quantification of the expression of marker genes for proliferation and cell cycle progression (Figure S3C). We also analyzed the effect of transfection of porcine preadipocytes with an miR-503-5p mimic, an inhibitor, and separate control reagents on preadipocyte differentiation. Surprisingly, miR-503-5p overexpression significantly inhibited the ability of preadipocytes to secrete lipid droplets, while cells transfected with the inhibitor had the opposite response (Figure S3D). Analysis of the expression of marker genes for lipogenic differentiation and fatty acid metabolism confirmed this result (Figure S3E). Taken together, these findings suggest that, in addition to enhancing the stability of *PLAAT3* by forming an RNA–RNA dimer [11], lnc*PLAAT3*-AS also helps to regulate adipogenesis and related molecules through miR-503-5p.

3.5. Differences in lncPLAAT3-AS and PLAAT3 Proliferation among Different Pig Breeds

To explore the role of lnc*PLAAT3*-AS and *PLAAT3* in fat deposition among different pig breeds, we downloaded the RNA-seq data for seven Chinese pig breeds, namely Chenghua, Neijiang, Qingyu, Yanan, Wujin, Yacha, and Tibetan, as well as one introduced Western breed, Yorkshire, with three biological replicates per breed. We analyzed *PLAAT3* and lnc*PLAAT3*-AS expression among the different breeds and found that both showed slight upregulation, though statistically insignificant, in six of the seven Chinese pig breeds (except for Wujin) compared with the Western breed, Yorkshire. This suggests that both *PLAAT3* and lnc*PLAAT3*-AS play an essential role in fat deposition in pigs and are potentially involved in the distinct adiposity phenotype between Chinese (relatively obese) and Western (relatively lean) breeds (Figure 4A,B).

3.6. SNPs of lncPLAAT3-AS and PLAAT3

Single nucleotide polymorphic SNPs may disrupt the structural features of many non-coding RNAs, interfere with their molecular function, and produce phenotypic effects [24]. We therefore searched for and identified a variety of SNP sites within lnc*PLAAT3*-AS (Figure S4). Although these SNPs do not occur in the region of lnc*PLAAT3*-AS that binds to miR-503-5p, they may change the secondary structure and/or expression of lnc*PLAAT3*-AS, which should be explored in future studies.

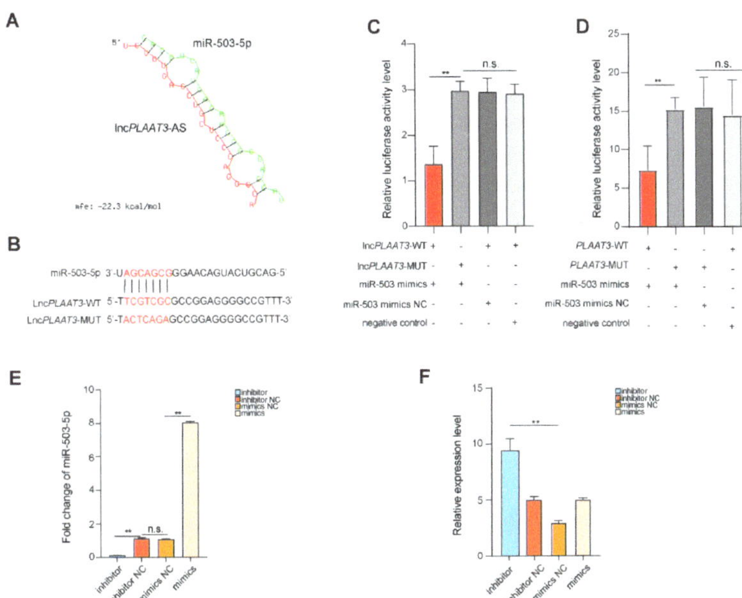

Figure 3. (**A**) Predicted miR-503-5p binding site in lnc*PLAAT3*-AS as determined by RNAhybrid; (**B**) binding capacity of lnc*PLAAT3*-AS and miR-503-5p as evaluated by RNAhybrid (Red sequences are miRNA seed region binding sequences); (**C**) luciferase assay revealed miR-503-5p-mediated inhibition of lnc*PLAAT3*-AS activity, miR-503-5p mimics NC is miR-503-5p mimics negative control ("+" indicates transfected, "−" indicates not transfected); (**D**) luciferase assay revealed miR-503-5p–mediated inhibition of *PLAAT3* activity; (**E**) Transfection efficiency as measured by qRT-PCR, miR-503-5p inhibitor NC is miR-503-5p inhibitor negative control; (**F**) qRT-PCR analysis of *PLAAT3* expression in transfected cells. All results are presented as means ± SEM; n = 3; ** $p < 0.01$; n.s., no significant difference.

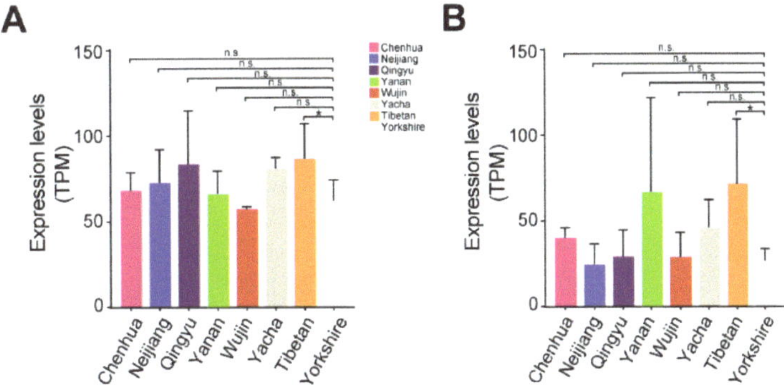

Figure 4. (**A**) Bar graph showing *PLAAT3* expression in eight pig breeds. (**B**) Bar graph showing lnc*PLAAT3*-AS expression in eight pig breeds. All results are presented as means ± SEM; n = 3; * $p < 0.05$; n.s., no significant difference.

4. Discussion

In this study, we found that lnc*PLAAT3*-AS promotes adipocyte lipogenic differentiation and mediates the miR-503-5p/*PLAAT3* interaction in adipocytes. The knockdown of lnc*PLAAT3*-AS in adipocytes decreases cell viability and inhibits adipocyte proliferation and lipogenic differentiation, whereas the knockdown of miR-503-5p had the opposite effect [25,26]. Our results suggest that lnc*PLAAT3*-AS and miR-503-5p could serve as targets for manipulating adipocyte lipogenic differentiation.

Based on previous studies of lnc*PLAAT3*-AS, we investigated the role of lnc*PLAAT3*-AS in regulating porcine primary preadipocyte development by overexpressing or disrupting this lncRNA in porcine primary preadipocytes and analyzing the effects with RNA-seq technology. In addition to enhancing *PLAAT3* stability by forming RNA–RNA dimers, it appears that lnc*PLAAT3*-AS functions within a second regulatory pathway to regulate adipogenesis and related molecules. One of the most well-accepted models for lncRNAs function is their role as ceRNAs which act as sponges to absorb free miRNAs via sequence complementarity, thereby inhibiting target miRNA function [27]. Li et al. reported that *UBE2CP3* promotes gastric cancer development mainly through the miR-138-5p/*ITGA2* axis. In addition, their study showed that *UBE2CP3*/*IGFBP7* can form an RNA duplex that interacts directly with the ILF3 protein. ILF3-mediated RNA–RNA interactions between *IGFBP7* mRNA and *UBE2CP3* in turn play an important role in protecting the stability of *UBE2CP3* mRNA [23]. We found that lnc*PLAAT3*-AS significantly promotes preadipocyte proliferation and differentiation by targeting miR-503-5p. Therefore, our findings suggest that lnc*PLAAT3*-AS plays an essential role in lipid deposition in pigs.

Adipocyte differentiation is essential for lipid deposition in mammals [28]. It has been shown that the normal expression of the *PLAAT3* gene promotes lipid deposition in mice. However, the rate of lipolysis was significantly higher in *PLAAT3*-deficient mice due to the significantly lower expression of adipose prostaglandin E2 bound to the Gai-coupled receptor EP3, increased cyclic AMP expression, and significantly lower adipose tissue mass [10,11]. In our study, we performed RNA-seq on porcine preadipocytes, followed by both GO and KEGG enrichment analysis, and found that lnc*PLAAT3*-AS is closely related to cellular metabolism, as is its host gene, *PLAAT3*. Further analysis showed that the expression changes of many critical genes involved in fat metabolism mirrored those of *PLAAT3*. The cluster of differentiation 36 (*CD36*) is the upstream gene of *PLAAT3*, whose main function is to release arachidonic acid (AA) from the cell membrane via cytoplasmic phospholipase A(2)α (cPLA(2)α), which is one of the products of *PLAAT3*. In addition to this, *CD36* contributes to the production of pro-inflammatory eicosanoids. Compared to control cells, CHO cells normally expressing human *CD36* released significantly more AA and prostaglandin E(2) (PGE(2)) in response to thapsigargin-induced ER stress [29]. The results in this study are worthy of further testing in a more significant number of samples.

British Yorkshire and local Chinese pigs differ significantly in meat production traits and have complementary characteristics. Local Chinese pigs have tender, flavorful, and juicy meat but grow slowly and produce little lean meat [30] Although Yorkshire pigs grow faster and produce more lean meat, the quality of the meat is not as high as that of local Chinese breeds. The differences in *PLAAT3* and lnc*PLAAT3*-AS expression among the eight pig breeds are also consistent with the abovementioned differences in phenotypic traits, demonstrating that *PLAAT3* and lnc*PLAAT3*-AS play an important role in pig fat deposition [15,16].

Increasingly, high-throughput sequencing has been used to identify non-coding RNAs involved in adipogenesis [31,32]. Non-coding RNAs, which were once considered nonsense transcripts, play a rich variety of essential roles in various biological processes. Therefore, it is crucial to continue to explore the diversity of non-coding RNAs and validate their functions to advance our understanding of epigenetic regulation and gain a deeper appreciation of the range of genetic information carried by these molecules [33].

5. Conclusions

In summary, we report the role of the *PLAAT3*–lnc*PLAAT3*-AS–miR-503-5p regulatory axis in preadipocyte development. lnc*PLAAT3*-AS promotes the expression of the host gene *PLAAT3* by adsorbing miR-503-5p, thereby promoting the lipogenic differentiation of porcine primary preadipocytes. Furthermore, we identified related genes that may be involved in adipogenesis by performing RNA-seq on preadipocytes in which lnc*PLAAT3*-AS was overexpressed or inhibited and explored the differences in *PLAAT3* and lnc*PLAAT3*-AS expression among different pig breeds. Our study provides a theoretical basis for other researchers to explore the molecular mechanisms of non-coding RNA-mediated regulation of lipid deposition.

Supplementary Materials: The following supporting information can be downloaded at: https://www.mdpi.com/article/10.3390/genes14010161/s1, Figure S1: RNA-seq data statistics; Figure S2: Functional enrichment analysis of differentially expressed genes; Figure S3: miR-503-5p cell validation experiments; Figure S4: Mutational mapping of the lnc*PLAAT3*-AS SNP locus; Table S1: Differentially expressed genes. Table S2: Differentially Expressed Genes (DEGs).

Author Contributions: Z.W., J.C., L.J. and M.L. designed the research. Z.W., K.L., Y.G. and Y.W. collected samples and extracted the RNA for sequencing. Z.W. and J.C. analyzed the RNA-seq data. Z.W., J.C. and Y.W. wrote the manuscript. L.J. and M.L. revised the manuscript. All authors have read and agreed to the published version of the manuscript.

Funding: This work was supported by grants from the National Key R&D Program of China (2021YFA0805903 and 2020YFA0509500), the Sichuan Science and Technology Program (2021ZDZX0008 and 2022JDJQ0054, 2021YFYZ0009), the Sichuan Science and Technology Innovation Talents (2022JDRC0037), and the National Natural Science Foundation of China (32272837).

Institutional Review Board Statement: All research involving animals was conducted according to Regulations for the Administration of Affairs Concerning Experimental Animals (Ministry of Science and Technology, China, revised in March 2017) and approved by the animal ethics and welfare committee (AEWC) of Sichuan Agricultural University under permit no. 20210162. This study was carried out in compliance with the ARRIVE guidelines.

Informed Consent Statement: Not applicable.

Data Availability Statement: The RNA-seq data is available in the NCBI Gene Expression Omnibus (GEO; https://www.ncbi.nlm.nih.gov/geo/, accessed on 26 September 2022) under series numbers GSE213924.

Conflicts of Interest: The authors declare no conflict of interest.

References

1. Liao, Q.; Liu, C.; Yuan, X.; Kang, S.; Miao, R.; Xiao, H.; Zhao, G.; Luo, H.; Bu, D.; Zhao, H.; et al. Large-scale prediction of long non-coding RNA functions in a coding–non-coding gene co-expression network. *Nucleic Acids Res.* **2011**, *39*, 3864–3878. [CrossRef] [PubMed]
2. He, H.; Wang, Y.; Ye, P.; Yi, D.; Cheng, Y.; Tang, H.; Zhu, Z.; Wang, X.; Jin, S. Long noncoding RNA ZFPM2-AS1 acts as a miRNA sponge and promotes cell invasion through regulation of miR-139/GDF10 in hepatocellular carcinoma. *J. Exp. Clin. Cancer Res.* **2020**, *39*, 159. [CrossRef] [PubMed]
3. Guo, X.; Li, H.; Zhang, M.; Li, R. LncRNA GAS6 antisense RNA 1 facilitates the tumorigenesis of clear cell renal cell carcinoma by regulating the AMP-activated protein kinase/mTOR signaling pathway. *Oncol. Lett.* **2021**, *22*, 727. [CrossRef]
4. Zhang, H.; Guan, R.; Zhang, Z.; Li, D.; Xu, J.; Gong, Y.; Chen, X.; Lu, W. LncRNA Nqo1-AS1 Attenuates Cigarette Smoke-Induced Oxidative Stress by Upregulating its Natural Antisense Transcript Nqo1. *Front. Pharmacol.* **2021**, *12*, 729062. [CrossRef] [PubMed]
5. Kuo, F.-C.; Neville, M.J.; Sabaratnam, R.; Wesolowska-Andersen, A.; Phillips, D.; Wittemans, L.B.; van Dam, A.D.; Loh, N.Y.; Todorčević, M.; Denton, N.; et al. HOTAIR interacts with PRC2 complex regulating the regional preadipocyte transcriptome and human fat distribution. *Cell Rep.* **2022**, *40*, 111136. [CrossRef]
6. Meurens, F.; Summerfield, A.; Nauwynck, H.; Saif, L.; Gerdts, V. The pig: A model for human infectious diseases. *Trends Microbiol.* **2012**, *20*, 50–57. [CrossRef]
7. Rosen, E.D.; MacDougald, O.A. Adipocyte differentiation from the inside out. *Nat. Rev. Mol. Cell Biol.* **2006**, *7*, 885–896. [CrossRef]
8. Groenen, M.A.M.; Archibald, A.L.; Uenishi, H.; Tuggle, C.K.; Takeuchi, Y.; Rothschild, M.F.; Rogel-Gaillard, C.; Park, C.; Milan, D.; Megens, H.-J.; et al. Analyses of pig genomes provide insight into porcine demography and evolution. *Nature* **2012**, *491*, 393–398. [CrossRef]
9. Pabst, R. The pig as a model for immunology research. *Cell Tissue Res.* **2020**, *380*, 287–304. [CrossRef]
10. Jaworski, K.; Ahmadian, M.; Duncan, R.E.; Sarkadi-Nagy, E.; Varady, K.A.; Hellerstein, M.K.; Lee, H.-Y.; Samuel, V.T.; Shulman, G.I.; Kim, K.-H.; et al. AdPLA ablation increases lipolysis and prevents obesity induced by high-fat feeding or leptin deficiency. *Nat. Med.* **2009**, *15*, 159–168. [CrossRef]

11. Liu, P.; Jin, L.; Zhao, L.; Long, K.; Song, Y.; Tang, Q.; Ma, J.; Wang, X.; Tang, G.; Jiang, Y.; et al. Identification of a novel antisense long non-coding RNA PLA2G16-AS that regulates the expression of PLA2G16 in pigs. *Gene* **2018**, *671*, 78–84. [CrossRef]

12. Du, J.; Xu, Y.; Zhang, P.; Zhao, X.; Gan, M.; Li, Q.; Ma, J.; Tang, G.; Jiang, Y.; Wang, J.; et al. MicroRNA-125a-5p Affects Adipocytes Proliferation, Differentiation and Fatty Acid Composition of Porcine Intramuscular Fat. *Int. J. Mol. Sci.* **2018**, *19*, 501. [CrossRef]

13. Mills, J.D.; Kawahara, Y.; Janitz, M. Strand-Specific RNA-Seq Provides Greater Resolution of Transcriptome Profiling. *Curr. Genom.* **2013**, *14*, 173–181. [CrossRef]

14. Papasavva, P.; Papaioannou, N.; Patsali, P.; Kurita, R.; Nakamura, Y.; Sitarou, M.; Christou, S.; Kleanthous, M.; Lederer, C. Distinct miRNA Signatures and Networks Discern Fetal from Adult Erythroid Differentiation and Primary from Immortalized Erythroid Cells. *Int. J. Mol. Sci.* **2021**, *22*, 3626. [CrossRef]

15. Tao, X.; Liang, Y.; Yang, X.; Pang, J.; Zhong, Z.; Chen, X.; Yang, Y.; Zeng, K.; Kang, R.; Lei, Y.; et al. Transcriptomic profiling in muscle and adipose tissue identifies genes related to growth and lipid deposition. *PLoS ONE* **2017**, *12*, e0184120. [CrossRef]

16. Shang, P.; Li, W.; Liu, G.; Zhang, J.; Li, M.; Wu, L.; Wang, K.; Chamba, Y. Identification of lncRNAs and Genes Responsible for Fatness and Fatty Acid Composition Traits between the Tibetan and Yorkshire Pigs. *J. Genom.* **2019**, *2019*, 5070975-12. [CrossRef]

17. Garin-Shkolnik, T.; Rudich, A.; Hotamisligil, G.S.; Rubinstein, M. FABP4 Attenuates PPARγ and Adipogenesis and Is Inversely Correlated With PPARγ in Adipose Tissues. *Diabetes* **2014**, *63*, 900–911. [CrossRef]

18. Gondret, F.; Riquet, J.; Tacher, S.; Demars, J.; Sanchez, M.; Billon, Y.; Robic, A.; Bidanel, J.; Milan, D. Towards candidate genes affecting body fatness at the SSC7 QTL by expression analyses. *J. Anim. Breed. Genet.* **2011**, *129*, 316–324. [CrossRef]

19. Pepino, M.Y.; Kuda, O.; Samovski, D.; Abumrad, N.A. Structure-function of CD36 and importance of fatty acid signal transduction in fat metabolism. *Annu. Rev. Nutr.* **2014**, *34*, 281–303. [CrossRef]

20. Guo, C.; Yuan, L.; Liu, X.; Du, A.; Huang, Y.; Zhang, L. Effect of ARB on expression of CD68 and MCP-1 in adipose tissue of rats on long-term high fat diet. *J. Huazhong Univ. Sci. Technol.* **2008**, *28*, 257–260. [CrossRef]

21. Chan, P.-C.; Liao, M.-T.; Lu, C.-H.; Tian, Y.-F.; Hsieh, P.-S. Targeting inhibition of CCR5 on improving obesity-associated insulin resistance and impairment of pancreatic insulin secretion in high fat-fed rodent models. *Eur. J. Pharmacol.* **2021**, *891*, 173703. [CrossRef] [PubMed]

22. Liu, C.; Li, P.; Li, H.; Wang, S.; Ding, L.; Wang, H.; Ye, H.; Jin, Y.; Hou, J.; Fang, X.; et al. TREM2 regulates obesity-induced insulin resistance via adipose tissue remodeling in mice of high-fat feeding. *J. Transl. Med.* **2019**, *17*, 300. [CrossRef] [PubMed]

23. Li, D.; She, J.; Hu, X.; Zhang, M.; Sun, R.; Qin, S. The ELF3-regulated lncRNA UBE2CP3 is over-stabilized by RNA-RNA interactions and drives gastric cancer metastasis via miR-138-5p/ITGA2 axis. *Oncogene* **2021**, *40*, 5403–5415. [CrossRef] [PubMed]

24. Sabarinathan, R.; Tafer, H.; Seemann, S.E.; Hofacker, I.L.; Stadler, P.F.; Gorodkin, J. RNAsnp: Efficient detection of local RNA secondary structure changes induced by SNPs. *Hum. Mutat.* **2013**, *34*, 546–556. [CrossRef] [PubMed]

25. Shen, L.; He, J.; Zhao, Y.; Niu, L.; Chen, L.; Tang, G.; Jiang, Y.; Hao, X.; Bai, L.; Li, X.; et al. MicroRNA-126b-5p Exacerbates Development of Adipose Tissue and Diet-Induced Obesity. *Int. J. Mol. Sci.* **2021**, *22*, 10261. [CrossRef]

26. Hu, M.; Kuang, R.; Guo, Y.; Ma, R.; Hou, Y.; Xu, Y.; Qi, X.; Wang, D.; Zhou, H.; Xiong, Y.; et al. Epigenomics analysis of miRNA cis-regulatory elements in pig muscle and fat tissues. *Genomics* **2022**, *114*, 110276. [CrossRef]

27. Klingenberg, M.; Matsuda, A.; Diederichs, S.; Patel, T. Non-coding RNA in hepatocellular carcinoma: Mechanisms, biomarkers and therapeutic targets. *J. Hepatol.* **2017**, *67*, 603–618. [CrossRef]

28. Lang, Y.-Y.; Xu, X.-Y.; Liu, Y.-L.; Ye, C.-F.; Hu, N.; Yao, Q.; Cheng, W.-S.; Cheng, Z.-G.; Liu, Y. Ghrelin Relieves Obesity-Induced Myocardial Injury by Regulating the Epigenetic Suppression of miR-196b Mediated by lncRNA HOTAIR. *Obes. Facts* **2022**, *15*, 540–549. [CrossRef]

29. Kuda, O.; Jenkins, C.M.; Skinner, J.R.; Moon, S.H.; Su, X.; Gross, R.W.; Abumrad, N.A. CD36 protein is involved in store-operated calcium flux, phospholipase A2 activation, and production of prostaglandin E2. *J. Biol. Chem.* **2011**, *286*, 17785–17795. [CrossRef]

30. Li, M.; Chen, L.; Tian, S.; Lin, Y.; Tang, Q.; Zhou, X.; Li, D.; Yeung, C.K.; Che, T.; Jin, L.; et al. Comprehensive variation discovery and recovery of missing sequence in the pig genome using multiple de novo assemblies. *Genome Res.* **2016**, *27*, 865–874. [CrossRef]

31. Squillaro, T.; Peluso, G.; Galderisi, U.; Di Bernardo, G. Long non-coding RNAs in regulation of adipogenesis and adipose tissue function. *eLife* **2020**, *9*, e59053. [CrossRef]

32. Kosinska-Selbi, B.; Mielczarek, M.; Szyda, J. Review: Long non-coding RNA in livestock. *Animal* **2020**, *14*, 2003–2013. [CrossRef]

33. Bridges, M.C.; Daulagala, A.C.; Kourtidis, A. LNCcation: lncRNA localization and function. *J. Cell Biol.* **2021**, *220*, 110276. [CrossRef]

Review

Current Analytical Methods and Research Trends Are Used to Identify Domestic Pig and Wild Boar DNA in Meat and Meat Products

Małgorzata Natonek-Wiśniewska *, Agata Piestrzynska-Kajtoch, Anna Koseniuk and Piotr Krzyścin

Department of Animal Molecular Biology, National Research Institute of Animal Production, 32-083 Balice, Poland
* Correspondence: malgorzata.natonek@iz.edu.pl

Abstract: The pig, one of the most important livestock species, is a meaningful source of global meat production. It is necessary, however, to prove whether a food product that a discerning customer selects in a store is actually made from pork or venison, or does not contain it at all. The problem of food authenticity is widespread worldwide, and cases of meat adulteration have accelerated the development of food and the identification methods of feed species. It is worth noting that several different molecular biology techniques can identify a porcine component. However, the precise differentiation between wild boar and a domestic pig in meat products is still challenging. This paper presents the current state of knowledge concerning the species identification of the domestic pig and wild boar DNA in meat and its products.

Keywords: pig; wild boar; species identification; genetic markers; genetic methods

Citation: Natonek-Wiśniewska, M.; Piestrzynska-Kajtoch, A.; Koseniuk, A.; Krzyścin, P. Current Analytical Methods and Research Trends Are Used to Identify Domestic Pig and Wild Boar DNA in Meat and Meat Products. *Genes* **2022**, *13*, 1825. https://doi.org/10.3390/genes13101825

Academic Editors: Katarzyna Piórkowska and Katarzyna Ropka-Molik

Received: 11 September 2022
Accepted: 5 October 2022
Published: 9 October 2022

Publisher's Note: MDPI stays neutral with regard to jurisdictional claims in published maps and institutional affiliations.

1. Introduction

The pig is believed to be one of the most important livestock species for humans. It is an excellent animal model for studying the molecular background of several human diseases [1], and above all, it is a meaningful source of global meat production. Pork is widely used in the food industry, but one problem is determining the food's authenticity, especially in terms of the species. Meat adulteration involves replacing or partially replacing high-value meat with cheaper quality meat [2], which involves economic, quality, safety, and socio-religious issues [3]. Together with pork, products may introduce veterinary drugs banned from the food chain, such as ractopamine, which is forbidden in many countries [4,5]. Moreover, undeclared pork may be contaminated with harmful microorganisms, such as the endoparasites Toxoplasma and Trichinella [6]. The presence of pork in food may also conflict with religious and cultural practices. Thus, the fraud of undeclared pork in food can affect consumer trust in the meat industry. Because of such illegal food practices, many governments have enacted laws prohibiting similar fraud. To ensure existing regulations are followed, laboratories worldwide have designed and developed various methods to identify animal species in food products, including methods to detect porcine components. According to scientific reports, efforts have also been made to provide a cost-efficient diagnostic tool for distinguishing wild boar (*Sus scrofa*), domestic pig (*Sus scrofa domestica*), and their hybrids (Figure 1). This is not easy, as the pig and the wild boar are evolutionarily closely related [7], and thus share many genetic markers. However, the identification methods have changed over time, accompanied by the development of innovative techniques that are now based on DNA testing. This work aims to present the current state of knowledge concerning the species identification of the domestic pig and wild boar DNA in meat and its products.

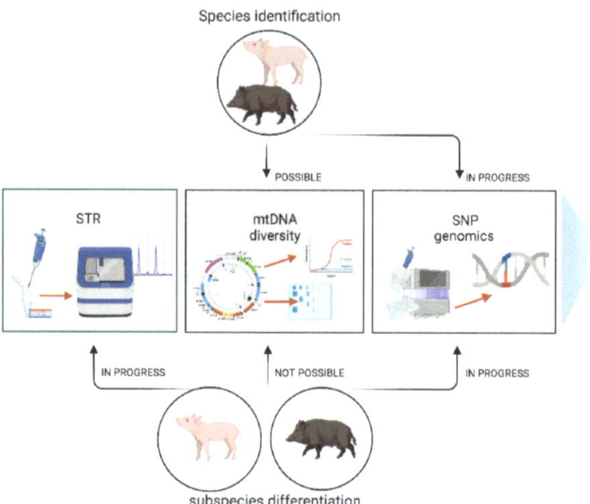

Figure 1. Methods of species identification of domestic pigs and wild boars and distinguishing between both subspecies.

Possibilities of species identification of pigs and wild boars (top of the figure) and distinguishing between both subspecies (bottom of the figure) using different molecular techniques.

2. Pig and Wild Boar Phylogenetics

A considerable number of animal breeds on Earth originated from diverse wild ancestors which, after being domesticated, exhibit a large phenotypic variety [8]. In the study by Groenen et al. [7], a comprehensive analysis of the phylogenetic background of the subspecies structure of wild boars and pigs was provided. The genomes of wild boars and domestic pigs from Europe and Asia revealed separate European and Asian lineages. The study also gave an insight into the history of population size changes. Both lines of wild boars diverged about 1.6–0.8 Myr (million years) ago, resulting in alleles divergence in the millions of loci [7]. Wild boars are the ancestors of modern domestic pigs, and it is believed that multiple domestication events occurred separately for species in Asia and Europe [7,9].

After domestication, many incidents of the admixture of domestic populations with local wild boars occurred because both species are found in the same habitat [9]. Moreover, there are farms (e.g., in Bulgaria and Sardinia) where pigs are kept in semi-wild conditions. It is highly likely that domestic pigs sporadically crossed with wild boars [10]. The frequent hybridisation of these two breeds was found in the studies on the European population [10–13]. Studies of the *MC1R* (melanocortin 1 receptor) gene in the Polish population of wild boars also proved the existence of admixed genotypes [14,15]. Crossbreeding domestic pigs with wild boars is documented and has served hunting purposes, better meat production, better taste, and reduced aggressive behaviour [11,12]. However, crossbreds have a high risk of pathogen transmission (e.g., ASF, African Swine Fever), which is the primary reason crossbreeding farms are under strict legislation in the European Union [12]. It is highly likely that some wild boars farmed in Poland were released or escaped from the farms and became a potential source of genome admixture in the wild population. Other studies have also suggested such a scenario [11,12,14]. The domestication of the pig and the mixing of domestic pig and wild boar populations cannot be ignored, as they provide the background to the problem of distinguishing between these subspecies.

3. Authentication of Meat from Domestic Pig

The authentication of domestic pig meat is mainly based on mitochondrial DNA (mtDNA). The well-known sequences of the mitochondrial genome allow comparison of the porcine mtDNA with the genome of any species. The advantages of mitochondrial DNA over genomic DNA are due to its properties. The mitochondrial genome is well understood and described in many databases (GenBank, EST, GSS). It is also species-specific and is present in many copies in every organism cell. Moreover, mtDNA is resistant to unfavourable factors, such as high temperature or pressure. This last feature is essential because it allows the identification of DNA in preserved meat products, the production of which involves thermal processing. Properties of mtDNA translate into biological specificity and high sensitivity of pork identification methods, which is very important when investigating highly processed products such as sausages, canned pork, pork fat, or gelatine. However, to date, it is impossible to distinguish pork from venison based on mtDNA variability only.

The heat treatment of food causes fragmentation of the DNA. This may reduce the chance of extracting good quality DNA sufficient for further analysis. In the literature, we can find many reports on the influence of individual types of heat treatment on DNA quality. Traditional cooking, heating in a microwave [16], or processing pork for canned meal influence the amount of DNA obtained, and the isolate rate remains at a reasonable level. In contrast, DNA extracts from gelatine or meat meal had much worse properties. In the first case, the quantity and quality of DNA described by the absorbance ratio had the values of 400–600 ng/μL and 1.7–1.9, respectively. However, only 15–145 ng/μL were obtained for gelatine or meat meal, with a purity often below 1.6 or above 1.9. These are the difficulties any laboratory faces in food product species identification.

The obtained DNA can be analyzed by many methods. Most of them are based on the PCR reaction, which can be used in several variants such as classical PCR, real-time PCR, sequencing, DNA barcoding and NGS (next-generation sequencing). The advantages and limitations of these methods are presented in Table 1.

Table 1. Advantages and limitations of different methods used for Sus scrofa meat identification.

Technique/Method	Advantages	Limitations and Difficulties	References
PCR	Simple to develop and easy to conduct Can detect small amounts of DNA High specificity species identification Analysis of monoplexes or multiplexes depending on needs Low detection limit Ability to detect processed and heat-treated samples	No quantitation Lower specificity when amplicon length is short More difficult PCR optimisation in case of multiplex PCR Unequal amplification efficiency which causes variable sensitivity Contaminants could produce false-negative results Lower DNA yield for heat-treated samples	[17–21]
PCR RAPD	Detects multi-species, and no previous knowledge of DNA is required Can detect a high level of polymorphism (for example, between similar species) Costs per assay are low	Unable to differentiate species in mixtures Mistakes during the analysis of degraded or autoclaved materials Difficulties in obtaining reproducible results	[22–24]
PCR-RFLP	Species-specific restriction pattern	The results can be challenging to interpret in the case of entirely unknown species components of the studied object Inadequate for highly processed or meat mixtures	[25,26]

Table 1. *Cont.*

Technique/Method	Advantages	Limitations and Difficulties	References
Real-Time PCR (quantification and qualification)	Possible quantitative result Very high sensitivity High sample throughput	Probe-based methods are expensive and time-intensive Dye-based methods are less accurate Quantification PCR requires appropriate reference material Precision of measure is different for different product types qPCR instruments are very costly HRM analysis requires HRM-capable real-time PCR machines and specialised software algorithms	[27–29]
Digital PCR	Better detection limit and accuracy	Little sample volume per reaction Small dynamic range if the number of partitions is limited The risk of falsely low quantification due to molecular dropout Restricted multiplex detection	[30,31]
LAMP	Rapid High specificity High amplification efficiency No need for thermal cycling Low operational cost	Less versatile Little to no multiplexing Less sensitive to inhibitors than PCR in case of complex samples. The method is more complicated than classical PCR owing to the presence of more primers	[32–35]
DNA barcoding and NGS	High sensitivity and specificity Single-step DNA sequencing and quantification by NGS High throughput detection	High-quality DNA necessary Long analysis time from DNA extraction to the finish of bioinformatics analysis Complex library preparation protocol for NGS Trace quantity of foreign DNA may cause inaccurate estimation of species Costly equipment to analyse and very highly skilled personnel	[36,37]

4. Methods Used for Species Identification—Possibilities and Application

4.1. PCR

Methods based on classical PCR are used mainly for qualitative analysis of a specific species or group of different species. They are very useful when we want to check the potential presence of swine DNA. They provide the highest analysis sensitivity of all DNA-based methods. Qualitative identification of porcine DNA is a perennial research problem, yet it still enjoys unabated interest. However, research tendencies have changed over the years. Twenty years ago, all methods were based on the analysis of long DNA fragments. Now, short counterparts, often below 100 bp, are preferred. These changes are related to the more excellent suitability of short amplification products for analysing processed samples.

4.2. PCR RAPD

The random amplified polymorphic DNA (RAPD) technique is based on DNA amplification, with a short arbitrary primer that amplifies multiple fragments of the analyzed genome. It is followed by separating obtained DNA fragments (based on their sizes) using gel electrophoresis. PCR-RAPD for species identification is most useful for rapid qualitative analysis. In the literature, it was described as a helpful tool for differentiating domestic animals, including pigs [38]. The identification of particular species DNA in meat samples using a random amplified polymorphic DNA (RAPD) technique because of numerous obstacles is rarely used for commercial samples [38,39].

4.3. PCR-RFLP

The basis of the PCR-RFLP (Restriction Fragment Length Polymorphism) reaction is the amplification of a large amplicon homologous for several species using a standard pair of sequence-specific primers. The large amplicons are then digested using one or more restriction enzymes into shorter reaction products. The method is efficiently implemented to screen for several species' DNA in one sample [40,41], even for gelatine [42]. Because of its properties, the technique has been successfully implemented for the complex analyses of animal groups (mammals, birds, or being a potential ingredient of a product, even gelatine) and then their specific representatives. Restriction enzyme analysis to detect pork has been performed frequently over the past several years. It enables us to simultaneously distinguish different species, including farm animals [40,43], and sometimes in combination with wild or companion animals [44,45]. Thus, despite many limitations (Table 1) and newer techniques, the restriction enzyme method is still used to determine pork in food products.

4.4. Real-Time PCR (Quantification and Qualification)

Real-time PCR is an easy method that allows qualitative and quantitative measurement of the species composition in any product. This technique is probably the most widely used for pork identification of all PCR-based techniques. Depending on the probes or dyes used, results occur in a shorter time than the classic PCR. In quantitative analysis, more accurate (accuracy, precision) results are obtained using probes (Taqman, TAMRA, Dabcyl, BHQ). The analysis is useful in controlling the declaration of the species composition in gelatine [46] and processed foods [29]. In both cases, the standard deviation of C_T and DNA concentration below 4% were demonstrated. The dyes (SYBR Green, EvaGreen) can also be used for a quantitative reaction, but the obtained results have a more significant measurement error [47]. On the other hand, in the qualitative determinations, the method with dyes works perfectly [48,49]. Currently, much emphasis is placed on the possibility of declaring a product vegan or kosher. A natural consequence of such a declaration should be the ability to identify falsifications at an extremely low level. A recent study indicated that the sensitivity of the real-time PCR method is improving. It is already possible to identify adulteration at 10 pg [47] or a concentration of 0.001% [48].

4.5. Digital PCR

Digital PCR (dPCR) is an alternative, sensitive, and specific method but does not rely on a standard curve. The basis of dPCR is to quantify the absolute number of targets present in a sample using limiting dilutions, PCR, and Poisson statistics. Because the habitat of reaction is divided into many smaller ones and endpoint amplification is less susceptible to competition between targets, this technique can detect small numbers of targets. It is useful whenever we need to perform quantitative determinations. Currently, this technique is increasingly more accurate than qPCR, owing to the lack of dependence on matching the type of reference material to the tested samples. It is described as suitable for usage by competent food safety authorities to verify compliance with the labelling of meat products, and to ensure quality and safety throughout the meat supply chain [50]. The studies found that the sensitivity of the method was very high—the lowest concentration of pork DNA, for which at least 95% of the replicates were positive, was 0.1 pg/μL [50]. As mentioned, the method shows the absolute result expressed in DNA copies. However, for practical purposes, there are ways of converting the result into the classical concentration of a component of a given species. Based on tests of 20 sausage samples carried out in six laboratories, it has been shown that the method can be used to evaluate meat in mixed meat products, with the highest accuracy and precision compared to the results generated by real-time PCR [31].

4.6. LAMP

Loop-mediated isothermal amplification (LAMP) is a rapid, economical and species-specific DNA-based assay for the authentication of animal species. The technique amplifies the animal mitochondrial D-loop region by isothermal amplification. In recent years, many laboratories have been working on developing this technique because of the shorter and less expensive hardware requirements compared to the PCR test, and the parameters sufficient for its use in the testing of commercial samples. The assay can detect pork to the level of 100 fg admixture, and the DNA detection limit was 0.0001% [51].

Cross-amplification of related species like cattle, buffalo, sheep, goat, and chicken was excluded by incorporating their DNA in the reaction assay [51,52]. It can also be integrated into compact lab-on-a-chip devices to develop micro-total analysis systems [51,53]. An essential feature of the research is the ability to analyze processed meat [52]. This makes LAMP a serious competitor to real-time PCR methods in pork identification in typical food samples.

4.7. DNA Barcoding and NGS

DNA barcoding generally relies on the use of a specific genetic target. The conventional DNA barcoding target is usually a ~650 bp region of the mitochondrial COI gene. This method is one of the most basic and oldest methods of species identification. It is mainly used for the analysis of cooked food samples. A study revealed that the rate of mislabelling for pork sausage was 21% [36,54].

The recent and fast development of benchtop next-generation sequencers has made NGS technologies highly applicable for meat speciation, including pork [37]. Most genome-wide association methods have been introduced using Porcine SNP Beadchips (Illumina, Inc., San Diego, CA, USA) and whole genome resequencing. These methods allow for the analysis of several SNP markers, CNVs, indels and STRs in a single reaction run. Studies have indicated that the sensitivity of this method is 1% of the total DNA [55].

Connecting both techniques (barcoding and NGS) improves the performance of analysis in comparison to traditional barcoding. Therefore, they can be serious candidates for species identification in the near future. Recently, the utility was presented for extracts from muscle meat, and DNA admixtures from model sausages [56] and commercial lamb sausages [57]. The observed limit of detection (LOD) was on the level of 0.1% [56] and allowed for the detection of a few adulterations in porcine DNA from processed food [57,58]. The NGS-based methods can also distinguish domestic pigs and wild boars.

5. Genetic Markers Used for Pig and Wild-Boar Distinguishing

Molecular techniques to identify animal species components in human food and animal feed use different genetic markers based on genomic or mitochondrial DNA. There are cost-effective and easily performed tests for DNA identification of cattle, horse, and poultry in food and animal feed [17,59]. Some of the methods used for pigs were discussed above. A reliable method for the differentiation of both subspecies would impact biodiversity, food fraud cases, detection of illegal hunting procedures, and zoonosis prevention. However, whether genomic DNA markers can be useful for analysis performed on DNA extracted from highly processed food products remains a subject of study.

5.1. Mitochondrial DNA Genome Loci

Mitochondrial sequences have been widely studied in the context of phylogenetics in many species [60–62], but using mtDNA only for distinguishing wild boar from domestic pig is unreliable. A different approach is needed. Fajardo et al. [63] conducted studies on wild boars and commercial pig breeds differentiation based on combining both nucleus (melanocortin 1 receptor, *MC1R*) and mitochondrial DNA (*D-loop*) analysis. The analysis of the *MC1R* gene proved to be more effective for species identification based on the polymorphism of mitochondrial DNA [63].

5.2. Melanocortin 1 Receptor Gene (MC1R)

The animal's coat colour is a visual feature that reflects changes in environment and breeding selection. It is a trait that phenotypically differentiates pigs and wild boars.

Wild animal forms are characterised by a dark fur colour, while domestic species have pale coat colours [61]. Several genes have been identified as associated with skin and hair pigmentation. *MC1R*, along with the *ASIP* gene, regulates the synthesis of two dyes: eumelanin (brown to black) and pheomelanin (crema to red) in melanocytes [64]. Both genes express an epistatic interaction [65,66]. An *MC1R* allele (E + allele, also called "wild" allele) unique to the wild boar's population was identified by Kijas et al. [67]. Most likely, owing to selection during the breeding work, individuals with the "wild" allele were lost. Implementing PCR-RFLP and real-time PCR protocols to identify four polymorphic *MC1R* loci gave inconsistent results. In Greece, a population of pigs and wild boars was distinguished using this marker [68]. However, further studies revealed that the "wild" allele was also present in some domestic breeds, and "domestic" alleles were found in wild boar samples [11,14,69].

5.3. Nuclear Receptor Subfamily 6, Group A, Member 1 (NR6A1)

Owing to increasing meat efficiency, domestic pig breeds are characterised by more vertebrae. This is another trait that differentiates wild boars and European commercial pigs. In the wild boar, the number of vertebrae is 19 and in pig breeds it is 21–23. A proline to leucine substitution at codon 192 (*p.Pro192Leu*) in the nuclear receptor subfamily 6, group A, member 1 (*NR6A1*) gene was shown to be the most likely causative mutation underlying the QTL (Quantitative Trait Loci) [70].

Since neither *MC1R* nor NR6A1 studies as single markers proved to be an effective tool to differentiate both species, the combination of both genes was studied. In general, the combination of single nucleotide polymorphisms in *MC1R* and *NR6A1* genes was tested using real-time PCR [11,71] or multiplexed using the SNaPshot Multiplex System method [72]. The results were promising.

5.4. Short Tandem Repeats (STRs)

Short tandem repeats (STRs, also known as microsatellites) are genetic markers that are commonly used in parentage testing and verification [73]. These markers are specific for the species, so they seemed good candidates to distinguish wild boar and pig. The hybridisation between pigs and wild boars based on STRs and *MC1R* was studied by Nikolov et al. [74]. STRs combined with SNPs in *MC1R*, *NR6A1* genes or mitochondrial D-loop have also been analyzed [10,75].

Bayesian clustering based on 10 STRs and mitochondrial D-loop polymorphic loci indicates that the wild and domestic forms are not divergent. Wild boars and domestic pigs share the most common mtDNA haplotypes, and microsatellite alleles weakly resolve both groups when examining a phylogenetic tree [10]. Nikolov et al. [74] proved the introgression of wild boar to Balkanian domestic pig breeds by studying *MC1R* genotypes and 10 STRs. In the population from Balkanste, the introgression in wild boar is more apparent than the introgression in domestic pigs from Western Europe [74]. Lorenzini et al. [75] implemented the STRs protocol combined with *MC1R* and *NR6A1* to identify both subspecies for forensic purposes. In this case, STRs were introduced to group individuals into the parental population. However, this marker type does not efficiently identify recent hybrids. STRs alone identify lower hybrids than genotyping polymorphic loci, along with *MC1R* and *NR6A1* [75].

5.5. Other Potential Genetic Markers

As pigs and wild boars are closely related, and genetic introgression from domestic pigs into European wild boar has been proved [74,76], phylogenetically close taxa distinguishing requires introduction of high-throughput, SNP-based (SNP, Single Nucleotide Polymorphism) molecular techniques. A 70K SNP chip was used in a French study [13] to determine the timing of the hybridisation (s) and to check the domestic pig admixture in the wild boar population. The study revealed that among wild boars, 83% to 100% of animals had a genome of "wild" origin and local ancestry analyses showed adaptive introgression from a domestic pig. These results indicated that population admixture must be considered in studies on pig and wild boar distinguishing.

Some loci responsible for phenotypic features important in domestication were revealed by Rubin et al. [77] with the whole-genome resequencing method. They found 33 candidate selective sweeps representing 18 different loci. Among others were regions on chromosome 1 (*SSC1*) including the *NR6A1* gene (described above), on *SSC4* including the *PLAG1* gene, on *SSC8* including the *LCORL* gene, and *SSC13* overlapping the *OSTN* gene. The *PLAG1* gene was previously associated with human and cattle height [78,79], and the *LCORL* gene was connected to the body size of various species [80–82]. *OSTN* produce the osteocrin, which is an inhibitor of osteoblast differentiation [83]. Its expression levels differed between muscle fibre types [84] and were correlated with food intake [85]. In the same project, authors found 72 nonsynonymous substitutions in the coding sequences, with the difference in allele frequency between pig and wild boar (i.e., *NR6A1*, *CYCS*, *ACER1*, *HK2*, *KITLG*, *TCTE1*, *SERPINA6*, *SEMA3D*, and many others) three of which (in *NR6A1*, *CCT8L2* and *MLL3*) were colocalized with putative sweep regions [77]. All of these associations between putative sweeps and phenotypic differences could potentially be used to distinguish wild boars and domestic pigs.

Wilkinson et al. [86] used a Porcine 60K SNP chip to find signatures of diversifying selection between different pig breeds (traditional and commercial European pig breeds), and also compared the pig breeds with wild boar. Among others, the results showed high levels of differentiation between pig and wild boar genomic regions connected to bone formation (on *SSC1*), growth (on *SSC7*), and fat deposition (on *SSC7* and *SSCX*). The differentiated genomic region on chromosome 7 was the most extensive and contained such genes as *PPARD* and *CDKN1A* associated with fat deposition [87,88]. It was close to the *MHC* loci, which is known to be a crucial, vertebrate immunity complex. Wild boar-specific selection signatures in *SSC7*, with one of the regions in *SSC7* located close to the *PPARD* gene, were also observed by Muñoz et al. [89]. Fatness was one of the features under the strong selection in domestic pigs [90].

Seeking global patterns in pig domestication, Yang et al. [8] calculated the genome-wide fixation index (Fst) between domestic pigs and wild boars (from Asia and Europe, separately) based on a genome-wide single nucleotide polymorphism. Among genes located close to the top outlier, SNPs with the highest Fst values were ones associated with phenotypic features and traits that changed during domestication, such as growth (*SOX2-OT*), muscle development (*MSTN*), energy balance (*NMU*, *LEP*, *GSK3A*), social behaviour (*TBX19*, *PAFAH1B3*), reproduction (*ESR1*, *GNRHR*, *PATZ1*, *PLSCR4*), metabolism (*TMEM67*, *FOXA1*, *INSIG2*), central nervous system (*LRRC4*, *VEPH1*, *CDH9*), immune system (*LAIR1*), and even perception of smell (*Olfr466*) [8]. The polymorphisms linked to genes, candidates for domestication loci, could potentially be used as markers for pig and wild boar distinguishing.

Recently, Moretti et al. [91] proposed a method to select a reduced SNP panel for the traceability of pig breeds and wild boars. The panel allowed discriminating among all species and wild boars in their study, and even to distinguish hybrid crossing.

The methods and markers implemented to date for distinguishing both subspecies in meat and muscle are summarised in Table 2. Although some of the genetic markers seem promising, none of the above markers was tested on DNA isolated from meat products. Since the analysis of such samples requires a unique approach, as the DNA isolated from food products is fragmented, methods for genotyping mutations may generate false negative results. Further studies are needed to resolve that problem.

Table 2. The molecular techniques and markers implemented for identifying domestic pig and wild boars and their hybrids in meat or muscles.

Technique	Marker/Gene	Tested on Sample of		References
		Muscle	**Meat**	
PCR RFLP and real-time PCR	*MC1R, NR6A1,* mtDNA regions	yes	yes	[11,14,63,68,70–72]
PCR multiplex	STRs	yes	no	[10,75]

6. Conclusions

Pork identification in food is essential from an economic, quality, safety, and socio-religious points of view. Several methods can be used for this species identification, such as PCR, real-time PCR, PCR-RAPD, PCR-RFLP, Digital PCR, LAMP, DNA barcoding and NGS. Some methods (classical PCR and real-time PCR) have been used since the 1990s and are still being improved for the analysis of more processed food. There are also possibilities for distinguishing between pork and wild boar meat. However, further study is needed to establish reliably distinguishing methods.

According to us, the best hope for further development in this field of research is provided by simple and/or modern techniques. A simple method like PCR (classic or real-time) is sufficient for fast and cheap pork identification, while modern techniques are preferable owing to the undeniable advantages of accuracy (Digital PCR), low price (LAMP), high specificity (NGS), and high throughput (DNA barcoding and NGS). In our opinion, each of these techniques is meaningful and useful for strictly defined types of samples (raw/processed/canned food; one-species meat/mixed meat of different species) and problems we must resolve, such as species identification of pig or differentiation their subspecies.

Author Contributions: Conceptualisation: M.N.-W., A.P.-K. and A.K.; methodology: M.N.-W., A.P.-K. and A.K.; writing—original draft preparation: M.N.-W., A.P.-K. and A.K.; writing—review and editing M.N.-W., A.P.-K., A.K. and P.K. All authors have read and agreed to the published version of the manuscript.

Funding: This research was funded by The National Research Institute of Animal Production (Number 501-180-711).

Institutional Review Board Statement: Not applicable.

Informed Consent Statement: Not applicable.

Data Availability Statement: Not applicable.

Conflicts of Interest: The authors declare no conflict of interest.

References

1. Bassols, A.; Costa, C.; Eckersall, P.D.; Osada, J.; Sabrià, J.; Tibau, J. The pig as an animal model for human pathologies: A proteomics perspective. *Proteom.—Clin. Appl.* **2014**, *8*, 715–731. [CrossRef]
2. Ropodi, A.; Pavlidis, D.; Mohareb, F.; Panagou, E.; Nychas, G.-J. Multispectral image analysis approach to detect adulteration of beef and pork in raw meats. *Food Res. Int.* **2014**, *67*, 12–18. [CrossRef]
3. Alamprese, C.; Casale, M.; Sinelli, N.; Lanteri, S.; Casiraghi, E. Detection of minced beef adulteration with turkey meat by UV–vis, NIR and MIR spectroscopy. *LWT* **2013**, *53*, 225–232. [CrossRef]
4. Cai, S.; Kong, F.; Xu, S. Detection of porcine-derived ingredients from adulterated meat based on real-time loop-mediated isothermal amplification. *Mol. Cell Probes* **2020**, *53*, 101609. [CrossRef] [PubMed]
5. Niño, A.M.; Granja, R.H.; Wanschel, A.C.; Salerno, A.G. The challenges of ractopamine use in meat production for export to European Union and Russia. *Food Control* **2017**, *72*, 289–292. [CrossRef]
6. King, H.; Bedale, W. *Hazard Analysis and Risk-Based Preventive Controls: Improving Food Safety in Human Food Manufacturing for Food Businesses*; Academic Press: Cambridge, MA, USA, 2017; ISBN 978-0-12-809475-4.

7. Groenen, M.A.M.; Archibald, A.L.; Uenishi, H.; Tuggle, C.K.; Takeuchi, Y.; Rothschild, M.F.; Rogel-Gaillard, C.; Park, C.; Milan, D.; Megens, H.-J.; et al. Analyses of pig genomes provide insight into porcine demography and evolution. *Nature* **2012**, *491*, 393–398. [CrossRef] [PubMed]
8. Yang, B.; Cui, L.; Perez-Enciso, M.; Traspov, A.; Crooijmans, R.P.M.A.; Zinovieva, N.; Schook, L.B.; Archibald, A.; Gatphayak, K.; Knorr, C.; et al. Genome-wide SNP data unveils the globalization of domesticated pigs. *Genet. Sel. Evol.* **2017**, *49*, 1–15. [CrossRef] [PubMed]
9. Larson, G.; Burger, J. A population genetics view of animal domestication. *Trends Genet.* **2013**, *29*, 197–205. [CrossRef]
10. Scandura, M.; Iacolina, L.; Crestanello, B.; Pecchioli, E.; DI Benedetto, M.F.; Russo, V.; Davoli, R.; Apollonio, M.; Bertorelle, G. Ancient vs. recent processes as factors shaping the genetic variation of the European wild boar: Are the effects of the last glaciation still detectable? *Mol. Ecol.* **2008**, *17*, 1745–1762. [CrossRef]
11. Fontanesi, L.; Ribani, A.; Scotti, E.; Utzeri, V.; Veličković, N.; Dall'Olio, S. Differentiation of meat from European wild boars and domestic pigs using polymorphisms in the *MC1R* and NR6A1 genes. *Meat Sci.* **2014**, *98*, 781–784. [CrossRef]
12. Iacolina, L.; Pertoldi, C.; Amills, M.; Kusza, S.; Megens, H.-J.; Bâlteanu, V.A.; Bakan, J.; Cubric-Curik, V.; Oja, R.; Saarma, U.; et al. Hotspots of recent hybridization between pigs and wild boars in Europe. *Sci. Rep.* **2018**, *8*, 17372. [CrossRef] [PubMed]
13. Mary, N.; Iannuccelli, N.; Petit, G.; Bonnet, N.; Pinton, A.; Barasc, H.; Amélie, F.; Calgaro, A.; Grosbois, V.; Servin, B.; et al. Genome-wide analysis of hybridization in wild boar populations reveals adaptive introgression from domestic pig. *Evol. Appl.* **2022**, *15*, 1115–1128. [CrossRef] [PubMed]
14. Dzialuk, A.; Zastempowska, E.; Skórzewski, R.; Twaružek, M.; Grajewski, J. High domestic pig contribution to the local gene pool of free-living European wild boar: A case study in Poland. *Mammal Res.* **2017**, *63*, 65–71. [CrossRef]
15. Babicz, M.; Pastwa, M.; Skrzypczak, E.; Buczyński, J.T. Variability in themelanocortin 1 receptor(*MC1R*) gene in wild boars and local pig breeds in Poland. *Anim. Genet.* **2013**, *44*, 357–358. [CrossRef]
16. Musto, M. DNA Quality and Integrity of Nuclear and Mitochondrial Sequences from Beef Meat as Affected by Different Cooking Methods. *Food Technol. Biotechnol.* **2011**, *49*, 523–528.
17. Natonek-Wiśniewska, M.; Krzyścin, P.; Piestrzyńska-Kajtoch, A. The species identification of bovine, porcine, ovine and chicken components in animal meals, feeds and their ingredients, based on COX I analysis and ribosomal DNA sequences. *Food Control* **2013**, *34*, 69–78. [CrossRef]
18. Tan, L.L.; Ahmed, S.A.; Ng, S.K.; Citartan, M.; Raabe, C.A.; Rozhdestvensky, T.S.; Tang, T.H. Rapid detection of porcine DNA in processed food samples using a streamlined DNA extraction method combined with the SYBR Green real-time PCR assay. *Food Chem.* **2019**, *309*, 125654. [CrossRef]
19. Prusakova, O.V.; Glukhova, X.A.; Afanas'Eva, G.V.; Trizna, Y.A.; Nazarova, L.F.; Beletsky, I.P. A simple and sensitive two-tube multiplex PCR assay for simultaneous detection of ten meat species. *Meat Sci.* **2018**, *137*, 34–40. [CrossRef]
20. Natonek-Wiśniewska, M.; Krzyścin, P. Evaluation of the suitability of mitochondrial DNA for species identification of microtraces and forensic traces. *Acta Biochim. Pol.* **2017**, *64*, 705–708. [CrossRef]
21. Man-an, S.K.A.; Fadilah, A.R.; Mardhiyyah, S. *Contemporary Issues and Development in the Global Halal Industry*; Ab.Manan, S.K., Abd Rahman, F., Sahri, M., Eds.; Springer: Singapore, 2016; ISBN 978-981-10-1450-5.
22. Mane, B.G.; Tanwar, V.K.; Girish, P.S.; Sharma, D.; Dixit, V.P. RAPD Markers for Differentiation of Meat Species. *Indian J. Vet. Res.* **2008**, *17*, 9–13.
23. Nakyinsige, K.; Man, Y.B.C.; Sazili, A.Q. Halal authenticity issues in meat and meat products. *Meat Sci.* **2012**, *91*, 207–214. [CrossRef] [PubMed]
24. Chappal-war, A.M.; Pathak, V.; Goswami, M.; Sharma, B.; Singh, P.; Mishra, R. Recent Novel Techniques Applied for Detection of Meat Adulteration and Fraudulent Practices. *Indian J. Vet. Public Health* **2020**, *7*, 1–6.
25. Rahmati, S.; Julkapli, N.M.; Yehye, W.A.; Basirun, W.J. Identification of meat origin in food products–A review. *Food Control* **2016**, *68*, 379–390. [CrossRef]
26. Hossain, M.A.M.; Ali, E.; Hamid, S.B.A.; Asing; Mustafa, S.; Desa, M.N.M.; Zaidul, I.S.M. Double Gene Targeting Multiplex Polymerase Chain Reaction–Restriction Fragment Length Polymorphism Assay Discriminates Beef, Buffalo, and Pork Substitution in Frankfurter Products. *J. Agric. Food Chem.* **2016**, *64*, 6343–6354. [CrossRef] [PubMed]
27. Lopez-Oceja, A.; Nuñez, C.; Baeta, M.; Gamarra, D.; de Pancorbo, M. Species identification in meat products: A new screening method based on high resolution melting analysis of cyt b gene. *Food Chem.* **2017**, *237*, 701–706. [CrossRef]
28. Thanakiatkrai, P.; Kitpipit, T. Meat species identification by two direct-triplex real-time PCR assays using low resolution melting. *Food Chem.* **2017**, *233*, 144–150. [CrossRef]
29. Natonek-Wiśniewska, M.; Krzyścin, P. Development of Easy and Effective Real-Time PCR tests to Identify Bovine, Porcine, and Ovine Components in Food. *Zywnosc Nauk. Technol. Jakosc/Food Sci. Technol. Qual.* **2015**, *22*, 73–84. [CrossRef]
30. Shehata, H.R.; Li, J.; Chen, S.; Redda, H.; Cheng, S.; Tabujara, N.; Li, H.; Warriner, K.; Hanner, R. Droplet digital polymerase chain reaction (ddPCR) assays integrated with an internal control for quantification of bovine, porcine, chicken and turkey species in food and feed. *PLoS ONE* **2017**, *12*, e0182872. [CrossRef]
31. Köppel, R.; Ganeshan, A.; Weber, S.; Pietsch, K.; Graf, C.; Hochegger, R.; Griffiths, K.; Burkhardt, S. Duplex digital PCR for the determination of meat proportions of sausages containing meat from chicken, turkey, horse, cow, pig and sheep. *Eur. Food Res. Technol.* **2019**, *245*, 853–862. [CrossRef]

32. Lee, S.-Y.; Kim, M.-J.; Hong, Y.; Kim, H.-Y. Development of a rapid on-site detection method for pork in processed meat products using real-time loop-mediated isothermal amplification. *Food Control* **2016**, *66*, 53–61. [CrossRef]
33. Ran, G.; Ren, L.; Han, X.; Liu, X.; Li, Z.; Pang, D.; Ouyang, H.; Tang, X. Development of a Rapid Method for the Visible Detection of Pork DNA in Halal Products by Loop-Mediated Isothermal Amplification. *Food Anal. Methods* **2015**, *9*, 565–570. [CrossRef]
34. Riztyan; Takasaki, K.; Yamakoshi, Y.; Futo, S. Single-Laboratory Validation of Rapid and Easy DNA Strip for Porcine DNA Detection in Beef Meatballs. *J. AOAC Int.* **2018**, *101*, 1653–1656. [CrossRef] [PubMed]
35. Tasrip, N.A.; Mokhtar Khairil, N.F.; Hanapi, U.K.; Abdul Manaf, Y.N.; Ali, M.E.; Cheah, Y.K. No Title Loop Mediated Isothermal Amplification; a Review on Its Application and Strategy in Animal Species Authentication of Meat Based Food Products. *Int. Food Res. J.* **2019**, *26*, 1–10.
36. Naaum, A.M.; Shehata, H.R.; Chen, S.; Li, J.; Tabujara, N.; Awmack, D.; Lutze-Wallace, C.; Hanner, R. Complementary molecular methods detect undeclared species in sausage products at retail markets in Canada. *Food Control* **2018**, *84*, 339–344. [CrossRef]
37. Haynes, E.; Jimenez, E.; Pardo, M.A.; Helyar, S.J. The future of NGS (Next Generation Sequencing) analysis in testing food authenticity. *Food Control* **2019**, *101*, 134–143. [CrossRef]
38. Arslan, A.; Ilhak, I.; Calicioglu, M.; Karahan, M. Identification of meats using random amplified polymorphic dna (rapd) technique. *J. Muscle Foods* **2005**, *16*, 37–45. [CrossRef]
39. Jin, L.-G.; Cho, J.-Y.; Seong, K.-B.; Park, J.-Y.; Kong, I.-S.; Hong, Y.-K. 18S rRNA gene sequences and random amplified polymorphic DNA used in discriminating Manchurian trout from other freshwater salmonids. *Fish. Sci.* **2006**, *72*, 903–905. [CrossRef]
40. Natonek-Wiśniewska, M.; Ząbek, T.; Słota, E. Species Identification of Mammalian Mt DNA Using PCR-RFLP. *Ann. Anim. Sci.* **2007**, *7*, 305–311.
41. Natonek-Wiśniewska, M.; Słota, E. A New Method for Species Identification of Poultry Based on 12S-RRNA Fragment Polymorphism. *Ann. Anim. Sci.* **2009**, *9*, 127–132.
42. Sultana, S.; Hossain, M.A.M.; Naquiah, N.N.A.; Ali, E. Novel multiplex PCR-RFLP assay discriminates bovine, porcine and fish gelatin substitution in Asian pharmaceuticals capsule shells. *Food Addit. Contam. Part A* **2018**, *35*, 1662–1673. [CrossRef]
43. Sun, Y.-L.; Lin, C.-S. Establishment and Application of a Fluorescent Polymerase Chain Reaction—Restriction Fragment Length Polymorphism (PCR-RFLP) Method for Identifying Porcine, Caprine, and Bovine Meats. *J. Agric. Food Chem.* **2003**, *51*, 1771–1776. [CrossRef] [PubMed]
44. Guan, F.; Jin, Y.-T.; Zhao, J.; Xu, A.-C.; Luo, Y.-Y. A PCR Method That Can Be Further Developed into PCR-RFLP Assay for Eight Animal Species Identification. *J. Anal. Methods Chem.* **2018**, *2018*, 5890140. [CrossRef] [PubMed]
45. Gargouri, H.; Moalla, N.; Kacem, H.H. PCR–RFLP and species-specific PCR efficiency for the identification of adulteries in meat and meat products. *Eur. Food Res. Technol.* **2021**, *247*, 2183–2192. [CrossRef]
46. Sultana, S.; Hossain, M.M.; Azlan, A.; Johan, M.R.; Chowdhury, Z.Z.; Ali, E. TaqMan probe based multiplex quantitative PCR assay for determination of bovine, porcine and fish DNA in gelatin admixture, food products and dietary supplements. *Food Chem.* **2020**, *325*, 126756. [CrossRef] [PubMed]
47. Janudin, A.A.S.; Chin, N.A.; Ahmed, M.U. An Eva Green Real-Time PCR Assay for Porcine DNA Analysis in Raw and Processed Foods. *Malays. J. Halal Res.* **2022**, *5*, 33–39. [CrossRef]
48. Lubis, H.; Salihah, N.T.; Norizan, N.A.; Hossain, M.M.; Ahmed, M.U. Fast and Sensitive Real-time PCR-based Detection of Porcine DNA in Food Samples by Using EvaGreen Dye. *Food Sci. Technol. Res.* **2018**, *24*, 803–810. [CrossRef]
49. Munir, M.A.; Inayatullah, A. Comparison of real time PCR and conventional PCR by identifying genomic DNA of bovine and porcine. *J. Kim. Ter. Indones.* **2021**, *23*, 63–71. [CrossRef]
50. Basanisi, M.G.; La Bella, G.; Nobili, G.; Coppola, R.; Damato, A.M.; Cafiero, M.A.; La Salandra, G. Application of the novel Droplet digital PCR technology for identification of meat species. *Int. J. Food Sci. Technol.* **2019**, *55*, 1145–1150. [CrossRef]
51. Ahmed, M.U.; Hasan, Q.; Hossain, M.M.; Saito, M.; Tamiya, E. Meat species identification based on the loop mediated isothermal amplification and electrochemical DNA sensor. *Food Control* **2010**, *21*, 599–605. [CrossRef]
52. Girish, P.; Barbuddhe, S.; Kumari, A.; Rawool, D.B.; Karabasanavar, N.S.; Muthukumar, M.; Vaithiyanathan, S. Rapid detection of pork using alkaline lysis- Loop Mediated Isothermal Amplification (AL-LAMP) technique. *Food Control* **2019**, *110*, 107015. [CrossRef]
53. Kumar, Y.; Bansal, S.; Jaiswal, P. Loop-Mediated Isothermal Amplification (LAMP): A Rapid and Sensitive Tool for Quality Assessment of Meat Products. *Compr. Rev. Food Sci. Food Saf.* **2017**, *16*, 1359–1378. [CrossRef] [PubMed]
54. Kane, D.E.; Hellberg, R.S. Identification of species in ground meat products sold on the U.S. commercial market using DNA-based methods. *Food Control* **2016**, *59*, 158–163, Corrigendum in **2016**, *63*, 267. [CrossRef]
55. Tillmar, A.O.; Dell'Amico, B.; Welander, J.; Holmlund, G. A Universal Method for Species Identification of Mammals Utilizing Next Generation Sequencing for the Analysis of DNA Mixtures. *PLoS ONE* **2013**, *8*, e83761. [CrossRef]
56. Dobrovolny, S.; Blaschitz, M.; Weinmaier, T.; Pechatschek, J.; Cichna-Markl, M.; Indra, A.; Hufnagl, P.; Hochegger, R. Development of a DNA metabarcoding method for the identification of fifteen mammalian and six poultry species in food. *Food Chem.* **2018**, *272*, 354–361. [CrossRef]
57. Xing, R.-R.; Wang, N.; Hu, R.-R.; Zhang, J.-K.; Han, J.-X.; Chen, Y. Application of next generation sequencing for species identification in meat and poultry products: A DNA metabarcoding approach. *Food Control* **2019**, *101*, 173–179. [CrossRef]
58. Cottenet, G.; Blancpain, C.; Chuah, P.F.; Cavin, C. Evaluation and application of a next generation sequencing approach for meat species identification. *Food Control* **2019**, *110*, 107003. [CrossRef]

59. Safdar, M.; Junejo, Y.; Arman, K.; Abasıyanık, M. A highly sensitive and specific tetraplex PCR assay for soybean, poultry, horse and pork species identification in sausages: Development and validation. *Meat Sci.* **2014**, *98*, 296–300. [CrossRef] [PubMed]
60. Fang, M.; Andersson, L. Mitochondrial diversity in European and Chinese pigs is consistent with population expansions that occurred prior to domestication. *Proc. R. Soc. B Boil. Sci.* **2006**, *273*, 1803–1810. [CrossRef] [PubMed]
61. Bruford, M.W.; Bradley, D.G.; Luikart, G. DNA markers reveal the complexity of livestock domestication. *Nat. Rev. Genet.* **2003**, *4*, 900–910. [CrossRef] [PubMed]
62. Koseniuk, A.; Słota, E. Mitochondrial control region diversity in Polish sheep breeds. *Arch. Anim. Breed.* **2016**, *59*, 227–233. [CrossRef]
63. Fajardo, V.; González, I.; Martín, I.; Rojas, M.; Hernández, P.E.; Garcia, T.; Martín, R. Differentiation of European wild boar (*Sus scrofa scrofa*) and domestic swine (*Sus scrofa domestica*) meats by PCR analysis targeting the mitochondrial D-loop and the nuclear melanocortin receptor 1 (*MC1R*) genes. *Meat Sci.* **2008**, *78*, 314–322. [CrossRef] [PubMed]
64. Klungland, H.; Våge, D.I. Pigmentary Switces in Domestic Animal Species. *Ann. N. Y. Acad. Sci.* **2003**, *994*, 331–338. [CrossRef] [PubMed]
65. Lu, D.; Willard, D.; Patel, I.R.; Kadwell, S.; Overton, L.; Kost, T.; Luther, M.; Chen, W.; Woychik, R.P.; Wilkison, W.O.; et al. Agouti Protein Isan Antagonist of the Melanocyte-Stimulating-Hormone Receptor. *Nature* **1994**, *371*, 799–802. [CrossRef]
66. Ollmann, M.M.; Lamoreux, M.L.; Wilson, B.D.; Barsh, G.S. Interaction of Agouti protein with the melanocortin 1 receptor in vitro and in vivo. *Genes Dev.* **1998**, *12*, 316–330. [CrossRef]
67. Kijas, J.M.H.; Wales, R.; Törnsten, A.; Chardon, P.; Moller, M.; Andersson, L. Melanocortin Receptor 1 (*MC1R*) Mutations and Coat Color in Pigs. *Genetics* **1998**, *150*, 1177–1185. [CrossRef] [PubMed]
68. Koutsogiannouli, E.A.; Moutou, K.A.; Sarafidou, T.; Stamatis, C.; Mamuris, Z. Detection of hybrids between wild boars (*Sus scrofa scrofa*) and domestic pigs (*Sus scrofa* f. domestica) in Greece, using the PCR-RFLP method on melanocortin-1 receptor (*MC1R*) mutations. *Mamm. Biol.* **2010**, *75*, 69–73. [CrossRef]
69. Koseniuk, A.; Smołucha, G.; Natonek-Wiśniewska, M.; Radko, A.; Rubiś, D. Differentiating Pigs from Wild Boars Based on *NR6A1* and *MC1R* Gene Polymorphisms. *Animals* **2021**, *11*, 2123. [CrossRef] [PubMed]
70. Mikawa, S.; Morozumi, T.; Shimanuki, S.-I.; Hayashi, T.; Uenishi, H.; Domukai, M.; Okumura, N.; Awata, T. Fine mapping of a swine quantitative trait locus for number of vertebrae and analysis of an orphan nuclear receptor, germ cell nuclear factor (NR6A1). *Genome Res.* **2007**, *17*, 586–593. [CrossRef] [PubMed]
71. Kaltenbrunner, M.; Mayer, W.; Kerkhoff, K.; Epp, R.; Rüggeberg, H.; Hochegger, R.; Cichna-Markl, M. Differentiation between wild boar and domestic pig in food by targeting two gene loci by real-time PCR. *Sci. Rep.* **2019**, *9*, 1–11. [CrossRef]
72. Beugin, M.-P.; Baubet, E.; De Citres, C.D.; Kaerle, C.; Muselet, L.; Klein, F.; Queney, G. A set of 20 multiplexed single nucleotide polymorphism (SNP) markers specifically selected for the identification of the wild boar (*Sus scrofa scrofa*) and the domestic pig (*Sus scrofa domesticus*). *Conserv. Genet. Resour.* **2017**, *9*, 671–675. [CrossRef]
73. Radko, A.; Smołucha, G.; Koseniuk, A. Microsatellite DNA Analysis for Diversity Study, Individual Identification and Parentage Control in Pig Breeds in Poland. *Genes* **2021**, *12*, 595. [CrossRef] [PubMed]
74. Nikolov, I.; Stoeckle, B.; Markov, G.; Kuehn, R. Substantial hybridisation between wild boars (*Sus scrofa scrofa*) and East Balkan pigs (*Sus scrofa* f. domestica) in natural environment as a result of semi-wild rearing in Bulgaria. *Czech J. Anim. Sci.* **2017**, *62*, 1–8. [CrossRef]
75. Lorenzini, R.; Fanelli, R.; Tancredi, F.; Siclari, A.; Garofalo, L. Matching STR and SNP genotyping to discriminate between wild boar, domestic pigs and their recent hybrids for forensic purposes. *Sci. Rep.* **2020**, *10*, 1–10. [CrossRef]
76. Goedbloed, D.J.; Megens, H.; Van Hooft, P.; Herrero-Medrano, J.M.; Lutz, W.; Alexandri, P.; Crooijmans, R.P.M.A.; Groenen, M.; Van Wieren, S.E.; Ydenberg, R.C.; et al. Genome-wide single nucleotide polymorphism analysis reveals recent genetic introgression from domestic pigs into Northwest European wild boar populations. *Mol. Ecol.* **2012**, *22*, 856–866. [CrossRef] [PubMed]
77. Rubin, C.-J.; Megens, H.-J.; Barrio, A.M.; Maqbool, K.; Sayyab, S.; Schwochow, D.; Wang, C.; Carlborg, Ö.; Jern, P.; Jørgensen, C.B.; et al. Strong signatures of selection in the domestic pig genome. *Proc. Natl. Acad. Sci. USA* **2012**, *109*, 19529–19536. [CrossRef] [PubMed]
78. Karim, L.; Takeda, H.; Lin, L.; Druet, T.; Arias, J.A.C.; Baurain, D.; Cambisano, N.; Davis, S.R.; Farnir, F.; Grisart, B.; et al. Variants modulating the expression of a chromosome domain encompassing PLAG1 influence bovine stature. *Nat. Genet.* **2011**, *43*, 405–413. [CrossRef] [PubMed]
79. Gudbjartsson, D.F.; Walters, G.B.; Thorleifsson, G.; Stefansson, H.; Halldorsson, B.V.; Zusmanovich, P.; Sulem, P.; Thorlacius, S.; Gylfason, A.; Steinberg, S.; et al. Many sequence variants affecting diversity of adult human height. *Nat. Genet.* **2008**, *40*, 609–615. [CrossRef]
80. Vaysse, A.; Ratnakumar, A.; Derrien, T.; Axelsson, E.; Pielberg, G.R.; Sigurdsson, S.; Fall, T.; Seppälä, E.H.; Hansen, M.S.T.; Lawley, C.T.; et al. Identification of Genomic Regions Associated with Phenotypic Variation between Dog Breeds using Selection Mapping. *PLoS Genet.* **2011**, *7*, e1002316. [CrossRef]
81. Pryce, J.E.; Hayes, B.; Bolormaa, S.; Goddard, M. Polymorphic Regions Affecting Human Height Also Control Stature in Cattle. *Genetics* **2011**, *187*, 981–984. [CrossRef] [PubMed]
82. Signer-Hasler, H.; Flury, C.; Haase, B.; Burger, D.; Simianer, H.; Leeb, T.; Rieder, S. A Genome-Wide Association Study Reveals Loci Influencing Height and Other Conformation Traits in Horses. *PLoS ONE* **2012**, *7*, e37282. [CrossRef] [PubMed]

83. Thomas, G.; Moffatt, P.; Salois, P.; Gaumond, M.-H.; Gingras, R.; Godin, E.; Miao, D.; Goltzman, D.; Lanctôt, C. Osteocrin, a Novel Bone-specific Secreted Protein That Modulates the Osteoblast Phenotype. *J. Biol. Chem.* **2003**, *278*, 50563–50571. [CrossRef] [PubMed]

84. Banzet, S.; Koulmann, N.; Sanchez, H.; Serrurier, B.; Peinnequin, A.; Bigard, A. Musclin gene expression is strongly related to fast-glycolytic phenotype. *Biochem. Biophys. Res. Commun.* **2007**, *353*, 713–718. [CrossRef] [PubMed]

85. Nishizawa, H.; Matsuda, M.; Yamada, Y.; Kawai, K.; Suzuki, E.; Makishima, M.; Kitamura, T.; Shimomura, I. Musclin, a Novel Skeletal Muscle-derived Secretory Factor. *J. Biol. Chem.* **2004**, *279*, 19391–19395. [CrossRef] [PubMed]

86. Wilkinson, S.; Lu, Z.H.; Megens, H.-J.; Archibald, A.L.; Haley, C.; Jackson, I.J.; Groenen, M.A.M.; Crooijmans, R.P.M.A.; Ogden, R.; Wiener, P. Signatures of Diversifying Selection in European Pig Breeds. *PLoS Genet.* **2013**, *9*, e1003453. [CrossRef]

87. Gondret, F.; Riquet, J.; Tacher, S.; Demars, J.; Sanchez, M.; Billon, Y.; Robic, A.; Bidanel, J.; Milan, D. Towards candidate genes affecting body fatness at the SSC7 QTL by expression analyses. *J. Anim. Breed. Genet.* **2011**, *129*, 316–324. [CrossRef]

88. Zhang, Y.; Gao, T.; Hu, S.; Lin, B.; Yan, D.; Xu, Z.; Zhang, Z.; Mao, Y.; Mao, H.; Wang, L.; et al. The Functional SNPs in the 5′ Regulatory Region of the Porcine PPARD Gene Have Significant Association with Fat Deposition Traits. *PLoS ONE* **2015**, *10*, e0143734. [CrossRef]

89. Muñoz, M.; Bozzi, R.; García-Casco, J.; Núñez, Y.; Ribani, A.; Franci, O.; García, F.; Škrlep, M.; Schiavo, G.; Bovo, S.; et al. Genomic diversity, linkage disequilibrium and selection signatures in European local pig breeds assessed with a high density SNP chip. *Sci. Rep.* **2019**, *9*, 1–14. [CrossRef] [PubMed]

90. Groenen, M.A.M. A decade of pig genome sequencing: A window on pig domestication and evolution. *Genet. Sel. Evol.* **2016**, *48*, 1–9. [CrossRef]

91. Moretti, R.; Criscione, A.; Turri, F.; Bordonaro, S.; Marletta, D.; Castiglioni, B.; Chessa, S. A 20-SNP Panel as a Tool for Genetic Authentication and Traceability of Pig Breeds. *Animals* **2022**, *12*, 1335. [CrossRef] [PubMed]

Article

Meta-Analysis of SNPs Determining Litter Traits in Pigs

Ewa Sell-Kubiak [1,*], Jan Dobrzanski [1], Martijn F. L. Derks [2], Marcos S. Lopes [2,3] and Tomasz Szwaczkowski [1]

[1] Department of Genetics and Animal Breeding, Poznań University of Life Sciences, Wołyńska 33, 60-637 Poznań, Poland
[2] Topigs Norsvin Research Centre, Schoenaker 6, 6641 SZ Beuningen, The Netherlands
[3] Topigs Norsvin, R. Visc. do Rio Branco, 1310-Centro, Curitiba 80420-210, PR, Brazil
* Correspondence: ewa.sell-kubiak@puls.edu.pl

Abstract: Nearly 2000 SNPs associated with pig litter size traits have been reported based on genome-wide association studies (GWASs). The aims of this study were to gather and integrate previously reported associations between SNPs and five litter traits: total number born (TNB), number born alive (NBA), number of stillborn (SB), litter birth weight (LWT), and corpus luteum number (CLN), in order to evaluate their common genetic background and to perform a meta-analysis (MA) of GWASs for total number born (TNB) recorded for animals from five pig populations. In this study, the genes with the largest number of associations with evaluated litter traits were *GABRG3*, *RBP7*, *PRKD1*, and *STXBP6*. Only 21 genes out of 233 associated with the evaluated litter traits were reported in more than one population or for more than one trait. Based on this evaluation, the most interesting candidate gene is *PRKD1*, which has an association with SB and TNB traits. Based on GO term analysis, *PRKD1* was shown to be involved in angiogenesis as well. As a result of the MA, two new genomic regions, which have not been previously reported, were found to be associated with the TNB trait. One SNP was located on *Sus scrofa* chromosome (SSC) 14 in the intron of the *FAM13C* gene. The second SNP was located on SSC9 within the intron of the *AGMO* gene. Functional analysis revealed a strong candidate causal gene underlying the QTL on SSC9. The third best hit and the most promising candidate gene for litter size was found within the *SOSTDC1* gene, associated with lower male fertility in rats. We showed that litter traits studied across pig populations have only a few genomic regions in common based on candidate gene comparison. *PRKD1* could be an interesting candidate gene with a wider association with fertility. The MA identified new genomic regions on SSC9 and SSC14 associated with TNB. Further functional analysis indicated the most promising gene was *SOSTDC1*, which was confirmed to affect male fertility in other mammals. This is an important finding, as litter traits are by default linked with females rather than males.

Keywords: corpus luteum number; number of stillborn; gene ontology; genomic regions; gene network; pCADD; protein–protein interaction

check for
updates

Citation: Sell-Kubiak, E.; Dobrzanski, J.; Derks, M.F.L.; Lopes, M.S.; Szwaczkowski, T. Meta-Analysis of SNPs Determining Litter Traits in Pigs. *Genes* **2022**, *13*, 1730. https://doi.org/10.3390/genes13101730

Academic Editors: Katarzyna Piórkowska and Katarzyna Ropka-Molik

Received: 30 August 2022
Accepted: 20 September 2022
Published: 26 September 2022

1. Introduction

Since the discovery of the first genomic association between litter size and estrogen receptors in 1996 [1], there have been high hopes of finding more major genomic regions for reproduction traits [2]. Currently, one of the most popular methods for uncovering the associations between traits and genes is genome-wide association study (GWAS) using single nucleotide polymorphism (SNP) markers [2–4]. Genomic regions related to litter traits were revealed by GWAS in various farm animal species, including rabbits [5,6], goats [7,8], cattle [9], and pigs [10–12]. Although GWAS has presented some shortcomings resulting from the rigid structure of the chips and uneven distribution of markers across the chromosomes, those studies still provided the greatest insight into the genetic bases of many reproduction traits in a variety of pig populations [10].

Many results of GWAS and earlier QTL mapping are stored in the Pig QTL database [13]. Although it doesn't store all associations reported in the literature, this database is a highly important source of information and gathers 30,869 QTLs for 692 traits, with 1625 of those being QTLs for litter traits. Among those QTLs, 356 are related to total number born (TNB), 223 to number born alive (NBA), 137 to number of stillborn (SB), 52 to litter birth weight (LWT), and 130 to corpus luteum number (CLN). The reported QTL regions were in most cases identified using SNPs in GWASs.

Even though those studies were performed on different populations, they often used the same SNP chips and statistical methodology. However, the results of those studies are not repeatable across breeds or even within one breed. This lack of overlap among GWAS results for TNB and NBA was recently described in a review by Bakoev et al. [12]. Moreover, litter traits proved to be highly complex polygenic traits. This, in contrast to production traits such as backfat and growth rate that have several major genes confirmed [14–16], resulted in detection of only one major QTL, the previously mentioned estrogen receptor [1].

Despite the high number of studies and QTLs reported for litter traits, no research aimed at reviewing the existing results of GWASs for several litter traits in a more systematic manner. One of the main approaches to integrate and pool the results from single GWASs is a meta-analysis (MA). The GWAS MA is widely used in medical research and is becoming more popular in livestock studies [17–19]. Meta-analysis could also be used to increase the power of association studies by combining datasets from different sources and reducing false positive associations [20,21]. This also could help indicate new major QTLs for litter traits.

Therefore, the aims of this study were to (1) gather and integrate previously reported associations between SNPs and five litter traits: total number born (TNB), number born alive (NBA), number of stillborn (SB), litter birth weight (LWT), and corpus luteum number (CLN) to evaluate their common genetic background; (2) investigate the relationships among reported candidate genes for litter traits by searching for functional interactions among proteins encoded by those genes; and (3) combine the full GWAS results from several studies to identify the effects of gene sets in a meta-analysis.

2. Materials and Methods

This study's main objectives were to: (1) gather and integrate previously reported associations between SNPs and five litter traits: total number born (TNB), number born alive (NBA), number of stillborn (SB), litter birth weight (LWT), and corpus luteum number (CLN) to evaluate their common genetic background; (2) investigate the relationships among reported candidate genes for litter traits by searching for functional interactions among proteins encoded by those genes; and (3) combine the full GWAS results from several studies to identify the effects of gene sets in a meta-analysis. This was performed using two datasets. Dataset **A** (Tables S1–S5; Supplementary_Tables) refers to the data generated from significant SNPs reported in studies performed for five traits: total number born (TNB), number born alive (NBA), number of stillborn (SB), litter birth weight (LWT), and corpus luteum number (CLN). Dataset **B** refers to data created from full GWAS results for TNB generated in five studies: [10,22–25] (Table 1).

Table 1. Overview of phenotypic and genomic data available for meta-analysis of GWAS results for total number born (TNB) collected in dataset **B**.

Source of Data	Uimari et al. [22]	Sell-Kubiak et al. [23]	Ma et al. [25]	Zhang et al. [10]	Balogh et al. [24]
Population	Finnish Landrace	Large White	Erhualian	Duroc	Hungarian Large White
Phenotype	deEBV [1]	deEBV [1]	EBV [2]	deEBV [1]	TNB
BeadChip			Porcine SNP60 Bead Chip		
Individuals	328	2351	48	1067	290
Available SNPs	57,868	40,969	28,020	32,147	56,592

Table 1. *Cont.*

Source of Data	Uimari et al. [22]	Sell-Kubiak et al. [23]	Ma et al. [25]	Zhang et al. [10]	Balogh et al. [24]
Method	single-SNP mixed model with pedigree relationship matrix	single-SNP mixed model with genomic relationship matrix	single-SNP mixed model with pedigree relationship matrix	single-SNP mixed model with genomic relationship matrix	multi-SNP mixed model with genomic relationship matrix
Detected SNPs	5	0	5	7	3
Threshold	$-\log_{10}(p\text{-value}) \leq 5.7$ [3]	$-\log_{10}(p\text{-value}) \geq 5$	$-\log_{10}(p\text{-value}) \leq 5.75$ [3]	$-\log_{10}(p\text{-value}) \geq 4$	$-\log_{10}(p\text{-value}) \geq 5$

[1] deEBV, i.e., deregressed breeding values of total number born; used to remove parents' average. [2] EBV, i.e., direct breeding values of total number born. [3] Bonferroni corrected *p*-value.

2.1. Dataset A

To create dataset **A**, this part of the study was performed following the guidelines from "Genome-wide association studies meta-analysis" by Thompson et al. [26]. The selection of published GWAS results included in the analysis was conducted from March 2020 to December 2020 within the following databases: Web of Knowledge, Web of Science, PubMed, and Google Scholar. Studies reporting associations between SNPs and TNB, NBA, SB, LWT, or CLN were collected using various combinations of the following terms: "pig", "litter size", "total number born", "number born alive", "litter size", "SNP", "litter birth weight", "stillbirth", "born dead", "ovulation", "corpus luteum number", "polymorphism", and "GWAS". Later, the survey was supported by searching for QTL associations in the Pig QTL database and also by screening the references of retrieved papers. The complete data collected for dataset **A** are presented in Supplementary_Tables in Tables S1–S5, which include *Sus scrofa* chromosome (SSC) location, allele, candidate gene, and breed. The general overview of dataset **A** is presented in Table 2.

Table 2. Publications reporting SNPs and genes associated with a total number of piglets born (TNB), number born alive (NBA), number of stillborn (SB), or litter birth weight (LWT).

Trait	Number of SNPs	Number of Genes	Publication	
	1	1	An et al.	[27]
	13	7	Coster et al.	[28]
	145	83	He et al.	[29]
	3	0	Kumchoo and Mekchay	[30]
	1	1	Li et al.	[31]
	1	1	Liu et al.	[32]
	7	5	Ma et al.	[25]
TNB	4	4	Sato et al.	[33]
	10	7	Sell-Kubiak et al.	[23]
	5	2	Uimari et al.	[22]
	2	2	Uzzaman et al.	[34]
	10	5	H. Wang et al.	[35]
	1	1	Y. Wang et al.	[36]
	40	6	Wu et al.	[37]
	5	3	Wu et al.	[38]
	7	4	Zhang et al.	[10]

Table 2. *Cont.*

Trait	Number of SNPs	Number of Genes	Publication	
NBA	1	1	An et al.	[27]
	17	8	Bergfelder-Drüing et al.	[39]
	11	6	Chen et al.	[11]
	3	3	Coster et al.	[28]
	3	3	Kumchoo and Mekchay	[30]
	1	1	Li et al.	[31]
	9	7	Ma et al.	[25]
	2	2	Sato et al.	[33]
	27	11	Suwannasing et al.	[40]
	1	0	Uzzaman et al.	[34]
	5	3	Y. Wang et al.	[36]
	101	19	Wu et al.	[37]
	15	6	Wu et al.	[38]
SB	46	22	Chen et al.	[11]
	6	3	Onteru et al.	[41]
	13	11	Schneider et al.	[42]
	2	1	Uimari et al.	[22]
	22	15	Verardo et al.	[43]
LWT	1	1	Coster et al.	[28]
	1	1	Liu et al.	[32]
	10	2	Zhang et al.	[10]
CLN	24	12	Schneider et al.	[44]

2.2. Dataset B

Dataset **B** was created by merging complete results of GWAS for TNB from five publications: Uimari et al. [22], Sell-Kubiak et al. [23], Ma et al. [25], Balogh et al. [24], and Zhang et al. [10]. The dataset created after merging the results from the five GWASs consisted of 63,531 SNPs available for further analysis. A specific overview of data collected in dataset **B** is presented in Table 1. In short, the data came from five different populations: Finnish Landrace, Large White, Erhualian, Duroc, and Hungarian Large White, and different methods were applied to perform the GWASs: single-SNP mixed model with pedigree additive relationship matrix or with genomic relationship matrix, and multiple-SNP Bayesian approach (Table 1). All populations were genotyped with Porcine SNP60 Bead Chip. The GWAS data from Sell-Kubiak et al. [23] covered phenotypes presented in a published paper, whereas the *p*-values of SNPs are the result of the single-SNP GWAS with genomics relationship matrix (more details in Supplementary_Method) and were not published in the mentioned paper.

2.3. Gene Ontology Analysis

Analysis of gene ontology (GO) terms [45] assigned to candidate genes collected in dataset **A** was performed using the PANTHER classification system Version 16 [46] with the newest available Ensembl Sscrofa11.1 as a reference genome. The analysis was performed with the binomial test of overrepresentation to explore biological processes in which the candidate genes are involved [46]. Results from the PANTHER were considered statistically significant at a false discovery rate (FDR)-corrected *p*-value $\leq$ 0.05 [47]. This analysis was performed on all candidate genes for five litter traits simultaneously.

2.4. Gene Network Analysis

The gene network analysis was performed with GeneMANIA [48] as a plug-in to Cytoscape v3.8.2 [49], with the human gene annotation as a reference. Settings allowed the authors to evaluate co-expression, physical interaction, gene interactions, shared protein domains, and co-localization. This analysis was performed only on candidate genes from dataset **A** that either had at least two SNPs detected along their sequence (Table 3) or were associated with more than one trait or presented in more than one study (Table 4).

Table 3. Genes with at least two SNPs in their sequence associated with total number born (TNB), number born alive (NBA), or number of stillborn (SB).

Trait	Gene	Number of SNPs	Publication	
TNB	GABRG3	5	Coster et al.	[28]
	MSI2	3	He et al.	[29]
	BICC1	2	Zhang et al.	[26]
	ENTPD1	2	He et al.	[29]
	ENOX1	2	Sell-Kubiak et al.	[23]
	SUGCT	2	Sell-Kubiak et al.	[23]
	NR3C2	2	Wu et al.	[38]
	GABRB3	2	Coster et al.	[28]
NBA	RBP7	5	Suwannasing et al.	[40]
	ARID1A	3	Chen et al.	[11]
	LRRK1	2	Suwannasing et al.	[40]
	ZMYND12	2	Suwannasing et al.	[40]
	RIMKLA	2	Suwannasing et al.	[40]
	RTL4	2	Suwannasing et al.	[40]
	UBE4B	3	Suwannasing et al.	[40]
	ALDH1A2	2	Wu et al.	[37]
	GABRA5	2	Wu et al.	[37]
	INPP4B	2	Wu et al.	[37]
	DNAJC6	2	Wu et al.	[37]
	QKI	2	Wu et al.	[37]
	SOX6	2	Wu et al.	[37]
SB	PRKD1	5	Chen et al.	[11]
	STXBP6	4	Chen et al.	[11]
	PBX1	2	Chen et al.	[11]
	GRM1	2	Chen et al.	[11]
	CYP24A1	2	Verardo et al.	[43]

2.5. Protein–Protein Interactions

Interactions between proteins coded by candidate genes reported in dataset **A** were investigated using a protein–protein interactions network using STRING Genomics v.11 [50]. The level of confidence for observed protein–protein interactions was limited to medium confidence interactions with scores > 0.40 with remaining settings at default [51].

2.6. Meta-Analysis on Dataset B

To combine estimates of SNP associations for TNB obtained from different populations, MA was performed on dataset **B**. Based on Garrick et al. [52], the decision was made that despite different definitions of phenotypes for TNB, the five datasets can be combined into one. All available 63,531 SNPs were included in the MA based on the weighted Z-score model. This approach considers the *p*-value, direction of effect, and number of individuals present in each study and was performed using METAL software [53]. The weighted Z-score model was chosen in accordance with Van den Berg et al. [54], who indicated this method as the most preferable when combining GWAS results with differences in the definition of the phenotypes, as is present in dataset **B**. Post-MA, the Bonferroni correction was applied to establish statistically significant associations. In addition, MA was performed in several runs with subsets of SNPs as follows: with all SNPs, with SNPs from at least 2 populations, with SNPs from at least 4 populations, and with SNPs present in all populations.

The evaluation of candidate genes found with the MA was performed with Bgee Version 14.2 (https://bgee.org/api/, accessed on 26 August 2022), GeneCards [55] and Ensembl BioMart.

Table 4. Genes with at least two SNPs in their sequence associated with more than one of the reproduction traits: total number born (TNB), number born alive (NBA), number of stillborn (SB), or litter birth weight (LWT), or reported in more than one population.

Candidate Gene	Publication and Population											
	An et al. [27] Berkshire	Chen et al. [11] Duroc	Coster et al. [28] Large White	He et al. [29] Erhualian	Li et al. [31] Yorkshire	Ma et al. [25] Erhualian	Sato et al. [33] Large White	Schneider et al. [44] Landrace x Duroc	Schneider et al. [42] Duroc, Yorkshire, Landrace	Verardo et al. [43] Large White	Wu et al. [37] Yorkshire	Y. Wang et al. [36] Large White
ASIC2				TNB							TNB	
CCL21											TNB, NBA	
CLSTN2				TNB, NBA								
COPG2			TNB, LBW									
DDAH1				TNB		TNB, NBA						
EIF3M			TNB, NBA									
FAT2											TNB, NBA	
GABRA5			TNB								TNB	
HECW1												TNB, NBA
IGFBP2	TNB											
INPP4B											TNB, NBA	
MAPK1IP1L									SB	SB		
NEK10				TNB							TNB	
PARD3						TNB, NBA						
PRKD1		SB	NBA									
RAD50							TNB, NBA					
SAMD4A								CLN		SB		
STXBP6		SB		TNB								
UBE3A			TNB, NBA									
UNC13C											TNB, NBA	
ZFYVE9					TNB, NBA			CLN				

2.7. Candidate Genes and Causal Variants

To assess possible candidate genes and variants underlying GWAS peaks, we also used the pCADD pipeline as described in [56]. In short, we extracted all sequence variants in linkage disequilibrium (LD) with the top SNP from the meta-GWAS. The candidate variants in LD with the top SNP were ranked according to their pCADD score. pCADD provides a per-base impact score [57] to distinguish between variants that likely have impact (positive or negative) and variants that are benign. The pipeline additionally provides gene annotation and functional genomic information to further fine-map the QTL region.

3. Results

In this study, two datasets were analyzed. First, dataset **A** (Tables S1–S5; Supplementary_Tables) refers to the data generated from significant SNPs reported in GWASs performed for five traits: total number born (TNB), number born alive (NBA), number of stillborn (SB), litter birth weight (LWT), and corpus luteum number (CLN). Second, dataset **B** refers to data created from full GWAS results for TNB generated in five studies [10,22–25] (Table 1).

3.1. SNPs and Candidate Genes from Dataset A

In total, 24 papers that studied at least one of the litter traits of interest were found (Tables S1–S5; Supplementary_Tables). Most of these studies focused on TNB and NBA, while studies for SB, LWT, and CLN were more limited (Table 2). The most common breeds evaluated in these studies were Large White, Yorkshire, Landrace, Duroc, and their crosses, with occasional occurrence of Erhualian and Berkshire. The definition of phenotypes varied across the publications and traits. While most of the publications used direct measurement of the trait, some authors chose deregressed breeding values as phenotypes [10,22,23]. In addition, in two studies, TNB phenotypes were divided into TNB at the first parity and the later parities [22,25].

The highest number of associations among SNPs and analyzed traits was reported as expected for TNB and NBA (Table 2), as those were the most-studied traits. Identified SNPs were rarely placed within the candidate gene. Most SNPs identified to be associated with the evaluated traits were located at least 50 Kbp away from the candidate gene. Only in the case of TNB, NBA, and SB were at least two SNPs with significant association with these traits in one population located within the candidate gene regions (Table 3). Genes with the largest number of associations along their sequences were *GABRG3*, *RBP7*, *PRKD1*, and *STXBP6*. Furthermore, only 21 genes out of 233 associated with the selected litter traits were reported in more than one population or for more than one trait (Table 4). For one gene out of this list (*DDAH1)* the overlap was expected, as the two studies reporting it were based on data from the same population [25,29]. This was not the case for the remaining populations presented in Table 4.

3.2. GO Term Analysis Results

In total 34 GO terms related to the biological processes were found (FDR < 0.05; Table 5). The most promising biological processes were "positive regulation of blood vessel endothelial cell migration", describing genes *NRP1*, *PIK3C2A*, *HDAC9*, *AKT3*, and *PRKD*, and "positive regulation of cell migration involved in sprouting angiogenesis", describing *NRP1*, *PIK3C2A*, *HDAC9*, and *AKT3*. Surprisingly, the majority of the remaining GO terms were involved with processes in nervous system development or regulation.

Table 5. Gene Ontology analysis for genes associated with total number born, number born alive, number of stillborn, litter birth weight, and corpus luteum number in pigs.

GO Term	Biological Processes (with Hierarchy When Applicable)	Observed Genes	Expected Genes	FDR
9987	cellular process	177	148.91	0.019
7399	→ nervous system development	32	15.2	0.027
21785	→ → branchiomotor neuron axon guidance	3	0.06	0.02
48518	positive regulation of the biological process	77	49.5	0.012
31344	→ regulation of cell projection organization	17	4.36	0.003
120035	→ → regulation of plasma membrane-bounded cell projection organization	17	4.29	0.003
51489	→ → → regulation of filopodium assembly	5	0.32	0.012
51491	→ → → → positive regulation of filopodium assembly	4	0.17	0.016
51239	regulation of multicellular organismal process	41	20.21	0.009
43536	→ positive regulation of blood vessel endothelial cell migration	5	0.45	0.04
90050	→ → positive regulation of cell migration involved in sprouting angiogenesis	4	0.2	0.026
8299	isoprenoid biosynthetic process	4	0.24	0.043

Table 5. *Cont.*

GO Term	Biological Processes (with Hierarchy When Applicable)				Observed Genes	Expected Genes	FDR
48813	dendrite morphogenesis				5	0.46	0.042
22603	regulation of anatomical structure morphogenesis				19	6.88	0.033
10975	↳	regulation of neuron projection development			14	2.84	0.002
50770		↳	regulation of axonogenesis		8	1.13	0.013
50771			↳	negative regulation of axonogenesis	5	0.48	0.048
34765	regulation of ion transmembrane transport				12	3.25	0.044
1904062	↳	regulation of cation transmembrane transport			12	2.25	0.004
48583	regulation of response to stimulus				54	30.14	0.009
50920	↳	regulation of chemotaxis			9	1.78	0.037
32409	regulation of transporter activity				9	1.8	0.039
1655	urogenital system development				12	2.55	0.01
72001	↳	renal system development			10	2.29	0.044
44057	regulation of system process				14	3.57	0.013

Table 5. *Cont.*

GO Term	Biological Processes (with Hierarchy When Applicable)	Observed Genes	Expected Genes	FDR
65008	regulation of biological quality	56	30.03	0.003
42592	⮡ homeostatic process	29	12.43	0.013
48878	⮡ chemical homeostasis	23	8.89	0.019
51240	positive regulation of multicellular organismal process	25	10.62	0.033
10646	regulation of cell communication	45	25.74	0.048
9966	⮡ regulation of signal transduction	42	22.78	0.033
51179	localization	74	44.32	0.003
51234	⮡ establishment of localization	60	33.77	0.004
6810	⮡ transport	57	32.4	0.009

3.3. Gene Network and Protein–Protein Interactions

The two gene network analyses indicated that most links among genes are based on co-expression and genetic interaction. The graphical representation of those results in presented in Figures 1 and 2.

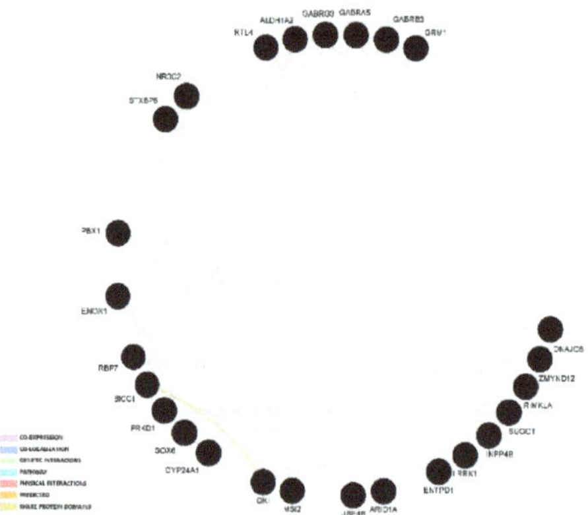

Figure 1. Gene network analysis of candidate genes that had at least two SNPs associated with reproduction traits.

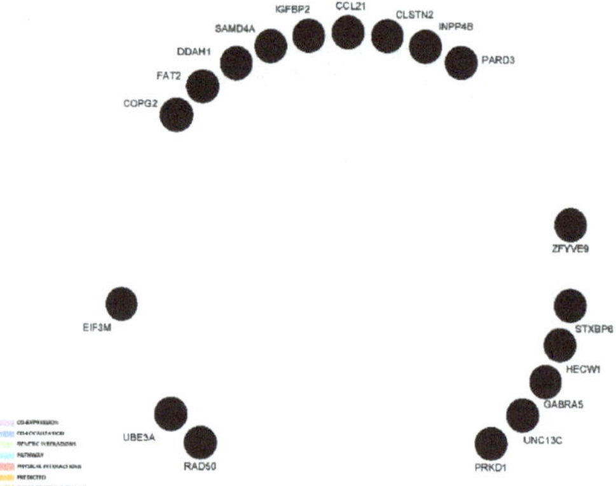

Figure 2. Gene network analysis of candidate genes that had an association reported with at least two reproduction traits or in two different populations.

Interactions among proteins (protein–protein interaction network; PPIN) coded by genes associated with the analyzed traits are presented in Figure 3. In total, 174 genes formed some pairs or chains of interactions with one evident cluster built of 138 proteins and 12 short chains that contained a maximum of 7 genes coding those proteins. The genes with the largest number of proven links among coded proteins were *CDC42*, *LRRK1*, and *AKT3*.

Overall, those two analyses did not provide strong evidence for interactions among candidate genes for the litter traits.

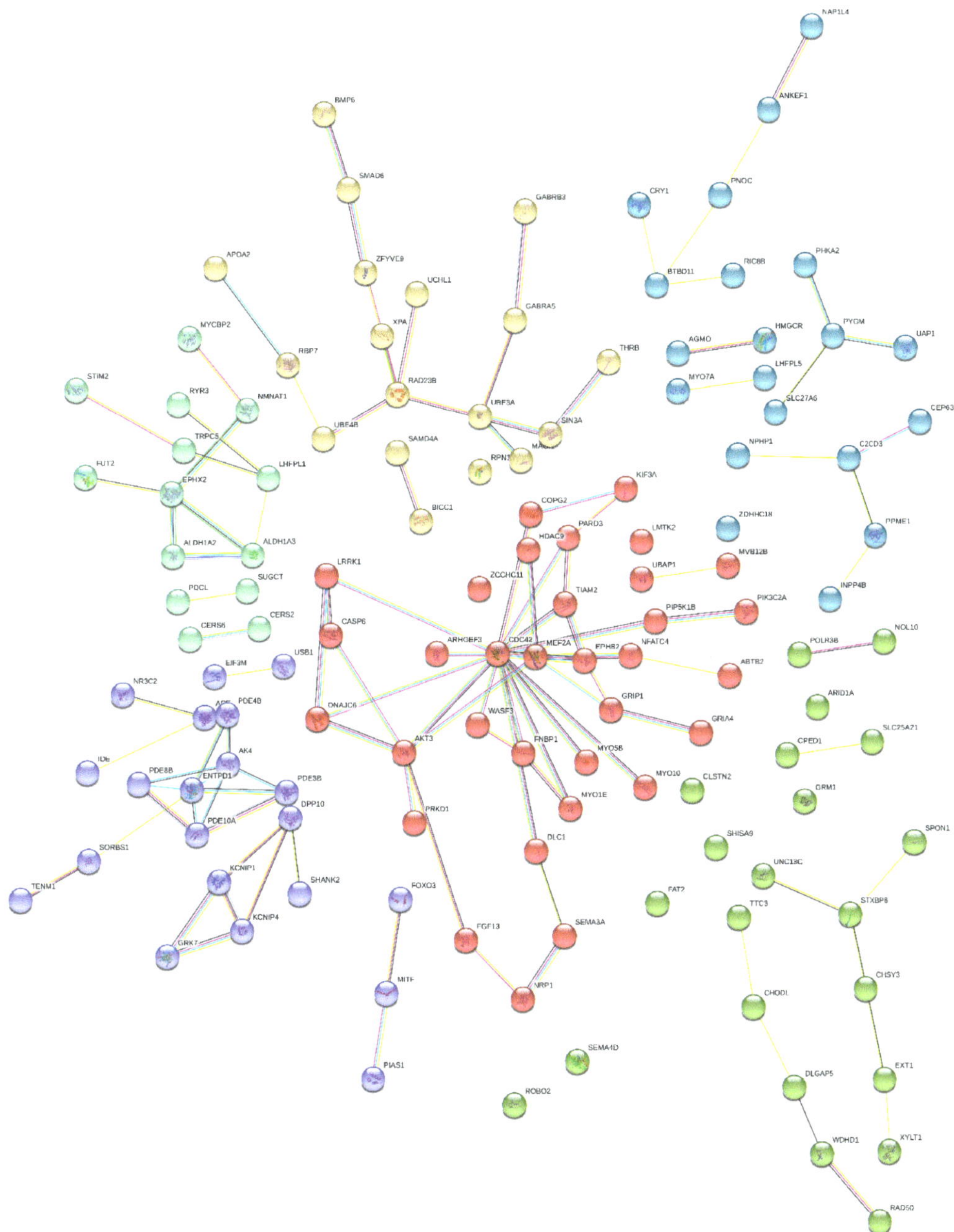

Figure 3. Protein–protein interaction network analysis of genes associated with total number born, number born alive, number of stillborn, litter birth weight, and corpus luteum number in pigs. Types of interactions: teal line—known interaction based on curated database, pink line—known interaction experimentally determined, green line—predicted interaction based on gene neighborhood, red line—predicted interaction based on gene fusion, blue line—predicted interaction based on a gene co-occurrence.

3.4. Meta-Analysis of Five GWA Studies

As a result of the MA performed on dataset B, two SNPs were found to be significantly associated with TNB. The first, rs80945731, located on SSC14: 62633073, was available in four out of five evaluated populations [10,22,24,25]. Based on Ensembl Sscrofa11.1, this SNP is located in the intron region of the *FAM13C* gene. The second, rs81300422, located on SSC9: 84696142, was unfortunately present in only one population [22]. According to Ensembl Sscrofa11.1, this gene is located within the intron of the gene *AGMO*. Importantly, these two SNPs found in the MA are reported for the first time to have an association with TNB. The last run of MA, which included only SNPs present in all populations, did not show any significant association with the evaluated traits.

3.5. Candidate Genes and Causal Variation

To assess candidate causal genes at the two QTL loci, we ran the pCADD-GWAS pipeline in four different breeds from Topigs Norsvin [45]. The rs81300422 SNP, which is located on SSC9, was segregated only in a synthetic boar line with a 7% minor allele frequency. The third best hit with the pCADD pipeline (after two intergenic SNPs) is upstream of the *SOSTDC1* gene. The *SOSTDC1* gene is a member of the sclerostin family functioning as a bone morphogenetic protein (BMP) antagonist, which is known to be associated with fertility. The descriptive statistics for TNB, NBA, SB, and mummies per genotype of rs81300422 is presented in Table 6. It can be observed that animals being homozygous for the alternative allele tend to lower SB (in sows and boars from a synthetic boar line and its crossbreeds) and higher rate of mummies (in sows from a synthetic boar line and in crossbreed sows) than homozygous animals for the reference allele. This is, however, not a statistically significant difference.

Table 6. Descriptive statistics of total number born (with SD), number born alive (with SD), number of stillborn, and mummies per genotype of rs81300422 for sows ♀and boars ♂from a synthetic boar line and its crosses.

Line	Genotype	N	TNB	NBA	SB	Mummies
	0/0	17,527	10.13 (3.10)	9.28 (3.10)	0.85	0.30
Synthetic boar line ♀	0/1	3247	10.07 (3.05)	9.29 (3.02)	0.78	0.32
	1/1	115	10.03 (2.95)	9.46 (2.85)	0.57	0.39
	0/0	25,500	10.08 (3.10)	9.23 (3.10)	0.85	0.33
Synthetic boar line ♂	0/1	4333	9.95 (3.18)	9.12 (3.17)	0.83	0.37
	1/1	231	9.64 (3.01)	8.86 (2.94)	0.78	0.34
	0/0	32,015	16.56 (3.89)	15.39 (3.74)	1.17	0.39
Crossbreed [1] ♀	0/1	5290	16.52 (3.86)	15.30 (3.75)	1.21	0.35
	1/1	206	16.41 (4.08)	15.37 (3.84)	1.04	0.43
	0/0	54,824	16.00 (3.93)	15.06 (3.85)	0.94	0.36
Crossbreed [2] ♂	0/1	9714	16.02 (3.80)	15.05 (3.78)	0.97	0.36
	1/1	172	15.83 (4.08)	15.02 (3.92)	0.80	0.34

[1] Synthetic boar line × (Large White × Landrace) [2] Synthetic boar line × (Landrace × Large White).

The rs80945731 SNP on SSC14 was common in all four commercial breeding populations. The results yielded SNPs within and close to the *PHYHIPL* and *FAM13C* genes, thus partially overlapping with the results of the candidate gene search applying the classical approach.

4. Discussion

This study intended to evaluate existing knowledge about the genomic regions associated with five litter traits in pigs: total number born (TNB), number born alive (NBA), number of stillborn (SB), litter birth weight (LWT), and corpus luteum number (CLN), and search for new candidate genes using bioinformatics analysis on combined results from previous studies. This study is the first one to evaluate the possibility of a common genetic background for those litter traits, as well as to present a meta-analysis for SNPs associated with TNB and combining data from five different populations.

4.1. Genetic Relationship between Litter Traits

One of the hypotheses at the beginning of this study was that there must be overlapping genomic regions and candidate genes among the evaluated litter traits. This hypothesis was based mostly on strong positive genetic correlations among TNB, NBA, CLN, and LWT [58,59]; however, it was not confirmed with the data collected in dataset **A**. Only one candidate gene, *ZFYVE9*, was reported for three litter traits (TNB, NBA, and CLN) and in two populations [31,42]. Another three candidate genes for SB (with a negative genetic correlation with the remaining litter traits [59]) were also found to be associated with TNB (*STXBP6* [11,29]), NBA (*PRKD1* [11,28]) or CLN (*SAMD4A* [42,43]). For LWT, only one candidate gene was in common with TNB (*COPG2* [28]), whereas only one candidate gene for SB overlapped between two populations (*MAPK1P1L* [43,44]). The abovementioned relationships among traits could suggest some pleiotropic effect of those genes on litter traits. In two populations three candidate genes for TNB were also reported: *ASIC2* [29,37], *GABRA5* [28,37], and *NEK10* [29,37]. Even though some connections across studies reporting candidate genes for litter traits can be observed, considering that we studied 233 candidate genes in total and only 21 of them covered more than one trait or population, one cannot assume the common genetic background of those traits. However, for production traits in pigs, except for major genes, such overlap between traits or populations is also not common [13]. It should be mentioned here that the older studies from PigQTLdb [13] reported very long QTLs because of high LD or access to SNP chips with low density. Thus, some of the reported candidate genes might in fact be very distant from the actual SNPs associated with the litter traits. Another reason might be the population stratification, which if not accounted for, can lead to false positives (e.g., [23]).

The lack of overlap between traits in terms of candidate genes was present not only among the studies for one trait but also within the same study if it focused on two or more traits. Moreover, more than one SNP was rarely reported within the candidate gene in the same population. Thus, the lack of overlap was not caused by the differences among the populations or the methodologies behind detecting associations, which are mentioned as two main reasons for differences among compared GWASs [12]. Even more so, the low repeatability of results across populations is surprising, because most of the studies were based on the same SNP chip and (in general) the most popular pig breeds. Another reason might have been the low heritability of the studied traits, which based on different studies varies from 0.05 to 0.2 [11,28,59,60]. The heritability level affects the ability to retrace the trait's heritability based on SNP associations [23].

Those results clearly show that the polygenic characteristics of the litter traits are very complex, and despite the undeniable relationship among traits their genetic background is mostly affected by different genomic regions often involved in nervous system development.

4.2. Connections among Candidate Genes

The gene network and protein–protein interaction analysis did not indicate clear clusters among candidate genes. It needs to be noted, however, that those analyses were based only on linking selected candidate genes with existing databases. This is why the current study did not produce gene expression or any other empirical data that could have been included as additional information for candidate genes. In addition, pig genomic databases are lacking information in comparison with human, mouse, or even cattle gene databases. Thus, the majority of the evidence suggesting functional links among analyzed genes were based on the co-expression of putative homologs present in other organisms (Ensembl Sscrofa11.1). In only a few cases were links between proteins experimentally determined or proteins involved in the same pathway [55].

Some similarities among litter traits can be seen in the GO terms analysis performed on dataset **A**. This analysis clearly showed that genes involved in neurodevelopment and the nervous system are overrepresented among candidate genes for the studied litter traits. This came as a surprising result as we expected to see more genes involved in vasculogenesis and angiogenesis, as blood supply to the uterus and developing fetus was indicated in the past as one of the most limiting factors for litter size [22,61]. That is why from all the possible biological processes that were significant in GO term analysis, we are certain that those involved in positive regulation of blood vessel endothelial cell migration and positive regulation of cell migration involved in sprouting angiogenesis require the most attention.

Interestingly, out of the five genes that have the aforementioned GO terms annotated to them, *PRKD1* is the gene with five SNPs along its sequence associated with number of stillborn [11] and one SNP with TNB [29]. *PRKD1* encodes a protein kinase involved in many cellular processes, including cell migration and differentiation, cell survival, and regulation of cell shape and adhesion [55]. Mutation in this gene causes congenital heart defects in humans [62,63]. In addition, it was also indicated as a candidate gene for age at puberty in maternal and terminal Landrace, Duroc, and Yorkshire lines [64]. Thus, it is associated with at least three traits related to reproduction in their different populations. This indicates that *PRKD1* should be one of the genes considered for molecular analysis involving gene expression or sequencing to confirm its association with litter traits in pigs.

4.3. The New Genomic Region Associated with Litter Size

The meta-analysis of GWAS results coming from five different populations indicated two new SNPs associated with litter size. Even though our MA produced far fewer significant results than each of the GWASs separately, this is the first time those SNPs are reported as associated with litter size.

The first of the significant SNPs on SSC9, rs81300422 (genotyped only in one population) is located in the intron of the gene *AGMO* encoding enzyme Alkylglycerol monooxygenase, not very well studied in swine, which is involved in the degradation of ether lipids [65] and the only enzyme that can breakdown alkylglycerols and lysol alkyl glycerophospholipids [66]. This gene was shown to be associated with neurodevelopmental disorders in humans [67], e.g., autism [68], type 2 diabetes [69–71], and immune defense [66]. In humans, this gene was expressed mostly in tissues of the digestive tract, especially in the liver, but also in the ovaries, uterus, prostate, and testes [55]. However, the literature does not report any direct links of *AGMO* with fertility.

Thus, a far more interesting and promising candidate gene is *SOSTDC1*, indicated by pCADD analysis, which has been found to affect fertility in male rats [72]. *SOSTDC1* is a negative regulator of spermatogenesis, and downregulation of this gene during puberty is essential for quantitatively and qualitatively normal spermatogenesis governing male fertility. The association between TNB and male fertility comes as a surprise, as litter traits are in general linked with females. However, in some pig breeds it is observed that male genetics plays a more important role in final litter size than originally expected [73,74]. Further comparison of litter traits among pigs with different genotypes for rs81300422 showed that the detected SNP could be affecting the number of stillborn and mummies in pig populations available for pCADD. This was, unfortunately, not confirmed by the statistical analysis.

The second significant SNP, rs80945731 (present in four populations), is located within the gene *FAM13C*, which does not have a clear function in pigs. Nonetheless, it was expressed in swine adipose tissue and tissues of the nervous system (e.g., amygdala or medulla oblongata). At the same time, in humans *FAM13C* was shown to be associated with endometrial mixed adenocarcinoma [55] as well as being used as a marker for prostate cancer [75] or as a rectal adenocarcinoma survival predictor [76]. In humans, this gene also was expressed (among others) in the tissue of the ovaries and uterus as well as the prostate and testes [55]. The second gene indicated by pCADD for rs80945731 was *PHYHIPL*, which is differentially expressed among early, full, and expanded blastocysts in bovine IVP embryos [77]. Both *FAM13C* and *PHYHIPL* in humans were expressed (among others) in the tissue of the ovaries and uterus as well as the prostate and testes [55].

5. Conclusions

We have shown that litter traits (total number born, number born alive, number of stillborn, litter birth weight, and corpus luteum number) studied across pig populations have only a few genomic regions in common based on candidate gene comparison. The most interesting candidate gene is *PRKD1*, which has an association with SB and TNB as well as being involved in angiogenesis based on GO term analysis. Our meta-analysis of GWAS results coming from five populations helped identify the new genomic regions on SSC9 and SSC14 associated with TNB. Further pCADD analysis indicated the most promising gene was *SOSTDC1*, which actually has a confirmed effect on male fertility. This is an important finding, as litter traits are by default linked with females rather than males.

Supplementary Materials: The following are available online at https://www.mdpi.com/article/10.3390/genes13101730/s1, Supplementary_Method: The description of not published GWAS [23], Table S1. SNPs associated with total number born in pigs with their location within pig genome (including chromosome - SSC, position and gene), alleles as well as breed in which the SNP was detected. Table S2. SNPs associated with number born alive in pigs with their location within pig genome (including chromosome - SSC, position and gene), alleles as well as breed in which the SNP was detected. Table S3. SNPs associated with number of stillborn in pigs with their location within pig genome (including chromosome - SSC, position and gene), alleles as well as breed in which the SNP was detected. Table S4. SNPs associated with litter birthweight in pigs with their location within pig genome (including chromosome, position and gene), alleles as well as breed in which the SNP was detected. Table S5. SNPs associated with corpus luteum number in pigs with their location within pig genome (including chromosome, position and gene), alleles as well as breed in which the SNP was detected.

Author Contributions: Conceptualization—E.S.-K.; methodology—E.S.-K., J.D., M.F.L.D., M.S.L. and T.S.; formal analysis—E.S.-K., J.D., M.F.L.D. and M.S.L.; data curation—J.D. and E.S.-K.; original draft preparation—E.S.-K. and J.D.; review and editing—E.S.-K., M.F.L.D., M.S.L. and T.S.; figures—E.S.-K.; tables in the main manuscript—E.S.-K. and J.D.; supplementary tables—J.D. and E.S.-K.; funding acquisition—E.S.-K. All authors have read and agreed to the published version of the manuscript.

Funding: This project was financed by the National Science Centre, Poland (NCN SONATA grant no. 2016/23/D/NZ9/00029; E.S.K. is the grant holder). E.S.K. also acknowledges the financial support of the Polish Ministry of Science and Higher Education (grant no. 1021/STYP/12/2017).

Institutional Review Board Statement: Not applicable.

Informed Consent Statement: Not applicable.

Data Availability Statement: Data of dataset **A** are contained within the article and supplementary materials. The data that support the findings of this study (combined in dataset **B**) are available from the National Engineering Research Center for Breeding Swine Industry and the Guangdong Provincial Key Lab of Agro-Animal Genomics and Molecular Breeding, College of Animal Science, South China Agricultural University (China); the Institute of Swine Science, Nanjing Agricultural University (China); Agrifood Research Finland, MTT, Biotechnology and Food Research (Finland); the TopigsNorsvin Research Centre, Beuningen (the Netherlands); and the NARIC Research Institute for Animal Breeding, Nutrition and Meat Science (Hungary), but restrictions apply to the availability of these data, which were used under license for the current study, and so are not publicly available. Data are however available from the authors upon reasonable request and with permission of: the National Engineering Research Center for Breeding Swine Industry and Guangdong Provincial Key Lab of Agro-Animal Genomics and Molecular Breeding, College of Animal Science, South China Agricultural University (China); the Institute of Swine Science, Nanjing Agricultural University (China); Agrifood Research Finland, MTT, Biotechnology and Food Research (Finland); the TopigsNorsvin Research Centre, Beuningen (the Netherlands); and the NARIC Research Institute for Animal Breeding, Nutrition and Meat Science (Hungary).

Acknowledgments: The authors would like to thank five scientific groups that generously provided GWAS results from their studies for meta-analysis: the National Engineering Research Center for Breeding Swine Industry and the Guangdong Provincial Key Lab of Agro-Animal Genomics and Molecular Breeding, College of Animal Science, South China Agricultural University (China); the Institute of Swine Science, Nanjing Agricultural University (China); Agrifood Research Finland, MTT, Biotechnology and Food Research (Finland); the TopigsNorsvin Research Centre, Beuningen (the Netherlands); and the NARIC Research Institute for Animal Breeding, Nutrition and Meat Science (Hungary).

Conflicts of Interest: The authors declare no conflict of interest.

References

1. Rothschild, M.; Jacobson, C.; Vaske, D.; Tuggle, C.; Wang, L.; Short, T.; Eckardt, G.; Sasaki, S.; Vincent, A.; McLaren, D.; et al. The estrogen receptor locus is associated with a major gene influencing litter size in pigs. *Proc. Natl. Acad. Sci. USA* **1996**, *93*, 201–205. [CrossRef]
2. Ernst, C.W.; Steibel, J.P. Molecular advances in QTL discovery and application in pig breeding. *Trends Genet.* **2013**, *29*, 215–224. [CrossRef] [PubMed]
3. Blaj, I.; Tetens, J.; Preuss, S.; Bennewitz, J.; Thaller, G. Genome-wide association studies and meta-analysis uncovers new candidate genes for growth and carcass traits in pigs. *PLoS ONE* **2018**, *13*, e0205576. [CrossRef]
4. Rothschild, M.F.; Hu, Z.-L.; Jiang, Z. Advances in QTL Mapping in Pigs. *Int. J. Biol. Sci.* **2007**, *3*, 192–197. [CrossRef]
5. Sosa-Madrid, B.S.; Santacreu, M.A.; Blasco, A.; Fontanesi, L.; Pena, R.N.; Ibáñez-Escriche, N. A genomewide association study in divergently selected lines in rabbits reveals novel genomic regions associated with litter size traits. *J. Anim. Breed. Genet.* **2020**, *137*, 123–138. [CrossRef]
6. Casto-Rebollo, C.; Argente, M.J.; Garciá, M.L.; Pena, R.; Ibáñez-Escriche, N. Identification of functional mutations associated with environmental variance of litter size in rabbits. *Genet. Sel. Evol.* **2020**, *52*, 22. [CrossRef]
7. Tao, L.; He, X.Y.; Jiang, Y.T.; Lan, R.; Li, M.; Li, Z.M.; Yang, W.F.; Hong, Q.H.; Chu, M.X. Combined approaches to reveal genes associated with litter size in Yunshang black goats. *Anim. Genet.* **2020**, *51*, 924–934. [CrossRef] [PubMed]
8. Islam, R.; Liu, X.; Gebreselassie, G.; Abied, A.; Ma, Q.; Ma, Y. Genome-wide association analysis reveals the genetic locus for high reproduction trait in Chinese Arbas Cashmere goat. *Genes Genom.* **2020**, *42*, 893–899. [CrossRef] [PubMed]
9. Jiang, J.; Ma, L.; Prakapenka, D.; VanRaden, P.M.; Cole, J.B.; Da, Y. A Large-Scale Genome-Wide Association Study in U.S. Holstein Cattle. *Front. Genet.* **2019**, *10*, 412. [CrossRef]
10. Zhang, Z.; Chen, Z.; Ye, S.; He, Y.; Huang, S.; Yuan, X.; Li, J. Genome-Wide Association Study for Reproductive Traits in a Duroc Pig Population. *Animals* **2019**, *9*, 732. [CrossRef]
11. Chen, Z.; Ye, S.; Teng, J.; Diao, S.; Yuan, X.; Chen, Z.; Zhang, H.; Li, J.; Zhang, Z. Genome-wide association studies for the number of animals born alive and dead in duroc pigs. *Theriogenology* **2019**, *139*, 36–42. [CrossRef] [PubMed]
12. Bakoev, S.; Getmantseva, L.; Bakoev, F.; Kolosova, M.; Gabova, V.; Kolosov, A.; Kostyunina, O. Survey of SNPs Associated with Total Number Born and Total Number Born Alive in Pig. *Genes* **2020**, *11*, 491. [CrossRef] [PubMed]
13. Animal Genome Project. Pig QTL Data Base. Available online: http//www.animalgenome.org/QTLdb/pig.html (accessed on 21 March 2021).
14. Van Son, M.; Enger, E.G.; Grove, H.; Ros-Freixedes, R.; Kent, M.P.; Lien, S.; Grindflek, E. Genome-wide association study confirm major QTL for backfat fatty acid composition on SSC14 in Duroc pigs. *BMC Genom.* **2017**, *18*, 369. [CrossRef]
15. Jungerius, B.J.; Van Laere, A.S.; Te Pas, M.F.W.; Van Oost, B.A.; Andersson, L.; Groenen, M.A.M. The IGF2-intron3-G3072A substitution explains a major imprinted QTL effect on backfat thickness in a Meishan x European white pig intercross. *Genet. Res.* **2004**, *84*, 95–101. [CrossRef]
16. Van Laere, A.-S.; Nguyen, M.; Braunschweig, M.; Nezer, C.; Collette, C.; Moreau, L.; Archibald, A.; Haley, C.; Buys, N.; Tally, M.; et al. A regulatory mutation in IGF2 causes a major QTL effect on muscle growth in the pig. *Nature* **2003**, *425*, 832–836. [CrossRef] [PubMed]
17. Alfonso, L. Use of meta-analysis to combine candidate gene association studies: Application to study the relationship between the ESR PvuII polymorphism and sow litter size. *Genet. Sel. Evol.* **2005**, *37*, 417–435. [CrossRef]
18. Bouwman, A.C.; Daetwyler, H.D.; Chamberlain, A.J.; Ponce, C.H.; Sargolzaei, M.; Schenkel, F.S.; Sahana, G.; Govignon-Gion, A.; Boitard, S.; Dolezal, M.; et al. Meta-analysis of genome-wide association studies for cattle stature identifies common genes that regulate body size in mammals. *Nat. Genet.* **2018**, *50*, 362–367. [CrossRef]
19. Duarte, D.A.S.; Newbold, C.J.; Detmann, E.; Silva, F.F.; Freitas, P.H.F.; Veroneze, R.; Duarte, M.S. Genome-wide association studies pathway-based meta-analysis for residual feed intake in beef cattle. *Anim. Genet.* **2019**, *50*, 150–153. [CrossRef] [PubMed]
20. Tropf, F.C.; Lee, S.H.; Verweij, R.M.; Stulp, G.; van der Most, P.J.; de Vlaming, R.; Bakshi, A.; Briley, D.A.; Rahal, C.; Hellpap, R.; et al. Hidden heritability due to heterogeneity across seven populations. *Nat. Hum. Behav.* **2017**, *1*, 757–765. [CrossRef]
21. Sauvant, D.; Letourneau-Montminy, M.P.; Schmidely, P.; Boval, M.; Loncke, C.; Daniel, J.B. Review: Use and misuse of meta-analysis in Animal Science. *Animal* **2020**, *14*, s207–s222. [CrossRef]

22. Uimari, P.; Sironen, A.; Sevón-Aimonen, M.-L. Whole-genome SNP association analysis of reproduction traits in the Finnish Landrace pig breed. *Genet. Sel. Evol.* **2011**, *43*, 42. [CrossRef] [PubMed]
23. Sell-Kubiak, E.; Duijvesteijn, N.; Lopes, M.S.; Janss, L.L.G.; Knol, E.F.; Bijma, P.; Mulder, H.A. Genome-wide association study reveals novel loci for litter size and its variability in a Large White pig population. *BMC Genom.* **2015**, *16*, 1049. [CrossRef]
24. Balogh, E.E.; Gábor, G.; Bodó, S.; Rózsa, L.; Rátky, J.; Zsolnai, A.; Anton, I. Effect of single-nucleotide polymorphisms on specific reproduction parameters in Hungarian Large White sows. *Acta Vet. Hung.* **2019**, *67*, 256–273. [CrossRef]
25. Ma, X.; Li, P.H.; Zhu, M.X.; He, L.C.; Sui, S.P.; Gao, S.; Su, G.S.; Ding, N.S.; Huang, Y.; Lu, Z.Q.; et al. Genome-wide association analysis reveals genomic regions on Chromosome 13 affecting litter size and candidate genes for uterine horn length in Erhualian pigs. *Animal* **2018**, *12*, 2453–2461. [CrossRef]
26. Thompson, J.R.; Attia, J.; Minelli, C. The meta-analysis of genome-wide association studies. *Brief. Bioinform.* **2011**, *12*, 259–269. [CrossRef]
27. An, S.M.; Hwang, J.H.; Kwon, S.; Yu, G.E.; Park, D.H.; Kang, D.G.; Kim, T.W.; Park, H.C.; Ha, J.; Kim, C.W. Effect of Single Nucleotide Polymorphisms in *IGFBP2* and *IGFBP3* Genes on Litter Size Traits in Berkshire Pigs. *Anim. Biotechnol.* **2018**, *29*, 301–308. [CrossRef]
28. Coster, A.; Madsen, O.; Heuven, H.C.M.; Dibbits, B.; Groenen, M.A.M.; van Arendonk, J.A.M.; Bovenhuis, H. The Imprinted Gene DIO3 Is a Candidate Gene for Litter Size in Pigs. *PLoS ONE* **2012**, *7*, e31825. [CrossRef]
29. He, L.C.; Li, P.H.; Ma, X.; Sui, S.P.; Gao, S.; Kim, S.W.; Gu, Y.Q.; Huang, Y.; Ding, N.S.; Huang, R.H. Identification of new single nucleotide polymorphisms affecting total number born and candidate genes related to ovulation rate in Chinese Erhualian pigs. *Anim. Genet.* **2017**, *48*, 48–54. [CrossRef]
30. Kumchoo, T.; Mekchay, S. Association of non-synonymous SNPs of OPN gene with litter size traits in pigs. *Arch. Anim. Breed.* **2015**, *58*, 317–323. [CrossRef]
31. Li, X.; Ye, J.; Han, X.; Qiao, R.; Li, X.; Lv, G.; Wang, K. Whole-genome sequencing identifies potential candidate genes for reproductive traits in pigs. *Genomics* **2020**, *112*, 199–206. [CrossRef]
32. Liu, C.; Ran, X.; Yu, C.; Xu, Q.; Niu, X.; Zhao, P.; Wang, J. Whole-genome analysis of structural variations between Xiang pigs with larger litter sizes and those with smaller litter sizes. *Genomics* **2018**, *111*, 310–319. [CrossRef]
33. Sato, S.; Kikuchi, T.; Uemoto, Y.; Mikawa, S.; Suzuki, K. Effect of candidate gene polymorphisms on reproductive traits in a Large White pig population. *Anim. Sci. J.* **2016**, *87*, 1455–1463. [CrossRef]
34. Uzzaman, M.R.; Park, J.E.; Lee, K.T.; Cho, E.S.; Choi, B.H.; Kim, T.H. A genome-wide association study of reproductive traits in a Yorkshire pig population. *Livest. Sci.* **2018**, *209*, 67–72. [CrossRef]
35. Wang, H.; Wu, S.; Wu, J.; Sun, S.; Wu, S.; Bao, W. Association analysis of the SNP (rs345476947) in the FUT2 gene with the production and reproductive traits in pigs. *Genes Genom.* **2018**, *40*, 199–206. [CrossRef]
36. Wang, Y.; Ding, X.; Tan, Z.; Xing, K.; Yang, T.; Pan, Y.; Mi, S.; Sun, D.; Wang, C. Genome-wide association study for reproductive traits in a Large White pig population. *Anim. Genet.* **2018**, *49*, 127–131. [CrossRef]
37. Wu, P.; Wang, K.; Yang, Q.; Zhou, J.; Chen, D.; Ma, J.; Tang, Q.; Jin, L.; Xiao, W.; Jiang, A.; et al. Identifying SNPs and candidate genes for three litter traits using single-step GWAS across six parities in Landrace and Large White pigs. *Physiol. Genom.* **2018**, *50*, 1026–1035. [CrossRef]
38. Wu, P.; Yang, Q.; Wang, K.; Zhou, J.; Ma, J.; Tang, Q.; Jin, L.; Xiao, W.; Jiang, A.; Jiang, Y.; et al. Single step genome-wide association studies based on genotyping by sequence data reveals novel loci for the litter traits of domestic pigs. *Genomics* **2018**, *110*, 171–179. [CrossRef]
39. Bergfelder-Drüing, S.; Grosse-Brinkhaus, C.; Lind, B.; Erbe, M.; Schellander, K.; Simianer, H.; Tholen, E. A Genome-Wide Association Study in Large White and Landrace Pig Populations for Number Piglets Born Alive. *PLoS ONE* **2015**, *10*, e0117468. [CrossRef]
40. Suwannasing, R.; Duangjinda, M.; Boonkum, W.; Taharnklaew, R.; Tuangsithtanon, K. The identification of novel regions for reproduction trait in Landrace and Large White pigs using a single step genome-wide association study. *Asian-Australas. J. Anim. Sci.* **2018**, *31*, 1852–1862. [CrossRef]
41. Onteru, S.K.; Fan, B.; Du, Z.-Q.; Garrick, D.J.; Stalder, K.J.; Rothschild, M.F. A whole-genome association study for pig reproductive traits. *Anim. Genet.* **2011**, *43*, 18–26. [CrossRef]
42. Schneider, J.F.; Miles, J.R.; Brown-Brandl, T.M.; Nienaber, J.A.; Rohrer, G.A.; Vallet, J.L. Genomewide association analysis for average birth interval and stillbirth in swine. *J. Anim. Sci.* **2015**, *93*, 529–540. [CrossRef]
43. Verardo, L.L.; Silva, F.F.; Lopes, M.S.; Madsen, O.; Bastiaansen, J.W.M.; Knol, E.F.; Kelly, M.; Varona, L.; Lopes, P.S.; Guimarães, S.E.F. Revealing new candidate genes for reproductive traits in pigs: Combining Bayesian GWAS and functional pathways. *Genet. Sel. Evol.* **2016**, *48*, 9. [CrossRef]
44. Schneider, J.F.; Nonneman, D.J.; Wiedmann, R.T.; Vallet, J.L.; Rohrer, G.A. Genomewide association and identification of candidate genes for ovulation rate in swine. *J. Anim. Sci.* **2014**, *92*, 3792–3803. [CrossRef]
45. Carbon, S.; Mungall, C. Gene Ontology Data Archive; Zenodo, Maryland, United States of America: 2021. Available online: https://doi.org/10.5281/zenodo.5228828 (accessed on 29 August 2022).
46. Mi, H.; Ebert, D.; Muruganujan, A.; Mills, C.; Albou, L.-P.; Mushayamaha, T.; Thomas, P.D. PANTHER version 16: A revised family classification, tree-based classification tool, enhancer regions and extensive API. *Nucleic Acids Res.* **2020**, *49*, D394–D403. [CrossRef]

47. Keel, B.N.; Snelling, W.M.; Lindholm-Perry, A.K.; Oliver, W.T.; Kuehn, L.A.; Rohrer, G. Using SNP Weights Derived From Gene Expression Modules to Improve GWAS Power for Feed Efficiency in Pigs. *Front. Genet.* **2020**, *10*, 1339. [CrossRef]
48. Warde-Farley, D.; Donaldson, S.L.; Comes, O.; Zuberi, K.; Badrawi, R.; Chao, P.; Franz, M.; Grouios, C.; Kazi, F.; Lopes, C.T.; et al. The GeneMANIA prediction server: Biological network integration for gene prioritization and predicting gene function. *Nucleic Acids Res.* **2010**, *38* (Suppl. 2), 214–220. [CrossRef]
49. Shannon, P.; Markiel, A.; Ozier, O.; Baliga, N.S.; Wang, J.T.; Ramage, D.; Amin, N.; Schwikowski, B.; Ideker, T. Cytoscape: A software environment for integrated models of Biomolecular Interaction Networks. *Genome Res.* **2003**, *13*, 2498–2504. [CrossRef]
50. Szklarczyk, D.; Morris, J.H.; Cook, H.; Kuhn, M.; Wyder, S.; Simonovic, M.; Santos, A.; Doncheva, N.T.; Roth, A.; Bork, P.; et al. The STRING database in 2017: Quality-controlled protein-protein association networks.; made broadly accessible. *Nucleic Acids Res.* **2017**, *45*, D362–D368. [CrossRef]
51. Bozhilova, L.V.; Whitmore, A.V.; Wray, J.; Reinert, G.; Deane, C.M. Measuring rank robustness in scored protein interaction networks. *BMC Bioinform.* **2019**, *20*, 446. [CrossRef]
52. Garrick, D.J.; Taylor, J.F.; Fernando, R.L. Deregressing estimated breeding values and weighting information for genomic regression analyses. *Genet. Sel. Evol.* **2009**, *41*, 55. [CrossRef]
53. Willer, C.; Li, Y.; Abecasis, G.R. METAL: Fast and efficient meta-analysis of genomewide association scans. *Bioinformatics* **2010**, *26*, 2190–2191. [CrossRef] [PubMed]
54. Berg, I.V.D.; Boichard, D.; Lund, M.S. Comparing power and precision of within-breed and multibreed genome-wide association studies of production traits using whole-genome sequence data for 5 French and Danish dairy cattle breeds. *J. Dairy Sci.* **2016**, *99*, 8932–8945. [CrossRef] [PubMed]
55. Stelzer, G.; Rosen, N.; Plaschkes, I.; Zimmerman, S.; Twik, M.; Fishilevich, S.; Stein, T.I.; Nudel, R.; Lieder, I.; Mazor, Y.; et al. The GeneCards Suite: From Gene Data Mining to Disease Genome Sequence Analysis. *Curr. Protoc. Bioinform.* **2016**, *54*, 1.30.1–1.30.33. [CrossRef] [PubMed]
56. Derks, M.F.; Gross, C.; Lopes, M.S.; Reinders, M.J.; Bosse, M.; Gjuvsland, A.B.; de Ridder, D.; Megens, H.-J.; Groenen, M.A. Accelerated discovery of functional genomic variation in pigs. *Genomics* **2021**, *113*, 2229–2239. [CrossRef] [PubMed]
57. Groß, C.; Derks, M.; Megens, H.-J.; Bosse, M.; Groenen, M.A.M.; Reinders, M.; De Ridder, D. pCADD: SNV prioritisation in Sus scrofa. *Genet. Sel. Evol.* **2020**, *52*, 4. [CrossRef]
58. Holm, B.; Bakken, M.; Klemetsdal, G.; Vangen, O. Genetic correlations between reproduction and production traits in swine. *J. Anim. Sci.* **2004**, *82*, 3458–3464. [CrossRef]
59. Sell-Kubiak, E. Selection for litter size and litter birthweight in Large White pigs: Maximum, mean and variability of reproduction traits. *Animal* **2021**, *15*, 100352. [CrossRef] [PubMed]
60. Li, H.; Wang, M.; Li, W.; He, L.; Zhou, Y.; Zhu, J.; Che, R.; Warburton, M.L.; Yang, X.; Yan, J. Genetic variants and underlying mechanisms influencing variance heterogeneity in maize. *Plant J.* **2020**, *103*, 1089–1102. [CrossRef]
61. Lu, S.; Liu, S.; Wietelmann, A.; Kojonazarov, B.; Atzberger, A.; Tang, C.; Schermuly, R.; Gröne, H.-J.; Offermanns, S. Developmental vascular remodeling defects and postnatal kidney failure in mice lacking Gpr116 (Adgrf5) and Eltd1 (Adgrl4). *PLoS ONE* **2017**, *12*, e0183166. [CrossRef]
62. Shaheen, R.; al Hashem, A.; Alghamdi, M.H.; Seidahmad, M.Z.; Wakil, S.M.; Dagriri, K.; Keavney, B.; Goodship, J.; Alyousif, S.; Al-Habshan, F.M.; et al. Positional mapping of PRKD1.; NRP1 and PRDM1 as novel candidate disease genes in truncus arteriosus. *J. Med. Genet.* **2015**, *52*, 322–329. [CrossRef]
63. Massadeh, S.; Albeladi, M.; Albesher, N.; Alhabshan, F.; Kampe, K.; Chaikhouni, F.; Kabbani, M.; Beetz, C.; Alaamery, M. Novel Autosomal Recessive Splice-Altering Variant in *PRKD1* Is Associated with Congenital Heart Disease. *Genes* **2021**, *12*, 612. [CrossRef]
64. Nonneman, D.; Lents, C.; Rohrer, G.; Rempel, L.; Vallet, J. Genome-wide association with delayed puberty in swine. *Anim. Genet.* **2013**, *45*, 130–132. [CrossRef] [PubMed]
65. Watschinger, K.; Werner, E.R. Alkylglycerol monooxygenase. *IUBMB Life* **2013**, *65*, 366–372. [CrossRef]
66. Sailer, S.; Keller, M.A.; Werner, E.R.; Watschinger, K. The Emerging Physiological Role of AGMO 10 Years after Its Gene Identification. *Life* **2021**, *11*, 88. [CrossRef] [PubMed]
67. Okur, V.; Watschinger, K.; Niyazov, D.; McCarrier, J.; Basel, D.; Hermann, M.; Werner, E.R.; Chung, W.K. Biallelic variants in AGMO with diminished enzyme activity are associated with a neurodevelopmental disorder. *Hum. Genet.* **2019**, *138*, 1259–1266. [CrossRef]
68. Sebat, J.; Lakshmi, B.; Malhotra, D.; Troge, J.; Lese-Martin, C.; Walsh, T.; Yamrom, B.; Yoon, S.; Krasnitz, A.; Kendall, J.; et al. Strong Association of De Novo Copy Number Mutations with Autism. *Science* **2007**, *316*, 445–449. [CrossRef]
69. Fontaine-Bisson, B.; The MAGIC investigators; Renström, F.; Rolandsson, O.; Payne, F.; Hallmans, G.; Barroso, I.; Franks, P.W. Evaluating the discriminative power of multi-trait genetic risk scores for type 2 diabetes in a northern Swedish population. *Diabetologia* **2010**, *53*, 2155–2162. [CrossRef] [PubMed]
70. Boesgaard, T.W.; Grarup, N.; Jørgensen, T.; Borch-Johnsen, K.; Hansen, T.; Pedersen, O. Variants at DGKB/TMEM195.; ADRA2A.; GLIS3 and C2CD4B loci are associated with reduced glucose-stimulated beta cell function in middle-aged Danish people. *Diabetologia* **2010**, *53*, 1647–1655. [CrossRef] [PubMed]

71. Dupuis, J.; Langenberg, C.; Prokopenko, I.; Saxena, R.; Soranzo, N.; Jackson, A.U.; Wheeler, E.; Glazer, N.L.; Bouatia-Naji, N.; Gloyn, A.L.; et al. Erratum: New genetic loci implicated in fasting glucose homeostasis and their impact on type 2 diabetes risk. *Nat. Genet.* **2010**, *42*, 464. [CrossRef]

72. Pradhan, B.; Bhattacharya, I.; Sarkar, R.; Majumdar, S.S. Downregulation of Sostdc1 in Testicular Sertoli Cells is Prerequisite for Onset of Robust Spermatogenesis at Puberty. *Sci. Rep.* **2019**, *9*, 11458. [CrossRef]

73. Myromslien, F.D.; Tremoen, N.H.; Andersen-Ranberg, I.; Fransplass, R.; Stenseth, E.-B.; Zeremichael, T.T.; Van Son, M.; Grindflek, E.; Gaustad, A.H. Sperm DNA integrity in Landrace and Duroc boar semen and its relationship to litter size. *Reprod. Domest. Anim.* **2018**, *54*, 160–166. [CrossRef]

74. Tremoen, N.H.; Van Son, M.; Andersen-Ranberg, I.; Grindflek, E.; Myromslien, F.D.; Gaustad, A.H.; Våge, D.I. Association between single-nucleotide polymorphisms within candidate genes and fertility in Landrace and Duroc pigs. *Acta Vet. Scand.* **2019**, *61*, 58. [CrossRef] [PubMed]

75. Burdelski, C.; Borcherding, L.; Kluth, M.; Hube-Magg, C.; Melling, N.; Simon, R.; Möller-Koop, C.; Weigand, P.; Minner, S.; Haese, A.; et al. Family with sequence similarity 13C (FAM13C) overexpression is an independent prognostic marker in prostate cancer. *Oncotarget* **2017**, *8*, 31494–31508. [CrossRef] [PubMed]

76. Huang, H.; Cao, J.; Guo, G.; Li, X.; Wang, Y.; Yu, Y.; Zhang, S.; Zhang, Q.; Zhang, Y. Genome-wide association study identifies QTLs for displacement of abomasum in Chinese Holstein cattle1. *J. Anim. Sci.* **2019**, *97*, 1133–1142. [CrossRef]

77. Georges, H.; Bishop, J.; Van Campen, H.; Barfield, J.; Hansen, T. 102 A delay in maternal zygotic transition may lead to early embryonic loss in poor-quality bovine blastocysts. *Reprod. Fertil. Dev.* **2020**, *32*, 177. [CrossRef]

Article

Variations in Fibrinogen-like 1 (*FGL1*) Gene Locus as a Genetic Marker Related to Fat Deposition Based on Pig Model and Liver RNA-Seq Data

Katarzyna Piórkowska [1,*], Kacper Żukowski [2], Katarzyna Ropka-Molik [1] and Mirosław Tyra [3]

1   Department of Animal Molecular Biology, National Research Institute of Animal Production, Krakowska 1, 32-083 Balice, Poland
2   Department of Cattle Breeding, National Research Institute of Animal Production, Krakowska 1, 32-083 Balice, Poland
3   Department of Pig Breeding, National Research Institute of Animal Production, Krakowska 1, 32-083 Balice, Poland
*   Correspondence: katarzyna.piorkowska@iz.edu.pl; Tel.: +48-66-608-1316

Abstract: The goal of this study was to evaluate the effects of mutations in the *FGL1* gene associated with pig productive traits to enrich the genetic marker pool for further selection and to support the studies on *FGL1* in the context of the fat deposition (FD) process. The variant calling and χ^2 analyses of liver RNA-seq data were used to indicate genetic markers. *FGL1* mutations were genotyped in the Złotnicka White (n = 72), Polish Large White (n = 208), Duroc (n = 72), Polish Landrace (PL) (n = 292), and Puławska (n = 178) pig breeds. An association study was performed using a general linear model (GLM) implemented in SAS® software. More than 50 crucial mutations were identified in the *FGL1* gene. The association study showed a significant effect of the *FGL1* on intramuscular fat (IMF), loin eye area, backfat thickness at the lumbar, ham mass ($p = 0.0374$), meat percentage ($p = 0.0205$), and loin fat ($p = 0.0003$). Alternate homozygotes and heterozygotes were found in the PL and Duroc, confirming the selective potential for these populations. Our study supports the theory that liver *FGL1* is involved in the FD process. Moreover, since fat is the major determinant of flavor development in meat, the *FGL1* rs340465447_A allele can be used as a target in pig selection focused on elevated fat levels.

Keywords: fibrinogen-like 1; *FGL1*; pig; fat deposition; selective markers

check for updates

Citation: Piórkowska, K.; Żukowski, K.; Ropka-Molik, K.; Tyra, M. Variations in Fibrinogen-like 1 (*FGL1*) Gene Locus as a Genetic Marker Related to Fat Deposition Based on Pig Model and Liver RNA-Seq Data. *Genes* **2022**, *13*, 1419. https://doi.org/10.3390/genes13081419

Academic Editor: Bo Zuo

Received: 12 July 2022
Accepted: 8 August 2022
Published: 9 August 2022

Publisher's Note: MDPI stays neutral with regard to jurisdictional claims in published maps and institutional affiliations.

1. Introduction

Next-generation sequencing (NGS) technology is currently widely applied to predict processes associated with fat deposition and obesity-related events [1,2]. Besides whole-genome sequencing (WGS) and genome-wide association studies (GWASs) [3], RNA sequencing delivers valuable information about polymorphic sites located in transcripts [4] to develop selective markers related to given traits. However, the main target of RNA-seq analysis is to estimate transcript levels, including ncRNA molecules such as lncRNA and miRNA [5]. Variant calling analysis based on transcriptome sequencing results could be an alternative way to identify gene mutations associated with the fat deposition (FD) process. Although, this type of analysis requires more detailed calculations than SNP identification-based WGS data [6]. Nevertheless, identifying mutations within RNA-seq data is the additive value because it requires only an extra in silico analysis.

Fibrinogen-like protein 1 (*FGL1*) is a member of the fibrinogen family mainly synthesized in the liver [7]. It has been demonstrated that increased *FGL1* expression is related to the regeneration of the liver organ by stimulating 3H-thymidine uptake leading to hepatocyte proliferation [8]. *FGL1* is also expressed in brown adipose tissue (BAT) and during liver injury; expression is also upregulated in this tissue, which suggests cross-talk between an injured liver and bAT. Consistently, a study of Fgl1-deficient mice showed global metabolic disorders, namely *FGL1*-null mice were heavier, had abnormal plasma lipid profiles, fasting hyperglycemia with enhanced gluconeogenesis, and exhibited differences in white and BAT morphology regarding wild types [9]. The authors implied that the structural similarity of Fgl1 to angiopoietin factors regulating lipid metabolism showed that *FGL1* likely plays a crucial role in lipid metabolism and liver regeneration. On the other hand, *FGL1* expression has been examined in gastric cancer (GC) tissue [7], where *FGL1* promotes GC proliferation, and patient overall survival time with increased *FGL1* expression was significantly shorter. Therefore, the *FGL1* gene has been proposed as a predictor in GC patients and a target for treating this cancer type. Wang et al. [10] showed that Fgl1 is a major ligand for immune inhibitory receptor lymphocyte-activation gene 3 (LAG-3) and inhibits antigen-specific T cell activation. According to these findings, elevated *FGL1* expression in the plasma of cancer patients is related to poor prognosis.

Pigs are a highly suited animal model for predicting FD events due to many convergences with the human body, such as the size of particular organs [11], body fat distribution, and similar metabolism [12].

In the present study, Złotnicka White (ZW) pigs were used as an animal model in the context of FD processes. ZW is a Polish indigenous breed included in the Polish Animal Genetic Resource Conservation Program [13] and has never been under selection pressure, thus retaining a high meat quality [14]. In addition, ZW pigs are differentiated in terms of fat deposition in the carcass; therefore, it was possible to select two groups to represent low and high-FD values. Variant calling analysis based on liver RNA-seq data identified, in a cheaper manner than GWAS, numerous polymorphic sites in transcripts, and the χ^2 test pinpointed mutations characterized by significantly different allele distributions between the FD pig groups. Following this work scheme, we found *FGL1* variants highly interesting regarding the possible application in the selection process.

2. Materials and Methods

Samples were collected after slathering in the Pig Test Station (PTS, National Research Institute of Animal Production). Following carcass evaluation, the meat was intended for sale and consumption. Ethical review and approval were waived for this study due to a non-invasive method of collecting. All conducted research was approved by the Approving Experiment Committee of the National Research Institute of Animal Production (Krakow, Poland) according to the Polish Act on the Protection of Animals Used for Scientific or Educational Purposes of 15 January 2015, which implements Directive 2010/63/EU of the European Parliament and the Council of 22 September 2010 on the protection of animals used for scientific purposes. Moreover, all procedures followed the guidelines and regulations of the Local Krakow Ethics Committee for Experiments with Animals.

2.1. Animals and Methods

Female pigs of Złotnicka White were used in RNA-seq (n = 16) and association (n = 72) analyses. The animals with an initial weight of 30 kg were delivered to the Pig Test Station (PTS) of the National Research Institute of Animal Production from different farms. The animals were maintained under the same environmental and diet conditions according to a local procedure [15], feeding *ad libitum* to the final weight of 100 ± 2.5 kg. During pig maintenance at the PTS, growth performance such as feed intake, feed conversion, and daily gain were measured. After slaughter and detailed dissection, body composition traits were measured, including meat percentage, carcass yield, weight of the most significant cuts such as ham and loin, backfat thickness, and numerous fat-related traits, including visceral and subcutaneous fat content and meat quality traits such as intramuscular fat (IMF), pH, meat color, and myowater exudation [16].

Liver samples (n = 72) were collected within 20 min after dissection and stabilized in RNAlater™ Stabilization Solution (Invitrogen, Waltham, MA, USA), and then frozen at −20 °C. Next, two pig groups were selected based on fat-level measurements performed

24 h after slaughter. The RNA-seq experiment included 16 ZW pigs, with 8 representing high-fat deposition (HFD, n = 8) and low (LFD, n = 8) values.

Moreover, *FGL1* variant frequency was estimated in Polish Large White (PLW) (n = 208), Polish Landrace (PL) (n = 292), Puławska (n = 178), and Duroc (n = 72) pigs that are still active in the Polish breeding. Animal samples for DNA isolation were collected as blood or hair follicles and were banked in the National Research Institute of Animal Production as part of other research projects.

2.2. Library Construction, Sequencing, and Aligning Raw Reads to the Pig Transcriptome

As described previously by Piórkowska et al. [17], RNA was isolated using a Pure-Link™ RNA Mini Kit (Invitrogen, Waltham, MA, USA). Its quality and quantity were calculated using TapeStatio2200 (RNA tape, Agilent, Santa Clara, CA, USA), and the RIN parameters were over 7.5 value. The cDNA libraries were prepared using a TruSeq RNA Sample Preparation Kit v2 (Illumina, San Diego, CA, USA), with a unique barcode for each sample. The quality and quantity of the cDNA libraries were assessed by the Qubit Fluorimeter (Invitrogen, Waltham, MA, USA) and TapeStation 2200 system (D1000 tape, Agilent, Santa Clara, CA, USA). The final concentration of cDNA libraries' was normalized to 10 nM, and they were pooled. Transcriptome sequencing was performed in 150PE cycles on the HiSeq 3000 System (Illumina, San Diego, CA, USA), employing commissioned sequencing in Admera Health Biopharma Services. Aligning of raw reads to the pig transcriptome was performed according to Piórkowska et al. [17].

2.3. Transcript Variant Identification

Picard tools and GATK v. 4.1.9 [18] were used to split reads containing Ns in their CIGAR string and for the base quality score recalibration, indel realignment, removal of duplicates, and finally to identify SNPs and indels. The filtering parameters were selected according to the Best Practices workflow [6,19]. Mutation sites for annotation and prediction were analyzed by SnpEff v. 4.1b [20] and the Variant Effect Predictor (Ensembl) [21]. Differences in allele distribution of mutations identified by RNA-seq using the χ^2 test (corrected p-value at false discovery rate (FDR) $\leq$ 0.05) between the HFD and LFD pigs were estimated. The FDR correction was conducted using the R_{stats} base procedure and default parameters. In the χ^2 test, it was expected that allele/genotypes would have normal according to Gaussian distribution. The procedure in R was based on Benjamin and Hochberg [22]. Functional analysis for proteins encoded by identified genes was performed by the STRING 11.0b tool (https://string-db.org/).

Four of the most significant *FGL1* variants located in the 3′UTR region, according to the χ^2 test, were validated by Sanger sequencing of DNA. The primers for sequencing are shown in Table S1. New indel mutations identified in the *FGL1* gene were submitted to GenBank with access number MW827172 (NCBI database). miRBase v.22 and mirPath v3.0 (Diana) tools we used to analyze significant SNPs and indels identified in the 3′UTR regions for miRNA binding and miRNA functional analysis, respectively.

Linkage disequilibrium was estimated with analysis based on Barrett et al. [23] method.

2.4. FGL1 Genotyping, Frequency Estimation, and Statistical Analyses

For *FGL1* 3′UTR genotyping, PCR-RFLP, Sanger sequencing, and PCR-ACRS methods were designed. The primers and restriction enzymes are shown in Table S1. *FGL1* frequency was estimated for ZW (sows), PLW (boar and sows), PL (boars), Puławska (boars and sows), and Duroc (boars) pigs. Besides ZW, all individuals used in the study are still active in Polish breeding. The examination of *FGL1* variants frequency allowed us to establish the selective potential targeted to increase organoleptic qualities of pork lost in the previous intense breeding.

The differences in fat deposition traits and pig phenotypes, including 16 ZW pigs used previously in RNA sequencing analysis, were calculated between HFD and LFD using the ANOVA procedure (SAS v. 8.02) with a post-hoc Duncan test (SAS Enterprise).

Association analysis was performed, including 72 ZW pigs using the GLM procedure (SAS v. 8.02). Additionally, severely affected pig phenotype *RYR1* mutation was genotyped in the ZW population. All individuals included in the analysis were free of the *RYR1* alternate allele [24].

The general linear model (GLM) (SAS Enterprise v. 7.1 with default settings; SAS Institute, Cary, NC, USA) was used for statistical analyses. The linear model for fixed analysis was:

$$Y_{ijkl} = \mu + d_i + b_j + \alpha(x_{ijk}) + e_{ijkl},$$

where: Y_{ijk}—observation, μ—overall mean, d_i—fixed effect of genotype group, b_j—fixed effect of the breed, $\alpha(x_{ijk})$—covariate for weight of the right side of the carcass, and e_{ijkl}—random error. The GLM model for analysis within breeds omitted b_j—fixed effect of breed. The ratio between investigated pigs and their mothers (sows) was 1.7, and the ratio between investigated pigs and their fathers (sires) was 2.28; therefore, these factors were not included in the statistical model. During the whole year, the animals were maintained in the pig house at the same temperature and humidity conditions and with the same feeding; therefore, the model did not account for the effect of slaughter season. The least-square means (LSM) method was used for the determination of significant differences between genotype groups. The differences in phenotype in particular genotype groups are presented as LSM $\pm$ SE.

Additive and dominance effects were calculated for a total of 72 pigs. Additive and dominance effects were calculated using the regression (REG) procedure (SAS v. 8.02). The additive effect was denoted as -1 and 1 for genotypes AA and GG, respectively. The dominance effects are represented as -1 for AG heterozygotes and 1 for both homozygotes.

3. Results

3.1. Animals Used in the RNA-Seq Method

HFD pigs used in the experiment showed over 40% higher fat deposition in the lumbar location and over 30% higher FD according to average backfat thickness and peritoneal and loin fat than in LFD individuals. Analysis of FGL1 expression between high and LFD pigs showed that in 180-day-old animals, FGL1 was not differentially expressed (Table 1). Table 1 presents details of the phenotypic characteristics of HFD and LFD pigs.

Table 1. Characteristics of Zlotnicka White pigs representing high and low-fat deposition traits; means $\pm$ SD.

Traits	HFD (n = 8)			LFD (n = 8)					All Pigs (n = 72)	
	Mean		SD	Mean		SD	*p*-Value	Group Diff *	Mean	SD
Daily gain (g)	641	a	34.5	718	b	84.2	0.03	11%	700	106
Backfat thickness (cm)	2.09	A	0.31	1.53	B	0.36	0.004	27%	1.90	0.39
Peritoneal fat (kg)	0.72	A	0.11	0.46	B	0.06	0.0001	36%	0.61	0.16
Backfat thickness in the K1 point (cm)	2.23	A	0.37	1.44	B	0.20	0.0002	35%	2.00	0.52
Ham fat mass with skin (kg)	2.49	A	0.21	1.80	B	0.23	1.86×10^{-5}	28%	2.15	0.37
Loin fat mass with skin (kg)	2.37	A	0.17	1.58	B	0.28	2.06×10^{-5}	33%	2.03	0.48
Fat over shoulder thickness (cm)	2.85	A	0.27	2.11	B	0.31	0.0002	26%	2.75	0.61
Lumbar fat I thickness (cm)	2.43	A	0.26	1.44	B	0.32	1.096×10^{-5}	41%	2.16	0.55
Lumbar fat II thickness (cm)	2.20	A	0.47	1.26	B	0.21	0.0004	43%	1.95	0.54
Lumbar fat III thickness (cm)	2.74	A	0.63	1.80	B	0.29	0.003	34%	2.40	0.59
Average backfat thickness (cm)	2.46	A	0.27	1.60	B	0.17	8.63×10^{-6}	35%	2.27	0.45
FGL1 expression level	3799.6		932.61	3731.6		1125.4	0.45			

Abbreviation: SD—standard deviation, HDF—high-fat deposition, LFD—low-fat deposition. * Group diff—differences between groups in percentage. Values with the same letters belong to the same statistical group (A, B = $p < 0.01$; a, b = $p < 0.05$).

3.2. Variant Calling and χ^2 Test Analyses

The average number of detected raw reads per sample was 62,276,008, and after filtration, it was 61,480,178. Mapping to the pig reference genome (Sscrofa11.1 GCA_000003025.6 assembly) showed that 70.8% of uniquely mapped reads matched the annotated exon regions, and 16.5% matched the introns. Over 100,000 mutations were identified in analyzed pigs, of which over 7000 showed significant differences in genotype distribution between HFD and LFD groups. It was found that 73.8% of identified mutations belonged to known variants deposited in the NCBI database. The number of insertions and deletions constituted a mere 7%, and significance according to the χ^2 test was 356. There were close to 2500 3′UTR significant variants. Significant downstream and intron mutations were numerous, over 2000 and 3500, respectively. Downstream and intron mutations are probably related to non-coding but functionally transcribed active genomic regions or unannotated exons (Table S2). However, they are considered an error or the result of poor genome annotation in most cases [25].

3.3. FGL1 Mutations Identified Using Variant Calling Analysis for 16 ZW Pigs

For further analysis, the *FGL1* variants were chosen because the *FGL1* plays an important role in the liver, the major organ for converting excess carbohydrates and proteins into fatty acids and triglycerides then exported and stored in adipose tissue [26]. On the other hand, String v. 11.0 showed that *FGL1* is often mentioned in publications as co-expressed (Figure S1) in the liver or plasma, with genes strongly associated with the regulation of fat deposition. We have identified 114 mutations in the *FGL1* gene region, including 3′UTR, intron, and synonymous variants, and the χ^2 test found 54 (3′UTR and intron) mutations as significant revealing differences in genotype distribution between the analyzed FD pig groups. The lowest *chicq* value (0.000335463) for 10 mutations in the 3′UTR region was observed. Haplotype analysis and visualization of linkage disequilibrium, including all SNPs, showed that they were clustered in a few blocks (Figure 1).

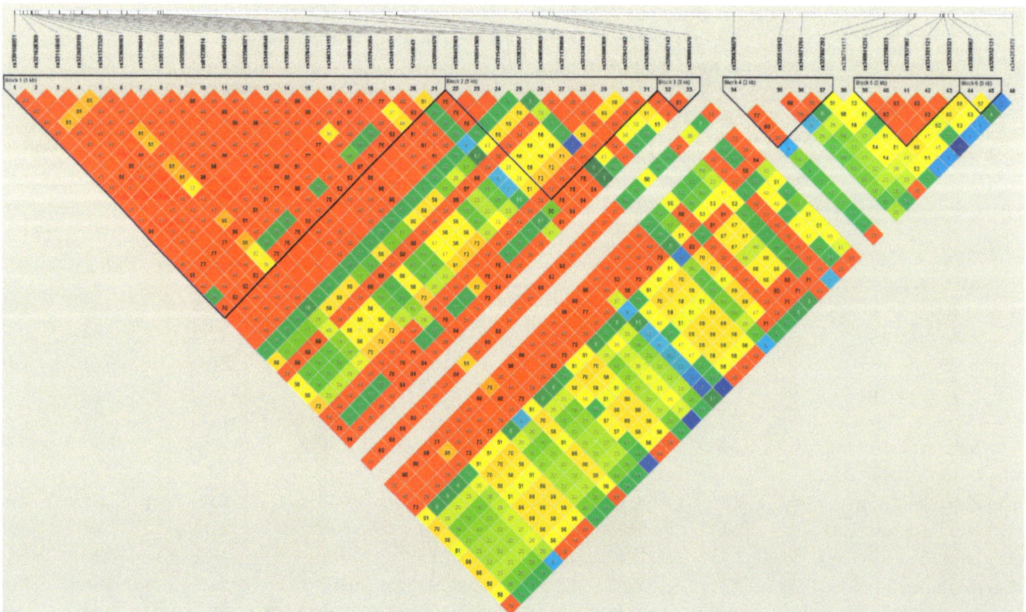

Figure 1. Linkage disequilibrium blocks estimated for SNPs identified across the FGL1 gene [23]. Red color shows high LD > 0.8, green and medium yellow 0.8 > LD > 0.5, and low blue LD < 0.5.

Haplotype analysis after variant calling calculation pinpointed that 10 of 3′UTR variants, including indels: NC_010459.5:g.5548991-5548992del and NC_010459.5:g. 5549161insCAGCA were 100% linked. Results of the variant calling for liver tissue are available as an Excel file containing a few sheets at shorturl.at/joxBT.

miRBase analysis showed that identified *FGL1* 3′UTR mutations may affect miRNA binding sites previously reported as correlated with obesity, including those involved in the insulin signaling pathway, maturity-onset diabetes of the young, type II diabetes mellitus, and adipocytokine signaling pathway (Table S3).

Four of the identified 3′UTR *FGL1* mutations were chosen to evaluate their effect on pig phenotypes (SNPs: rs340465447 and rs330493983; indels: NC_010459.5:g.5548991-5548992del and NC_010459.5:g. 5549161insCAGCA). Figure 2 presents the results of *FGL1* mutation genotyping. New indel mutations were submitted as new alleles of the *FGL1* gene to GenBank under accession number MW827172.

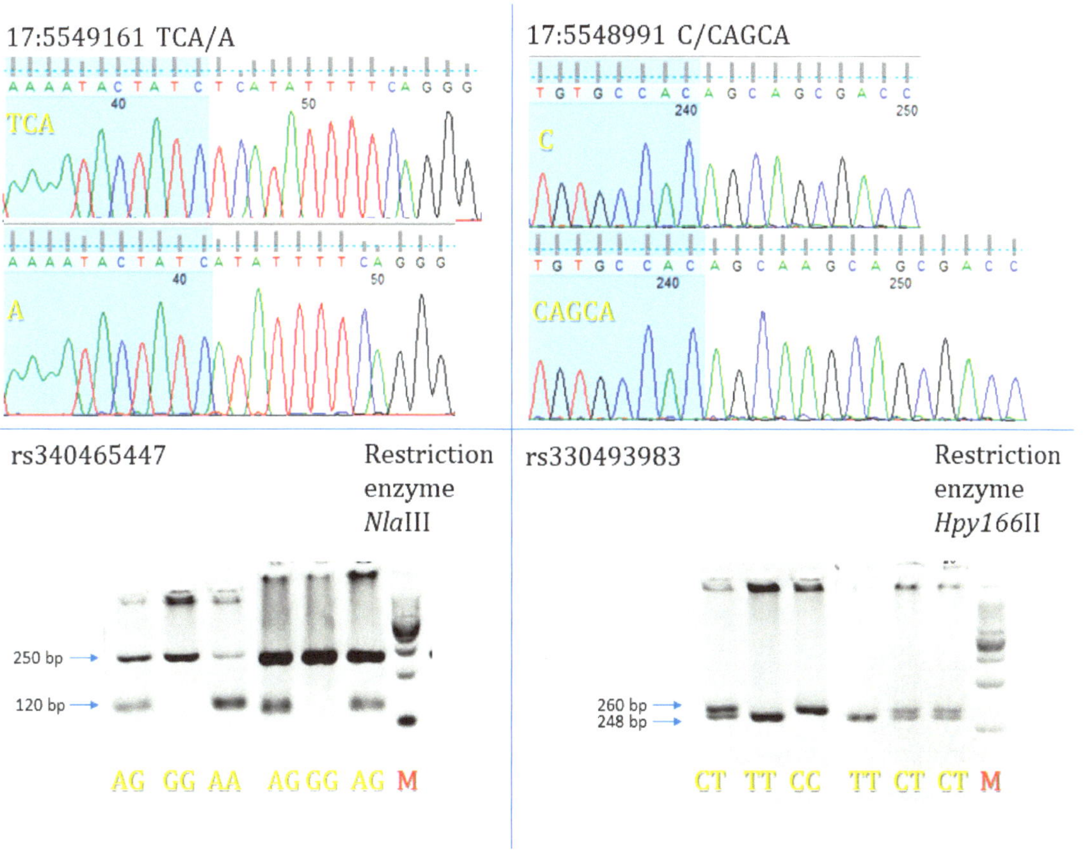

Figure 2. Showed genotyping results obtained for *FGL1* 3′UTR mutations two SNPs rs340465447, rs330493983 and two INDELs: NC_010459.5:g.5548991-5548992del and NC_010459.5:g.5549161insCAGCA. M—marker 100 bp DNA Ladder (New England Biolabs Inc., Ipswich, MA, USA). rs340465447 and rs330493983 SNPs were separated on 3.5% and 5% agarose gel, respectively.

Preliminary association analysis, including 16 pigs that differed in fat deposition (the same as in the RNA-seq method), showed that 10 fully linked 3′UTR polymorphisms were significantly associated with fat deposition. Although only two alternative homozygous animals were found (AA-rs340465447 or CC-rs330493983), the heterozygous animals were quite distinct in FD traits from reference homozygotes. They showed almost 1 cm thicker backfat ($p < 0.001$) at three measured points of lumbar, 800 and 730 g ($p < 0.001$) heavier loin and ham fats, respectively. The differences between the FGL1 genotypes are shown in Table 2.

Table 2. Means $\pm$ S.E. based on ANOVA test for important pig traits dependent on rs340465447, rs330493983, INDELs: NC_010459.5:g.5548991-5548992del and NC_010459.5:g.5549161insCAGCA genotypes in 16 ZW pigs differed in fat deposition.

Traits	FGL1 Genotype		
	GG TT	AG CT	AA CC
Loin weight (kg)	4.87 $\pm$ 0.40 [A]	4.38 $\pm$ 0.30 [B]	4.39 $\pm$ 0.30
Ham weight (kg)	8.06 $\pm$ 0.30 [a]	7.48 $\pm$ 0.30 [b]	7.07 $\pm$ 0.7 [b]
Feet mass (kg)	1.02 $\pm$ 0.05 [Aa]	0.95 $\pm$ 0.01 [b]	0.89 $\pm$ 0.07 [Bc]
Knuckle fat with skin (kg)	0.21 $\pm$ 0.01 [A]	0.25 $\pm$ 0.02 [B]	0.24 $\pm$ 0.01 [AB]
Loin eye height (cm)	6.80 $\pm$ 0.16 [A]	6.22 $\pm$ 0.27 [B]	6.45 $\pm$ 0.32 [AB]
Loin eye area (cm^2)	47.1 $\pm$ 2.75 [a]	43.5 $\pm$ 3.45 [a]	42.4 $\pm$ 3.48 [ab]
Meat percentage in primary cuts (kg)	64.2 $\pm$ 1.08 [A]	58.9 $\pm$ 1.21 [Ba]	58.3 $\pm$ 2.54 [b]
Meat percentage %	56.0 $\pm$ 1.05 [A]	50.5 $\pm$ 1.21 [Ba]	50.2 $\pm$ 2.51 [b]
Peritoneal fat (kg)	0.46 $\pm$ 1.99 [A]	0.70 $\pm$ 2.88 [B]	0.76 $\pm$ 4.89 [AB]
Loin fat with skin (kg)	1.58 $\pm$ 0.20 [Aa]	2.39 $\pm$ 0.20 [B]	2.28 $\pm$ 0.10 [Bb]
Ham fat with skin (kg)	1.80 $\pm$ 0.30 [A]	2.53 $\pm$ 0.01 [B]	2.38 $\pm$ 0.02 [B]
Backfat over shoulder (cm)	2.11 $\pm$ 0.31 [A]	2.83 $\pm$ 0.10 [B]	2.90 $\pm$ 0.31 [B]
Backfat over back (cm)	1.53 $\pm$ 0.36 [A]	2.10 $\pm$ 0.36 [B]	2.05 $\pm$ 0.05 [B]
Backfat over lumbar I	1.44 $\pm$ 0.32 [Aa]	2.43 $\pm$ 0.28 [B]	2.40 $\pm$ 0.20 [b]
Backfat over lumbar II	1.26 $\pm$ 0.21 [A]	2.18 $\pm$ 0.52 [B]	2.25 $\pm$ 0.25 [B]
Backfat over lumbar III	1.80 $\pm$ 0.71 [Aa]	2.78 $\pm$ 0.20 [B]	2.60 $\pm$ 0.29 [b]
Average backfat thickness (cm)	1.63 $\pm$ 0.18 [A]	2.47 $\pm$ 0.31 [B]	2.44 $\pm$ 0.10 [B]
Backfat in the point C1	1.44 $\pm$ 0.20 [Aa]	2.25 $\pm$ 0.15 [B]	2.15 $\pm$ 0.41 [Bb]
Backfat in the point K1	1.44 $\pm$ 0.20 [Aa]	2.25 $\pm$ 0.15 [B]	2.15 $\pm$ 0.41 [Bb]

Allele: G—wild, A—mutation. The probability was measured at least squares means for gene effect where Pr > |t| for H0: LSMean = LSMean (SAS v. 8.02). Values with the same superscripts belong to the same statistical group (A, B = $p < 0.01$; a, b = $p < 0.05$), p-value

3.4. GLM Analysis between Złotnicka White Phenotype and FGL1 Variants

The analysis of rs340465447, rs330493983, and indels NC_010459.5:g.5548991-5548992del (TCA/A) and NC_010459.5:g.5549161insCAGCA (C/CAGCA) FGL1 mutations of the 3′UTR region, including the more numerous Złotnicka White population ($n = 72$), showed that polymorphism linkage was not as strong as in the 16 individuals selected for RNA sequencing. Two SNPs were fully linked but revealed a slightly different allele distribution than observed for two indel mutations (Table S4). For NC_010459.5:g.5548991-5548992del(TCA/A), we identified a higher number of heterozygotes than for rs340465447 (G/A) and fewer alternate homozygotes. According to the SNPs, the ZW population was not in HWE.

In turn, indel frequency was consistent with the HWE. The association analysis was performed for rs340465447 (G/A) and NC_010459.5:g.5548991-5548992del (TCA/A) and showed that these SNP mutations are associated with IMF and subcutaneous fat deposition traits. Pigs with the AA genotype showed 24%, 23%, and 32% higher IMF, ABT, and loin fat, respectively. Additive effect analysis confirmed this observation and showed that a change A $\rightarrow$ G leads to -0.24 cm ($p < 0.01$) of ABT, $+2.90$ cm^2 ($p < 0.01$) of loin eye area and $+9$ day ($p < 0.05$) of slaughter age. The rs340465447 mutation leads to changes in daily gain and pH of loin and ham values measured 24 h after slaughter (Table 3).

The NC_010459.5:g.5548991-5548992del (TCA/A) polymorphism influenced fat and meat traits, such as loin and ham mass, meat percentage, loin eye area and primary cuts, and different kinds of subcutaneous fats, such as average backfat thickness and loin and ham fats. In turn, for IMF, only a value of trend was observed. The dominance effect comparing heterozygous with homozygous values showed that changes ins/del $\rightarrow$ ins/ins leads to $+2.04\%$ ($p < 0.01$) of meat percentage, $+280$ g ($p < 0.05$) of ham mass and -0.32 cm ($p < 0.01$) in backfat thickness in the K1 point—over the lateral edge of the longissimus dorsi muscle (Table 4). The differences between pig traits were shown even between ins/del heterozygotes and ins/ins homozygotes.

3.5. FGL1 Frequency in Pigs Active in Polish Breeding

SNP and indel mutation frequencies were tested in the boars and sows used in Polish breeding. Analysis showed that rs340465447/rs330493983 FGL1 polymorphisms were rare in the PLW and Puławska breeds. There were only heterozygous variants present. In turn, in the PL and Duroc, alternate homozygous AA was also observed. Deletion NC_010459.5:g.5548991-5548992del (TCA/A) or insertion NC_010459.5:g. 5549161insCAGCA(C/CAGCA) showed a lower frequency. Only heterozygotes were observed in the PL, and none of the alternate homozygote pigs were found (Table S4).

Table 3. Least mean square (LSM) $\pm$ S.E. for important pig traits dependent on ENSSSCT00000037614.2:c.*1662G>A *FGL1* genotypes (rs340465447).

Traits	*FGL1* Genotype							GLM Significance		Additive Effect	Dominance Effect
	AA (n = 26)		AG (n = 16)		GG (n = 30)						
	LSM	SE	LSM	SE	LSM	SE	*p*-Value	X2P	*FGL1*	A $\to$ G	Het $\to$ Hom
pH_{ham24}	5.57 AB	0.03	5.64 A	0.03	5.53 B	0.02	0.0080	x	**	ns	−0.04 *
pH_{loin24}	5.59 A	0.02	5.55 AB	0.03	5.46 B	0.02	0.0006	x	***	−0.06 ***	ns
IMF	1.64 a	0.10	1.32 b	0.09	1.49 ab	0.06	0.0462	x	*	ns	+0.13 *
Carcass yield (kg)	74.8 a	0.18	74.6 ab	0.22	74.1 b	0.16	0.0262	***	*	ns	ns
Average backfat thickness (cm)	2.61 a	0.14	2.28 ab	0.13	2.12 b	0.10	0.0234	ns	*	−0.24 **	ns
Loin eye height (cm)	5.62 A	0.19	6.08 AB	0.18	6.40 B	0.13	0.0046	ns	**	+0.41 **	ns
Loin eye area (cm^2)	40.1 a	1.53	44.3 ab	1.47	45.3 b	1.10	0.0236	**	*	+2.90 **	ns
Meat percentage (%)	50.9	1.13	52.3	1.08	53.5	0.80	0.1491	ns	ns	+1.33 *	ns
Primary cut (kg)	18.8	0.43	19.3	0.41	19.8	0.31	0.1919	ns	ns	+0.60 *	ns
Loin fat with skin (kg)	2.32 A	0.13	2.17 ABa	0.13	1.76 Bb	0.09	AB 0.0033 ab 0.0344	***	**	−0.25 *	ns
Backfat over back (cm)	2.35	0.16	1.90	0.15	1.87	0.12	0.0557	ns	*	−0.23 *	ns
Backfat over lumbar I (cm)	2.61 A	0.17	2.20 AB	0.16	1.97 B	0.12	0.0087	ns	**	−0.32 **	ns
Backfat over lumbar II (cm)	2.33 a	0.16	1.94 ab	0.15	1.76 b	0.11	0.0136	ns	**	−0.28 **	ns
Backfat over lumbar III (cm)	2.72	0.18	2.52	0.17	2.24	0.13	0.0939	ns	*	−0.24 *	ns
Daily gain (0–180 days) kg	530	16	521	15	493	11	0.1529	ns	ns	−18 *	ns
Slaughter age (day)	187	6.4	192	6.2	204	4.9	0.0831	x	*	+9 *	ns

Allele: G—wild, A—mutation. Mean and SE were estimated using GLM model, values with the same superscripts belong to the same statistical group (A, B = $p < 0.01$; a, b = $p < 0.05$), p-value in GLM significant * $p < 0.05$, ** $p < 0.01$, *** $p < 0.001$, ns—not significant, X2P—covariate for weight of the right side of the carcass.

Table 4. Least mean square (LSM) $\pm$ S.E. for important pig traits dependent on NC_010459.5:g.5548991-5548992del deletion (TCA/A).

| Traits | *FGL1* Genotype | | | | | | | GLM Significance | | Additive Effect | Dominance Effect |
| | TCA/TCA (n = 31) | | TCA/A (n = 32) | | A/A (n = 9) | | *p*-Value | X2P | FGL1 | Ins $\rightarrow$ Del | Het $\rightarrow$ Hom |
	LSM	SE	LSM	SE	LSM	SE					
IMF	1.49	0.06	1.40	0.06	1.90	0.21	0.0760	x	*	0.21 *	0.15 *
pH_{ham24}	5.53 [a]	0.02	5.62 [b]	0.02	5.58 [ab]	0.06	0.0354	x	**	ns	ns
pH_{loin24}	5.47 [a]	0.02	5.56 [b]	0.02	5.60 [ab]	0.04	0.0103	x	**	0.06 *	ns
Carcass yield (kg)	74.08 [A]	0.14	74.6 [B]	0.16	75.4 [AB]	0.35	0.0039	***	**	ns	ns
Average backfat thickness (cm)	2.11 [A]	0.10	2.56 [B]	0.10	1.99 [AB]	0.24	0.0093	ns	**	ns	−0.26 **
Loin eye area (cm^2)	45.5 *	1.10	41.6	1.20	43.7	2.70	0.0506	**	*	ns	ns
Loin eye height (cm)	6.42 [a]	0.14	5.81 [b]	0.15	5.82 [ab]	0.33	0.0107	ns	**	−0.32 *	ns
Meat percentage (%)	53.6 [a]	0.74	50.6 [b]	0.80	55.8 [ab]	1.81	0.0205	ns	**	ns	+2.04 **
Primary cut (kg)	19.8 [a]	0.28	18.6 [b]	0.31	20.7 [ab]	0.69	0.0227	***	**	ns	+0.70 *
Loin fat with skin (kg)	1.76 [A]	0.09	2.33 [B]	0.10	1.88 [AB]	0.22	0.0003	***	***	ns	−0.28 **
Loin mass without fat and skin (kg)	4.62 [a]	0.10	4.24 [b]	0.10	4.82 [ab]	0.24	0.0326	***	**	ns	+0.22 *
Ham fat mass (kg)	1.99 [a]	0.08	2.30 [b]	0.09	1.84 [ab]	0.19	0.0271	*	**	ns	−0.21 **
Ham mass (kg)	7.75 [a]	0.14	7.31 [b]	0.15	8.16 [ab]	0.26	0.0374	***	**	ns	+0.28 *
Backfat over back (cm)	1.85	0.11	2.23	0.12	1.72	0.3	0.0827	ns	*	ns	−0.22 *
Backfat over lumbar I (cm)	1.97 [a]	0.10	2.50 [b]	0.10	2.02 [ab]	0.29	0.0102	ns	*	ns	−0.25 **
Backfat over lumbar II (cm)	1.75 [A]	0.11	2.25 [B]	0.12	1.65 [AB]	0.26	0.0078	ns	**	ns	−0.28 **
Backfat over lumbar III (cm)	2.23 [A]	0.12	2.78 [B]	0.13	2.03 [AB]	0.29	0.0083	ns	**	ns	−0.33 **
Backfat in the point C1	1.86 [a]	0.12	2.38 [b]	0.13	1.55 [ab]	0.29	0.0130	ns	**	ns	−0.34 **
Backfat in the point K1	1.85 [a]	0.11	2.38 [b]	0.13	1.60 [ab]	0.28	0.0118	ns	**	ns	−0.32 **
Daily gain	729	20	662	22	714	50	0.0754	ns	*	ns	+30 *

Allele: TCA—wild, A—mutation. Mean and SE were estimated using the GLM model; values with the same superscripts belong to the same statistical group (A, B = $p < 0.01$; a, b = $p < 0.05$), *p*-value in GLM significant * $p < 0.05$, ** $p < 0.01$, *** $p < 0.001$, * in FGL1 genotypes showed trends, ns—not significant, K1 point—backfat thickness over the lateral edge of longissimus dorsi muscle, C1—backfat thickness in height extension of loin eye, X2P—covariate for weight of the right side of the carcass.

4. Discussion

4.1. Pig Potential as an Animal Model in Fat Deposition Research

At the end of the 20th century and the beginning of the 21st century, the rate of life significantly accelerated due to vast civilization leap, the development of science, and access to new technologies. The consequence of increased stress levels resulted in civilization diseases, such as obesity leading to substantial health disorders such as diabetes [27], hypertension [28], cancer [29], or premature death [30]. In the face of such a severe epidemic, various studies have determined the precise processes related to fat deposition [31], treatment of its effects, and analysis of environmental factors conducive to obesity [32]. The introduction of appropriate prevention through changing nutritional habits has also been studied and implemented [33].

The molecular process associated with fat accumulation is often challenging in humans; the specific clinical and physiological conditions related to obesity may be limiting due to costs or complexity, the difficulties in obtaining appropriate tissue samples, or ethical problems. Thus, researchers use animal models where molecular DNA/RNA/protein functions remain unchanged between species. Pig animal model better reflects the processes occurring in humans than small lab animals, as they show human-like body proportions [34], metabolism [35], and a similar distribution of adipose tissue and adipocyte size [36]. On the other hand, mutations identified in pigs leading to significant phenotype disorders deliver the information that may be applied in human research. The identified mutations may be helpful in finding a potential problem solution. For example, the melanocortin receptor 4 (*MC4R*) gene, the mutation described by Kim et al. [37], was found to be significantly associated with food consumption and increasing the fat content in pigs in different pig breeds and countries, including in Poland [38]. A few years later, Chagnon et al. [39] found several mutations significantly associated with morbid obesity in the study of human *MC4R*.

The present study identified mutations in the *FGL1* gene, which were significantly related to fat deposition in Złotnicka White (ZW) pigs using variant calling methods based on RNA-seq data. We found more than 50 polymorphism sites that, according to the χ^2 test, showed significantly different genotype/allele distribution when comparing low- and high-fat deposition pig groups. Most of them were located in the 3′UTR region, which controls gene expression and does not change the amino acid sequence. As previously mentioned, *FGL1* is expressed mainly in the liver and is related to the regeneration of this organ, being involved in the hepatocyte proliferation process [8]. Demchev et al. [9] suggested in their reports that a lack of the *FGL1* gene leads to increased body weight and an abnormal plasma lipid profile. Nevertheless, RNA-seq data used in the present study for gene variant identification showed that *FGL1* was not differentially expressed between high- and low-FD pig groups at 180 days old. However, these findings do not preclude prior differentiation at the beginning of ontogenetic growth or in an embryonic stage.

Moreover, our String v. 11.0 analysis showed that *FGL1* is often mentioned in the publications as co-expressed in the liver or plasma with apolipoprotein C3 (*APOC3*), apolipoprotein A1 (*APOA1*), and apolipoprotein A2 (*APOA2*), which are strongly associated with the regulation of very-low-density lipoproteins, cholesterol efflux, and transport, and triglyceride catabolic process with [40].

It is assumed that identified significant *FGL1* 3′UTR variants change binding sites for numerous miRNAs previously described as playing a role in regulating lipid processes related to diabetes, appearing as potentially regulating porcine *FGL1* gene expression, such as mir338-5p and mir338-3p. Their binding site was probably controlled by rs330493983 FGL1 SNP, which revealed a significant effect on fat deposition in pigs. The T variant of rs330493983 occurring in lean pigs favors miRNA-338 binding, whereas the C variant precluded interacting with this regulator. miR-338-3p is expressed in the liver and is related to cancer pathogenesis [41,42]. However, this miRNA was also examined for its role in metabolic disorders such as hepatic insulin resistance and mediation in glycogenesis by regulating the AKT/GSK3β signaling pathway. Moreover, Li et al. [41] showed that mir338-3p was down-regulated in db/db, HFD-fed, and TNFa-treated C57BL/6j mice. This observation suggested an adverse effect of mir-338-3p on fat deposition, similar to hepatic *FGL1*, which Demchev et al. [9] observed in *FGL1*-null mice. Thus, the binding site in the 3′UTR region significantly affecting FD in pigs indicates likely regulation of mir-338-3p-FGL1. However, it should be confirmed in further experimental research, including cell culture and luciferase assay.

4.2. FGL1 Variants as a Potential Selective Marker to Improve Fat Content in Pigs

Long-term selection toward lean meat in pigs, both in the breeds used as dam and sire lines, has led to decreased backfat thickness and IMF, which determine meat flavor and technological suitability [43]. The attempted recovery of higher fat content in pig carcasses by traditional breeding methods is time-consuming and expensive. The application of genetic markers enables the selection process acceleration due to the possibility of young animal evaluation and removing those with adverse gene variants. In Polish pig breeding, only one requirement is identifying a mutation in the *RYR1* gene and eliminating carriers and burdened individuals from the dam and sire populations. Moreover, the Polish Ministry of Agriculture and Rural Development presently funds the monitoring of *IGF2* and *MC4R* polymorphisms [37,44] in PL pigs as a maternal component to indicate the possibility of improving fat tissue levels. However, they still are searching for new potential genetic markers that could be useful in this case.

In the present study, the possibility of using *FGL1* 3′UTR variants was tested in the ZW breed, which is not under selection pressure, so within this population, a wide range of fat-related and meat traits is represented; thus, ZW is a suitable material for association study. Our research showed for interesting correlation of rs340465447_A, rs330493983_C, NC_010459.5:g.5548991-5548992del_A and NC_010459.5:g.5549161insCAGCA_CAGCA *FGL1* alleles that were positively associated with subcutaneous fat and also IMF. Identification of these alleles in Polish pig populations used as maternal and paternal components in breeding was carried out. Their presence in these populations was rather poor. However, in the PL and Duroc pigs, allele rs340465447_A occurred in the alternate homozygous form, which is promising if breeders would try to introduce this selection marker to increase fat levels, especially as only boars were analyzed in the PL and Duroc; therefore, its spread in the PL population would be easier.

5. Conclusions

Mice and human studies show that *FGL1* is indirectly related to the occurrence of obesity and diabetes. Our study supports this theory because we detected interesting *FGL1* variants in the 3′UTR region using liver RNA-seq data related to fat levels in the pig carcasses. We indicated that identified mutations are probably associated with miRNA binding, which regulates *FGL1* expression, and probably fat deposition in pigs. Moreover, association analysis in ZW pigs pinpointed the possibility of the *FGL1* variant, especially rs340465447_A utility, in the selection process toward increasing fat level in pigs since fat is the leading meat flavor carrier.

Supplementary Materials: The following supporting information can be downloaded at: https://www.mdpi.com/article/10.3390/genes13081419/s1. Table S1: Primers used in the study. Table S2: Potential changes in miRNA binding sites identified by miRBase dependent on mutations in 3′UTR region of FGL1 gene in pigs Table S3: Means ± S.E. based on ANOVA test for important pig traits dependent on rs340465447, rs330493983, INDELs: NC_010459.5:g.5548991-5548992del and NC_010459.5:g.5549161insCAGCA genotypes in 16 ZW pigs differed in fat deposition. Table S4: The frequencies of alleles and genotypes of SNP and INDEL in 3′UTR of FGL1 gene in Złotnicka White and pigs active in Polish breeding. Figure S1: Fgl1 and their interactions. By color are indicated the involvement of particular interactions in the biological process, molecular function and KEGG pathway associated with lipid metabolism. The protein interactions were generated by STRINGv.11b: functional protein association networks. https://string-db.org/.

Author Contributions: Conceptualization, K.P.; methodology, K.P., M.T. and K.Ż.; validation, M.T.; formal analysis, K.Ż.; investigation, M.T.; resources, K.R.-M.; data curation, K.P. and K.R.-M.; writing—original draft preparation, K.P. and K.Ż.; writing—review and editing, K.P. and K.R.-M.; supervision, K.P.; project administration, M.T.; funding acquisition, K.P. and M.T. All authors have read and agreed to the published version of the manuscript.

Funding: This study was supported by the statutory activity of the National Research Institute of Animal Production no. 01-18-05-21 and The National Center for Research and Development in Poland—BIOSTRATEG2/297267/14/NCBR/2016.

Institutional Review Board Statement: The biological material for this study was collected after slathering from pigs maintained and slaughtered in the Pig Test Station (PTS, National Research Institute of Animal Production). Following carcass evaluation, the meat was intended for sale and consumption. Ethical review and approval were waived for this study due to a non-invasive method of collecting. All conducted research was approved by the Approving Experiment Committee of the National Research Institute of Animal Production (Krakow, Poland) according to the Polish Act on the Protection of Animals Used for Scientific or Educational Purposes of 15 January 2015, which implements Directive 2010/63/EU of the European Parliament and the Council of 22 September 2010 on the protection of animals used for scientific purposes. Moreover, all procedures followed the guidelines and regulations of the Local Krakow Ethics Committee for Experiments with Animals.

Informed Consent Statement: Not applicable.

Data Availability Statement: New INDEL mutations identified in the FGL1 gene were submitted to GenBank with access number MW827172 (NCBI database). The sequence data for RNA libraries have been submitted to the Gene Expression Omnibus (GSE160436). Variant calling results are available at link shorturl.at/ksKU1 (accessed on 20 Oct 2021).

Conflicts of Interest: The authors declare no conflict of interest.

References

1. Bakhtiarizadeh, M.R.; Salehi, A.; Alamouti, A.A.; Abdollahi-Arpanahi, R.; Salami, S.A. Deep transcriptome analysis using RNA-Seq suggests novel insights into molecular aspects of fat-tail metabolism in sheep. *Sci. Rep.* **2019**, *9*, 9203. [CrossRef] [PubMed]
2. Xing, K.; Wang, K.; Ao, H.; Chen, S.; Tan, Z.; Wang, Y.; Xitong, Z.; Yang, T.; Zhang, F.; Liu, Y.; et al. Comparative adipose transcriptome analysis digs out genes related to fat deposition in two pig breeds. *Sci. Rep.* **2019**, *9*, 12925. [CrossRef] [PubMed]
3. Rask-Andersen, M.; Karlsson, T.; Ek, W.E.; Johansson, Å. Genome-wide association study of body fat distribution identifies adiposity loci and sex-specific genetic effects. *Nat. Commun.* **2019**, *10*, 339. [CrossRef] [PubMed]
4. Piórkowska, K.; Żukowski, K.; Ropka-Molik, K.; Tyra, M. Detection of genetic variants between different Polish Landrace and Puławska pigs by means of RNA-seq analysis. *Anim. Genet.* **2018**, *49*, 215–225. [CrossRef]
5. Xing, K.; Zhu, F.; Zhai, L.; Chen, S.; Tan, Z.; Sun, Y.; Hou, Z.; Wang, C. Identification of genes for controlling swine adipose deposition by integrating transcriptome, whole-genome resequencing, and quantitative trait loci data. *Sci. Rep.* **2016**, *6*, 23219. [CrossRef]
6. Piskol, R.; Ramaswami, G.; Li, J.B. Reliable identification of genomic variants from RNA-seq data. *Am. J. Hum. Genet.* **2013**, *93*, 641–651. [CrossRef]
7. Zhang, Y.; Qiao, H.X.; Zhou, Y.T.; Hong, L.; Chen, J.H. Fibrinogen-like-protein 1 promotes the invasion and metastasis of gastric cancer and is associated with poor prognosis. *Mol. Med. Rep.* **2018**, *18*, 1465–1472. [CrossRef]
8. Hara, H.; Yoshimura, H.; Uchida, S.; Toyoda, Y.; Aoki, M.; Sakai, Y.; Morimoto, S.; Shiokawa, K. Molecular cloning and functional expression analysis of a cDNA for human hepassocin, a liver-specific protein with hepatocyte mitogenic activity. *Biochim. Biophys. Acta-Gene Struct. Expr.* **2001**, *1520*, 45–53. [CrossRef]
9. Demchev, V.; Malana, G.; Vangala, D.; Stoll, J.; Desai, A.; Kang, H.W.; Li, Y.; Nayeb-Hashemi, H.; Niepel, M.; Cohen, D.E.; et al. Targeted Deletion of Fibrinogen Like Protein 1 Reveals a Novel Role in Energy Substrate Utilization. *PLoS ONE* **2013**, *8*, e58084. [CrossRef]
10. Wang, J.; Sanmamed, M.F.; Datar, I.; Su, T.T.; Ji, L.; Sun, J.; Chen, L.; Chen, Y.; Zhu, G.; Yin, W.; et al. Fibrinogen-like Protein 1 Is a Major Immune Inhibitory Ligand of LAG-3. *Cell* **2019**, *176*, 334–347.e12. [CrossRef]
11. Sykes, M.; Sachs, D.H. Transplanting organs from pigs to humans. *Sci. Immunol.* **2019**, *4*, 41. [CrossRef]
12. Houpt, K.A.; Houpt, T.R.; Pond, W.G. The pig as a model for the study of obesity and of control of food intake: A review. *Yale J. Biol. Med.* **1979**, *52*, 307–329.
13. Grzeskowiak, E.; Borys, A.; Borzuta, K.; Buczynski, J.T.; Lisiak, D. Slaughter value, meat quality and backfat fatty acid profile in Zlotnicka White and Zlotnicka Spotted fatteners. *Anim. Sci. Pap. Rep.* **2009**, *27*, 115–125.

14. Szyndler-Nędza, M.; Świątkiewicz, M.; Migdał, Ł.; Migdał, W. The Quality and Health-Promoting Value of Meat from Pigs of the Native Breed as the Effect of Extensive Feeding with Acorns. *Animals* **2021**, *11*, 789. [CrossRef]

15. Różycki, M.; Tyra, M. Rules at evaluating the pigs in Pig Slaughter Testing Station. State of pig breeding and pig evaluation results. *Inst. Zootech. Kraków* **2010**, 92–117.

16. Tyra, M.; Żak, G. Analysis of the possibility of improving the indicators of pork quality through selection with particular consideration of intramuscular fat (imf) conntenttent*. *Ann. Anim. Sci.* **2013**, *13*, 33–44. [CrossRef]

17. Piórkowska, K.; Żukowski, K.; Ropka-Molik, K.; Tyra, M. New long-non coding RNAs related to fat deposition based on pig model. *Ann. Anim. Sci.* **2022**. [CrossRef]

18. McKenna, A.; Hanna, M.; Banks, E.; Sivachenko, A.; Cibulskis, K.; Kernytsky, A.; Garimella, K.; Altshuler, D.; Gabriel, S.; Daly, M.; et al. The genome analysis toolkit: A MapReduce framework for analyzing next-generation DNA sequencing data. *Genome Res.* **2010**, *20*, 1297–1303. [CrossRef]

19. Depristo, M.A.; Banks, E.; Poplin, R.; Garimella, K.V.; Maguire, J.R.; Hartl, C.; Philippakis, A.A.; Del Angel, G.; Rivas, M.A.; Hanna, M.; et al. A framework for variation discovery and genotyping using next-generation DNA sequencing data. *Nat. Genet.* **2011**, *43*, 491–501. [CrossRef]

20. Cingolani, P.; Platts, A.; Wang, L.L.; Coon, M.; Nguyen, T.; Wang, L.; Land, S.J.; Lu, X.; Ruden, D.M. A program for annotating and predicting the effects of single nucleotide polymorphisms, SnpEff: SNPs in the genome of Drosophila melanogaster strain w1118; iso-2; iso-3. *Fly* **2012**, *6*, 80–92. [CrossRef]

21. McLaren, W.; Gil, L.; Hunt, S.E.; Riat, H.S.; Ritchie, G.R.S.; Thormann, A.; Flicek, P.; Cunningham, F. The Ensembl Variant Effect Predictor. *Genome Biol.* **2016**, *17*, 122. [CrossRef] [PubMed]

22. Benjamini, Y.; Hochberg, Y. Controlling the False Discovery Rate: A Practical and Powerful Approach to Multiple Testing. *J. R. Stat. Soc. B* **1995**, *57*, 289–300. [CrossRef]

23. Barrett, J.C.; Fry, B.; Maller, J.; Daly, M.J. Haploview: Analysis and visualization of LD and haplotype maps. *Bioinformatics* **2005**, *21*, 263–265. [CrossRef] [PubMed]

24. Škrlep, M.; Kavar, T.; Čandek-Potokar, M. Comparison of PRKAG3 and RYR1 gene effect on carcass traits and meat quality in Slovenian commercial pigs. *Czech J. Anim. Sci.* **2010**, *55*, 149–159. [CrossRef]

25. Collado-Torres, L.; Nellore, A.; Frazee, A.C.; Wilks, C.; Love, M.I.; Langmead, B.; Irizarry, R.A.; Leek, J.T.; Jaffe, A.E. Flexible expressed region analysis for RNA-seq with derfinder. *Nucleic Acids Res.* **2017**, *45*, e9. [CrossRef]

26. Frayn, K.N.; Arner, P.; Yki-Järvinen, H. Fatty acid metabolism in adipose tissue, muscle and liver in health and disease. *Essays Biochem.* **2006**, *42*, 89–103. [CrossRef]

27. Pillon, N.J.; Loos, R.J.F.; Marshall, S.M.; Zierath, J.R. Metabolic consequences of obesity and type 2 diabetes: Balancing genes and environment for personalized care. *Cell* **2021**, *184*, 1530–1544. [CrossRef]

28. Cohen, J.B. Hypertension in Obesity and the Impact of Weight Loss. *Curr. Cardiol. Rep.* **2017**, *19*, 98. [CrossRef]

29. Arnold, M.; Leitzmann, M.; Freisling, H.; Bray, F.; Romieu, I.; Renehan, A.; Soerjomataram, I. Obesity and cancer: An update of the global impact. *Cancer Epidemiol.* **2016**, *41*, 8–15. [CrossRef]

30. Abdelaal, M.; le Roux, C.W.; Docherty, N.G. Morbidity and mortality associated with obesity. *Ann. Transl. Med.* **2017**, *5*, 161. [CrossRef]

31. Jo, M.G.; Kim, M.W.; Jo, M.H.; bin Abid, N.; Kim, M.O. Adiponectin homolog osmotin, a potential anti-obesity compound, suppresses abdominal fat accumulation in C57BL/6 mice on high-fat diet and in 3T3-L1 adipocytes. *Int. J. Obes.* **2019**, *43*, 2422–2433. [CrossRef]

32. Wang, Y.; Hollis-Hansen, K.; Ren, X.; Qiu, Y.; Qu, W. Do environmental pollutants increase obesity risk in humans? *Obes. Rev.* **2016**, *17*, 1179–1197. [CrossRef]

33. Tuso, P.; Beattie, S. Nutrition reconciliation and nutrition prophylaxis: Toward total health. *Perm. J.* **2015**, *19*, 80–86. [CrossRef]

34. Oster, M.; Reyer, H.; Gerlinger, C.; Trakooljul, N.; Siengdee, P.; Keiler, J.; Ponsuksili, S.; Wolf, P.; Wimmers, K. mRNA Profiles of Porcine Parathyroid Glands Following Variable Phosphorus Supplies throughout Fetal and Postnatal Life. *Biomedicines* **2021**, *9*, 454. [CrossRef]

35. Nielsen, K.L.; Hartvigsen, M.L.; Hedemann, M.S.; Læke, H.N.; Hermansen, K.; Bach Knudsen, K.E. Similar metabolic responses in pigs and humans to breads with different contents and compositions of dietary fibers: A metabolomics study. *Am. J. Clin. Nutr.* **2014**, *99*, 941–949. [CrossRef]

36. Stachowiak, M.; Szczerbal, I.; Switonski, M. Genetics of Adiposity in Large Animal Models for Human Obesity—Studies on Pigs and Dogs. In *Progress in Molecular Biology and Translational Science*; Elsevier: Amsterdam, The Netherlands, 2016; Volume 140, pp. 233–270. ISBN 9780128046159.

37. Kim, K.S.; Larsen, N.; Short, T.; Plastow, G.; Rothschild, M.F. A missense variant of the porcine melanocortin-4 receptor (MC4R) gene is associated with fatness, growth, and feed intake traits. *Mamm. Genome* **2000**, *11*, 131–135. [CrossRef]

38. Piórkowska, K.; Tyra, M.; Rogoz, M.; Ropka-Molik, K.; Oczkowicz, M.; Rózycki, M. Association of the melanocortin-4 receptor (MC4R) with feed intake, growth, fatness and carcass composition in pigs raised in Poland. *Meat Sci.* **2010**, *85*, 297–301. [CrossRef]

39. Chagnon, Y.C.; Rankinen, T.; Snyder, E.E.; Weisnagel, S.J.; Pérusse, L.; Bouchard, C. The human obesity gene map: The 2002 update. *Obes. Res.* **2003**, *11*, 313–367. [CrossRef]

40. Sayeed, A.; Dalvano, B.E.; Kaplan, D.E.; Viswanathan, U.; Kulp, J.; Janneh, A.H.; Hwang, L.Y.; Ertel, A.; Doria, C.; Block, T. Research Paper Profiling the circulating mRNA transcriptome in human liver disease. *Oncotarget* **2020**, *11*, 2226–2242. [CrossRef]

41. Li, P.; Chen, X.; Su, L.; Li, C.; Zhi, Q.; Yu, B.; Sheng, H.; Wang, J.; Feng, R.; Cai, Q.; et al. Epigenetic Silencing of miR-338-3p Contributes to Tumorigenicity in Gastric Cancer by Targeting SSX2IP. *PLoS ONE* **2013**, *8*, e66782. [CrossRef]

42. Xu, H.; Zhao, L.; Fang, Q.; Sun, J.; Zhang, S.; Zhan, C.; Liu, S.; Zhang, Y. MiR-338-3p inhibits hepatocarcinoma cells and sensitizes these cells to sorafenib by targeting hypoxia-induced factor 1. *PLoS ONE* **2014**, *9*, e115565. [CrossRef]

43. Tyra, M.; Zak, G. Analysis of relationships between fattening and slaughter performance of pigs and the level of intramuscular fat (IMF) in longissimus dorsi muscle. *Ann. Anim. Sci.* **2012**, *12*, 169–178. [CrossRef]

44. Van Laere, A.S.; Nguyen, M.; Braunschweig, M.; Nezer, C.; Collette, C.; Moreau, L.; Archibald, A.L.; Haley, C.S.; Buys, N.; Tally, M.; et al. A regulatory mutation in IGF2 causes a major QTL effect on muscle growth in the pig. *Nature* **2003**, *425*, 832–836. [CrossRef] [PubMed]

Article

Alternative Splicing Isoforms of Porcine *CREB* Are Differentially Involved in Transcriptional Transactivation

Dongjie Zhang [1] , Qian Zhang [2], Liang Wang [1], Jiaxin Li [2], Wanjun Hao [2], Yuanlu Sun [2], Di Liu [1],*
and Xiuqin Yang [2],*

1   Institute of Animal Husbandry, Heilongjiang Academy of Agricultural Sciences, Harbin 150086, China;
     djzhang8109@163.com (D.Z.); wlwl1448@163.com (L.W.)
2   College of Animal Science and Technology, Northeast Agricultural University, Harbin 150030, China;
     15169615521@163.com (Q.Z.); ljxneau@163.com (J.L.); haowanjun1109@163.com (W.H.);
     sunyuanlu2021@163.com (Y.S.)
*   Correspondence: liudi1963@163.com (D.L.); xiuqinyang@neau.edu.cn (X.Y.); Tel.: +86-451-86677458 (D.L.);
     +86-451-55191738 (X.Y.)

Citation: Zhang, D.; Zhang, Q.; Wang, L.; Li, J.; Hao, W.; Sun, Y.; Liu, D.; Yang, X. Alternative Splicing Isoforms of Porcine *CREB* Are Differentially Involved in Transcriptional Transactivation. *Genes* **2022**, *13*, 1304. https://doi.org/10.3390/genes13081304

Academic Editors: Katarzyna Piórkowska and Katarzyna Ropka-Molik

Received: 3 July 2022
Accepted: 21 July 2022
Published: 22 July 2022

Publisher's Note: MDPI stays neutral with regard to jurisdictional claims in published maps and institutional affiliations.

Abstract: The cAMP response element-binding protein (CREB), a basic leucine zipper transcription factor, is involved in the activation of numerous genes in a variety of cell types. The *CREB* gene is rich in alternative splicing (AS) events. However, studies on the AS of *CREB* genes in pigs are limited, and few reports have compared the roles of isoforms in activating gene expression. Here, five AS transcripts, V1–5, were characterized by RT-PCR and two, V3 and V5, were new identifications. Both V1 and V2 have all the functional domains of the CREB protein, with similar tissue expression profiles and mRNA stability, suggesting that they have similar roles. The transcriptional transactivation activities of four isoforms encoding complete polypeptides were analyzed on the expression of the B-cell CLL/lymphoma 2-like protein 2 and the poly (A)-binding protein, nuclear 1 genes with a dual-luciferase reporter system, and differential activities were observed. Both V1 and V2 have promoting effects, but their roles are gene-specific. V3 has no effect on the promoter of the two genes, while V4 functions as a repressor. The mechanisms underlying the differential roles of V1 and V2 were analyzed with RNA-seq, and the genes specifically regulated by V1 and V2 were identified. These results will contribute to further revealing the role of CREB and to analyzing the significance of AS in genes.

Keywords: pig; CREB; alternative splicing variants; function; RNA-seq

1. Introduction

The cAMP response element (CRE)-binding protein (CREB), also known as CREB1, is a 43-kDa basic leucine zipper (bZIP) transcription factor (TF). It is involved in the activation of many cellular signals through the regulation of genes harboring the CRE motif. Genome-wide screening for CRE motifs indicated that more than 4000 genes might be activated by CREB. This suggests that CREB is a general TF and is important in many cellular processes [1]. *CREB* gene knockout mice had embryonal neuronal deficits and a reduced lifespan [2,3]. The dysregulation of the *CREB* gene results in aberrant signal transduction affecting apoptosis, proliferation and differentiation, immune surveillance, and metabolism [4–7].

The consensus sequence of the CRE motif is TGACGTCA, and it was first characterized in the neuropeptide somatostatin gene [8]. The CREB gene also recognizes the conserved half (TGACG or CGTCA) of the CRE sequence. The CGTCA sequence in the promoter of TAT or Sox 3 has been identified as a prototypical binding site of CREB [9–11]. Many genes containing TGACG are targets of CREB, including c-Fos [10], CYP11B [12], and Sox 9 [13]. A few nucleotide variations from the CRE consensus sequence in some genes have little effect on the binding of CREB, except for affinity changes [14–16]. Some genes with a nonpalindromic CRE motif, such as the half CRE element, TGAGC, are more efficient in the induction of CREB than the full ones [14]. This might be because the competitive TFs, such as ATF2 or ATF3, for CRE cannot bind to the half CRE element [17], or because the affinity is attenuated [18] and the binding of CREB is easier on the half sequences.

There are many alternative transcript variants encoding distinct polypeptides in the *CREB* gene [19–21]. Some rare isoforms identified in the testis have been implicated in spermatogenesis in rats [22] and humans [21]. This indicates that the alternative splicing (AS) of the *CREB* gene has functional significance and physiological importance. Among the CREB alternative protein isoforms, isoforms A (NM_004379) and B (NM_134442) are major forms and are universally expressed in tissues. Both of these isoforms contain all the functional domains of CREB, including the N-terminal glutamine-rich (Q) domains, the kinase-induced domain (KID), and the C-terminal bZIP domain that have similar expression profiles in mice [19]. It is unlikely that two isoforms with identical roles exist simultaneously in tissues. However, it is not well established whether these isoforms are different in their transcriptional transactivation potential and whether the specific signals regulated by each isoform exist. In this study, we aimed to analyze the differential roles of the CREB isoforms and characterize novel AS transcripts in pigs.

CREB plays important roles in many physiological and pathological processes in humans [23,24]. Pigs have many histological and physiological similarities to humans, and they have been used as an animal model for many human diseases. Our results will help to reveal the detailed roles of CREB and contribute to the improvement of human health. Additionally, pork accounts for more than 50% of the total meat consumption in China. The clarification of the roles of CREB, a general TF that is important for cellular processes, might help to improve the growth and productivity of pigs.

2. Results

2.1. Identification of Alternative Splicing Transcripts

A total of five AS variants (named V1–5) were obtained by RT-PCR from mixed RNAs isolated from the spleen, kidneys, muscles, lungs, stomach, liver, heart, fat, large intestine, and small intestine of pigs. V1, V2, and V4 have the same coding sequence (CDS) as those deposited in GenBank (NM_001361427, NM_001099929, and XM_021075309) and encode the polypeptides of 341, 327, and 278 aa, respectively. V1 and V2 correspond to isoforms B (NM_134442) and A (NM_004379) in humans [21,25], respectively. Both V1 and V2 show a similarity of 100% with the homologs from humans. Transcripts V3 and V5, encoding the polypeptides of 299 and 104 aa, respectively, were identified for the first time.

The porcine *CREB* gene is composed of 10 exons, and the CDS starts from the ninth base pair of exon 3 (Figure 1A). All the splice sites in transcript V1 follow the GT-AG rule. In contrast with V1, V2–4 are formed by in-frame deletions. Both V2 and V4 are formed by exon skipping. V2 is absent from exon 5, and V4 lacks exons 4 and 5. V3 is formed by an internal deletion of 84 bp in exon 4, spanning the sequence from 27 bp to 110 bp, indicating the presence of a cryptic intron. The splice sites of this cryptic intron are AT-CC. V5 contains a premature termination codon (PTC) owing to the retention of a partial sequence of intron 5 and encodes a truncated polypeptide (Figure 1B). Q1, KID, Q2, and bZIP domains exist in sequence in the polypeptide of variant V1. In contrast with V1, V2 lacks an α-helix between the Q1 and KID domains. Both V3 and V4 have lost domain Q1 (Figure 1C). V3 and V5 have been submitted to GenBank under the accession Nos. OL439245 and OL439247, respectively.

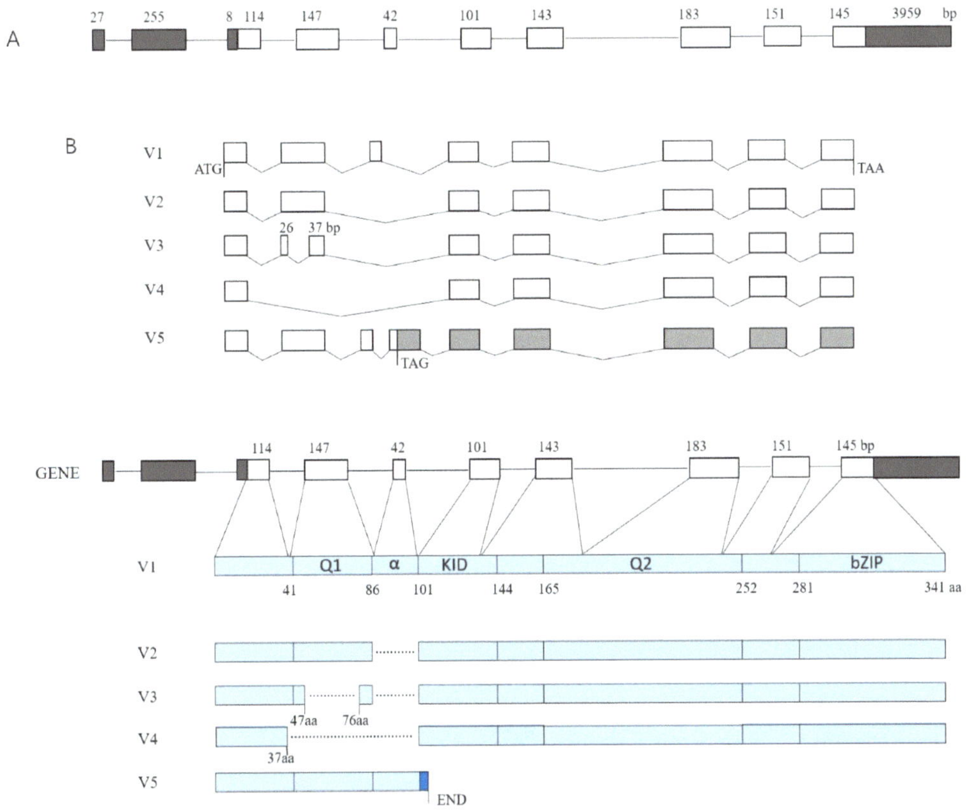

Figure 1. Schematic structure of the identified transcript variants of the porcine *CREB* gene. (**A**) Schematic diagram of the *CREB* gene (NM_001361427) structure. (**B**) Schematic structure of the alternative splicing transcripts. (**C**) Domain composition of the alternative splicing isoforms. Exons and introns are indicated with a box and a straight line, respectively. Oblique lines indicate that the sequences are spliced out. A blank box indicates a coding sequence, while a closed box indicates an untranslated region.

2.2. Tissue Expression Profile

Real-time PCR analysis showed that isoforms V1–3 were ubiquitously expressed in all the tissues studied. V4 was not detected in the heart, liver, stomach, and fat tissues. V5 was not found in the heart and fat tissues (Figure 2).

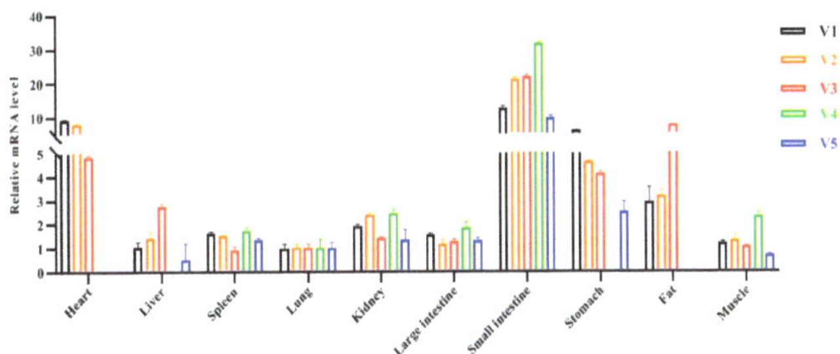

Figure 2. Tissue expression profile of five transcript variants of the porcine *CREB* gene.

2.3. Transcriptional Regulatory Effects of the CREB Variants

To analyze the transcriptional transactivation activity of the in-frame variants, V1–4, dual-luciferase reporter analysis was performed on the basis of bioinformatic analysis. The results demonstrated that there were potential CREB-binding sites in the promoters of the porcine B-cell CLL/lymphoma 2-like protein 2 (BCL2L2) and the poly (A)-binding protein, nuclear 1 (PABPN1) genes.

The reporter gene of PABPN1 has strong promoter activity, as revealed by the luciferase activity analysis. The deletion of the putative cis-element resulted in a significant decrease in the promoter activity ($p < 0.01$) (Figure 3A). The overexpression of CREB V1 or V2 promoted the activity significantly, while no promotive effects of exogenous V1 or V2 were found on the plasmids lacking the binding site of CREB (Figure 3B,C). The overexpression of CREB V3 had no significant effect on the promoter activity of the *PABPN1* gene ($p > 0.05$) (Figure 3D). The ectopic expression of V4 decreased the promoter activity significantly ($p < 0.05$) (Figure 3E).

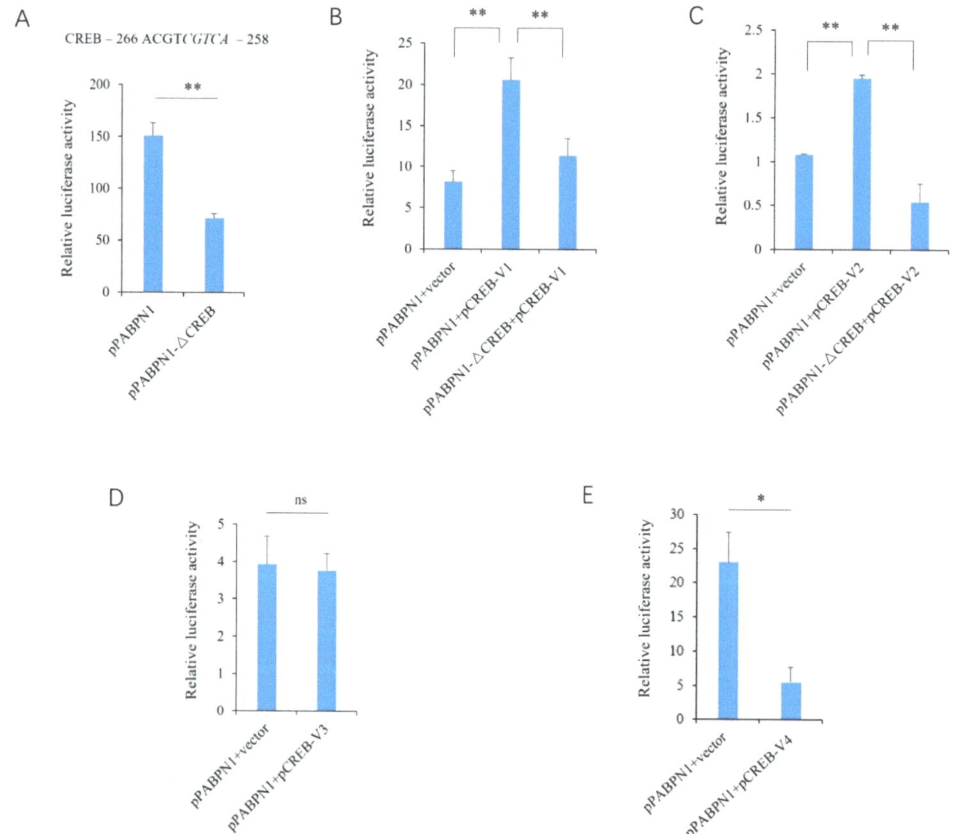

Figure 3. Effects of the transcript variants of CREB on the porcine PABPN1 promoter. (**A**) Effects of CREB motif deletion on promoter activity. The putative binding site is shown above, and the sequence deleted in the mutant-type reporter gene is indicated in italics. (**B**) Effects of V1 overexpression on promoter activity. (**C**) Effects of V2 overexpression on promoter activity. (**D**) Effects of V3 overexpression on promoter activity. (**E**) Effects of V4 overexpression on promoter activity. A Student's *t*-test was used to compare the differences between the two groups. * and ** indicate that the differences were significant, at $p < 0.05$ and $p < 0.01$ levels, respectively, while ns, not significant, indicates that the differences were not significant ($p > 0.05$).

The reporter gene of BCL2L2 has strong promoter activity, and the deletion of the putative binding site inhibits the promoter activity significantly ($p < 0.01$) (Figure 4A). The overexpression of CREB V1 had no effect on the luciferase activity, while exogenous V2 increased the promoter activity significantly ($p < 0.01$) (Figure 4B). The deletion of the CREB motif in the reporter gene abolished the effects of exogenous V2 on the promoter activity (Figure 4C). The effects of V3 and V4 on the porcine *BCL2L2* promoter were similar to that on the *PABPN1* promoter. V3 did not significantly affect the promoter activity (Figure 4D), while V4 decreased the activity (Figure 4E).

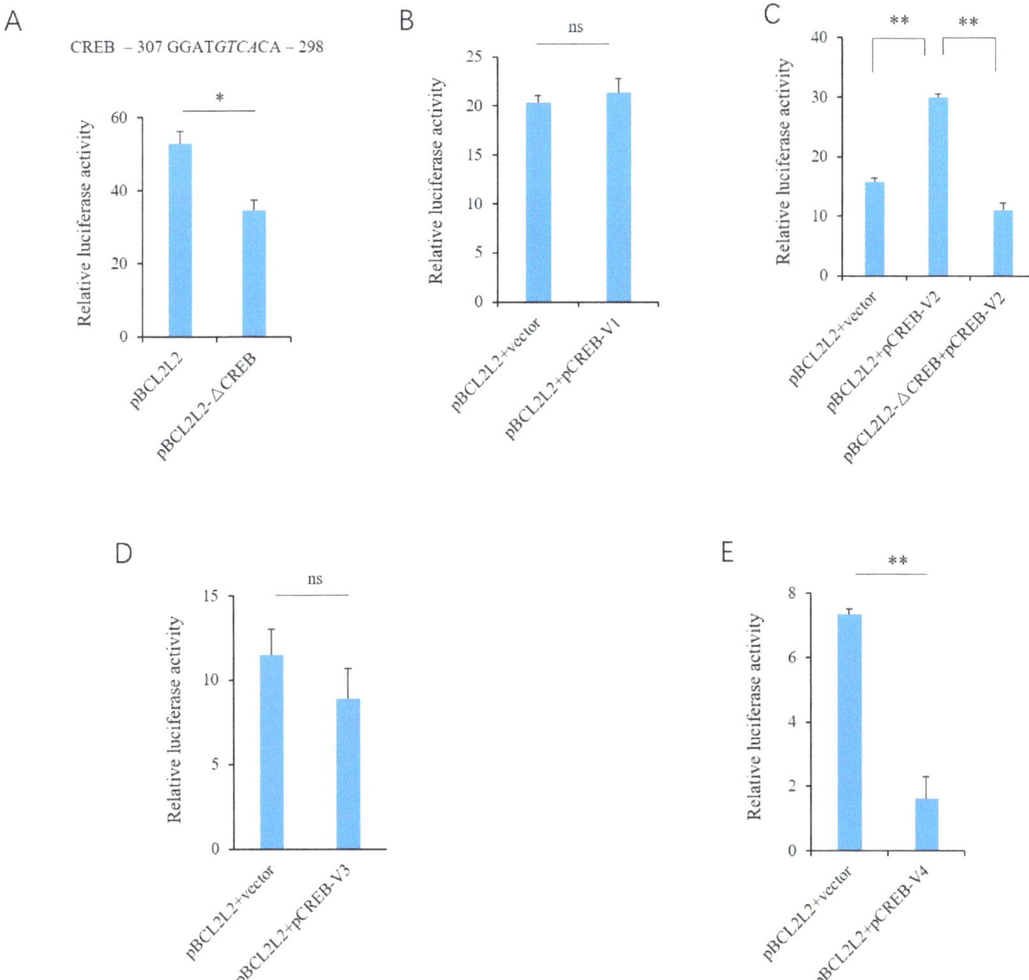

Figure 4. Effects of the transcript variants of CREB on the porcine BCL2L2 promoter. (**A**) Effects of CREB motif deletion on promoter activity. The putative binding site is shown above, and the sequence deleted in the mutant-type reporter gene is indicated in italics. (**B**) Effects of V1 overexpression on promoter activity. (**C**) Effects of V2 overexpression on promoter activity. (**D**) Effects of V3 overexpression on promoter activity. (**E**) Effects of V4 overexpression on promoter activity. A Student's *t*-test was used to compare the differences between the two groups. * and ** indicate that the differences were significant, at $p < 0.05$ and $p < 0.01$ levels, respectively, while ns, not significant, indicates that the differences were not significant ($p > 0.05$).

2.4. Effects of the CREB Variants on the Endogenous Expression of PABPN1 and BCL2L2

We measured the effects of the CREB transcript variants on the expression of endogenous *PABPN1* and *BCL2L2*. Real-time PCR analysis showed that the overexpression of V1 significantly increased the mRNA level of *PABPN1* in PK-15 cells ($p < 0.01$), while no change was observed in the expression of *BCL2L2* ($p > 0.05$) (Figure 5A). Exogenous V2 increased the expression of both *PABPN1* and *BCL2L2* significantly ($p < 0.01$) (Figure 5B). The overexpression of V4 inhibited the expression of *PABPN1*, but the change was not significant ($p > 0.05$), while it decreased the mRNA level of *BCL2L2* significantly ($p < 0.05$) (Figure 5C).

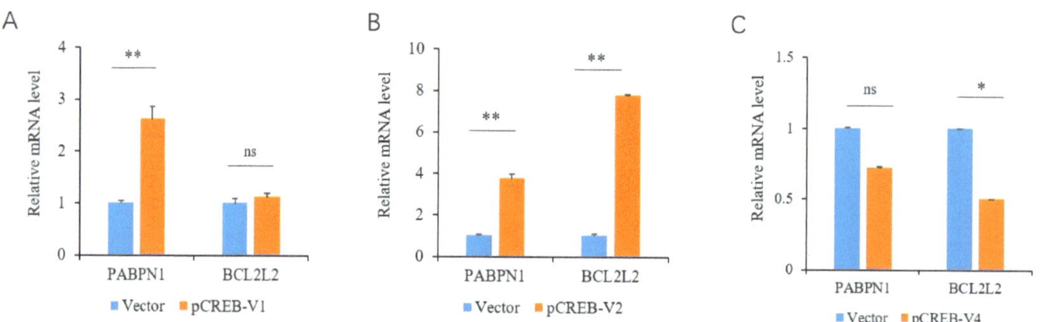

Figure 5. Effects of variants V1 (**A**), V2 (**B**), and V3 (**C**) of CREB on the expression of endogenous *PABPN1* and *BCL2L2* genes. A Student's *t*-test was used to compare the differences between the two groups. * and ** indicate that the differences were significant, at $p < 0.05$ and $p < 0.01$ levels, respectively, while ns, not significant, indicates that the differences were not significant ($p > 0.05$).

2.5. Comparison of Variants V1 and V2

The above analysis showed that there are differences between the transcriptional regulatory activities in variants V1 and V2, although both increased the transcription of target genes. We studied the mechanisms underlying these differences. Transcripts V1 and V2 are similar in stability, as revealed by half-time analysis (Figure 6A). Competitive RT-PCR was used to compare the expression levels of the two variants in the same tissues, and it revealed that V2 was expressed more abundantly than V1 in all the tissues examined (Figure 6B).

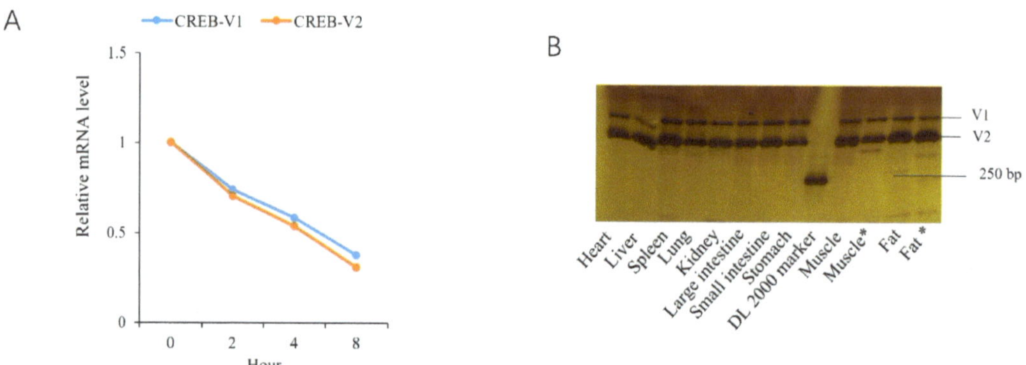

Figure 6. Comparison of isoforms V1 and V2. (**A**) mRNA stability analysis. The mRNA level is shown as the ratio of mRNA remaining to that at time zero. (**B**) Comparison of the mRNA levels in the same tissues. * indicates that the tissues were from 7-month-old pigs, while the other tissues were from 1-month-old pigs.

2.6. Overview of the RNA-Seq Data

RNA-seq analysis was used to reveal the mechanisms underlying the differential roles of variants V1 and V2 in regulating gene expression. A total of 179.43 million clean read pairs were obtained in nine cDNA libraries, with 19.24, 19.09, and 21.48 million on average for cells transfected with V1, V2, and an empty vector (negative control, NC), respectively, comprising at least 94.3% of Quality 20 (Q20) reads and 86.6% of Q30 reads. More than 93.12% of the clean reads in each sample were mapped to the reference genome (*Sus scrofa* 11.1).

When compared to the NC group, there were 1996 differentially expressed genes (DEGs), with 1046 upregulated genes and 950 downregulated genes in cells overexpressing V1. There were 165 DEGs, with 95 upregulated genes and 70 downregulated genes in cells overexpressing V2 (Figure 7A; Table S1A,B). Among the DEGs regulated by V1 and V2, 57 were common (Figure 7B). DEGs specifically regulated by V1 or V2 are shown in Table S1C. A total of 392 and 28 DEGs were identified as TFs in the V1- and V2-treated groups, respectively, of which 10 were common to the two groups (Figure 7C; Table S2). Nine DEGs, six from the NC-vs.-V1 group and three from the NC-vs.-V2 group, were selected to validate the RNA-seq data with real-time PCR analysis, and consistent results were obtained (Figure 7D).

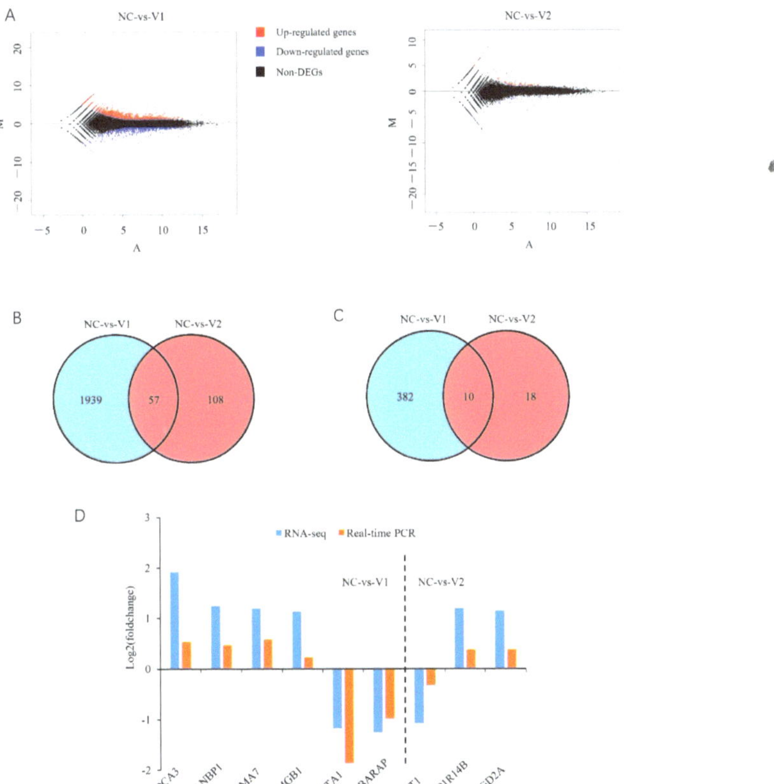

Figure 7. Overview of the RNA-seq data. (**A**) MA plot of the RNA-seq data. (**B**) Venn diagram of DEGs in cells overexpressing V1 and V2. (**C**) Venn diagram of differentially expressed transcription factors in cells overexpressing V1 and V2. (**D**) Validation of the RNA-seq data with real-time quantitative PCR.

2.7. Functional Characterization of the Differentially Expressed Genes

Gene Ontology (GO) and Kyoto Encyclopedia of Genes and Genomes (KEGG) analyses were performed on the NC-vs.-V1 and NC-vs.-V2 groups, and the results were compared to reveal the reason for their differential roles in the regulation of gene expression. Two GO terms were enriched in the cellular component category in each group, and there was no difference between the two groups (Figure 8A). Among the top 10 GO terms enriched in the biological pathway category, nine were common to the two groups. The terms developmental process and biological process involved in the interspecies interaction between organisms were specific to NC-vs.-V1 and NC-vs.-V2, respectively. In the molecular function category, a greater number of items differed between the two groups (Figure 8B). A total of six GO terms were differentially enriched in the two groups, of which the molecular transducer activity, structural molecule activity, antioxidant activity, and cargo receptor activity were specific to NC-vs.-V1, while the molecular function regulator and translation regulator activity were specific to NC-vs.-V2 (Figure 8C).

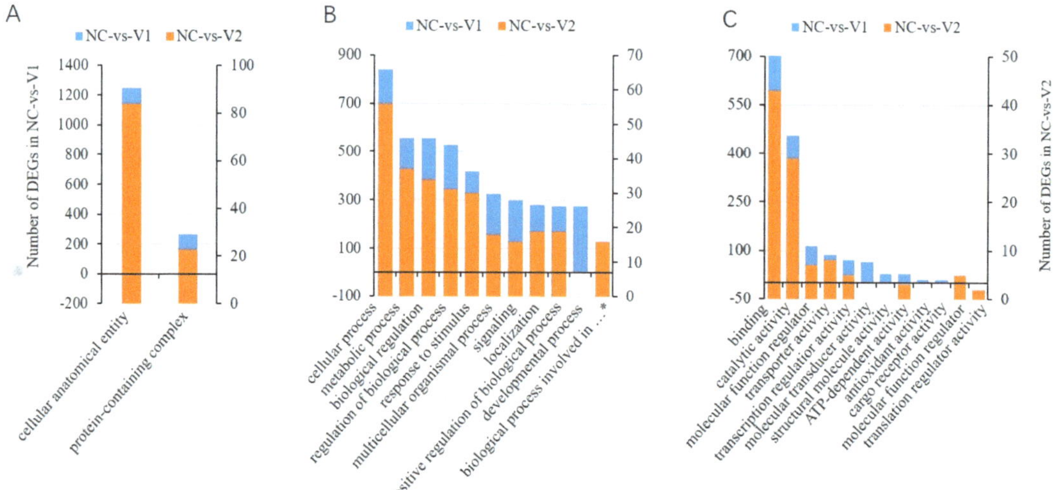

Figure 8. Top GO terms involved in the cellular component (**A**), biological process (**B**), and molecular function (**C**). * Biological process involved in the interspecies interaction between organisms.

DEGs were involved in various pathways in both the NC-vs.-V1 and NC-vs.-V2 groups. Pathways enriched in NC-vs.-V1 were very different from those enriched in the NC-vs.-V2 group. The most significantly enriched pathways were arachidonic acid metabolism and ferroptosis in the NC-vs.-V1 group, and the cytosolic DNA-sensing pathway and basal TFs in the NC-vs.-V2 group. Among all the significantly enriched pathways ($p < 0.05$), only three were common to both groups (Figure 9A,B). Both groups contained pathways related to adipogenesis, such as arachidonic acid metabolism and the PPAR signaling pathway in NC-vs.-V1, and the PI3K-Akt signaling pathway and focal adhesion in NC-vs.-V2. This suggests that the two variants regulate fat deposition via different mechanisms.

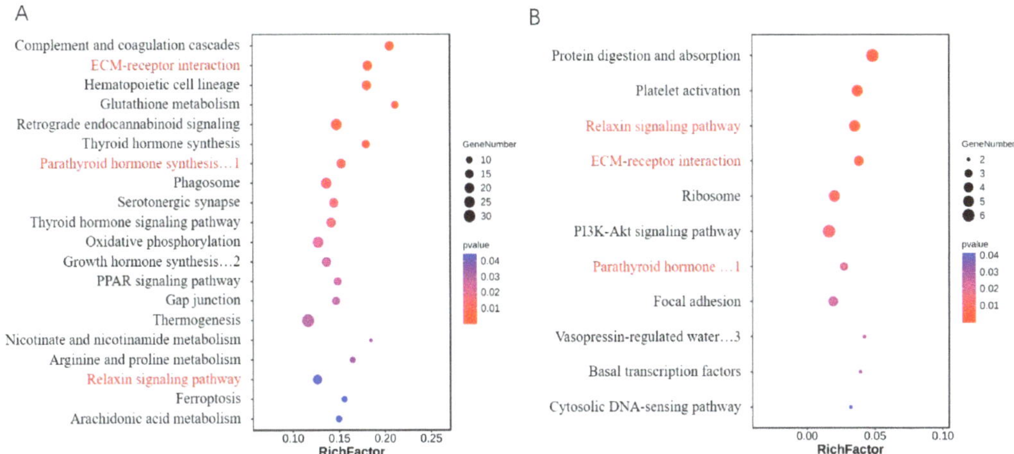

Figure 9. KEGG pathways significantly enriched by DEGs, regulated by V1 (**A**) and V2 (**B**). Pathways shared by the two groups are indicated in red. 1. Parathyroid hormone synthesis, secretion, and action. 2. Growth hormone synthesis, secretion, and action. 3. Vasopressin-regulated water reabsorption.

3. Discussion

CREB is a critical TF involved in various physiological processes. It is rich in AS transcripts, some of which play distinctive roles in cell growth and development, while some isoforms, such as α and β in humans and mice, have similar tissue expression profiles and identical domains [21], suggesting that they play the same roles. However, these isoforms may not be redundant in the transcriptome and the proteome, and little effort has been made to analyze the functional differences between them. In this study, AS transcript variants of CREB were characterized in pigs using molecular biology techniques. Through dual-luciferase reporter analysis, four of the isoforms containing in-frame variations were analyzed to compare their transcriptional transactivation activities in PK-15 cells. The mechanisms underlying the differential activities of two canonical isoforms, corresponding to α and β in humans and mice, were analyzed at the genome level. The results help to reveal the role of CREB and to analyze the significance of AS in genes.

Five transcripts of CREB, V1–5, were characterized in pigs, and two, V3 and V5, were identified for the first time. In contrast to V1, isoforms V2–4 are in-frame variations. V2 contains all the same functional domains of CREB as V1, while isoforms V3 and V4 have lost the Q1 domain because AS now occurs in that region. The CREB protein is composed of distinct domains with different functions. N-terminal domains, including Q1, Q2, and KID, function synergistically to induce the transcriptional transactivation of CREB. Q1 functions to induce transcriptional enhancement, and Q2 is essential for inducing CREB-mediated transcription. The C-terminal portion harbors the bZIP domain responsible for the DNA binding and dimerization of CREB. The KID domain, intermediated between Q1 and Q2, can be activated by a series of kinases, such as protein kinase A, which affects the dimerization and DNA binding of CREB [26]. Thus, V3 and V4 have domains that are essential for transcriptional regulation, suggesting that they have transcriptional transactivation activity.

The expression profiles of transcripts V3 and V4 in tissues indicate that they have distinctive roles. V1 and V2, composed of identical domains, are ubiquitously expressed with similar profiles, indicating their identical roles. However, it is impossible to express two isoforms with identical roles simultaneously. To reveal the differences between isoforms V1 and V2 and to verify the potential of V3 and V4 as TFs, dual-luciferase reporter analysis was performed using *PABPN1* and *BCL2L2* as representatives of the target genes. We found that both V1 and V2 can promote the transcription of *PABPN1*, while only V2 has transcriptional regulation activity in the promoter of *BCL2L2*. This indicates that there is a division of labor between V1 and V2 in activating gene expression. V3 has no effect on the promoter activity of the two genes, while V4 inhibits the expression of both *PABPN1* and *BCL2L2*, in contrast to the promoter-activating ability of V1 and V2. Conflicting roles in the regulation of gene expression have been observed in the AS isoforms of the *CREB* gene. It is predominantly a positive regulator of the cAMP-responsive genes, but in the human testis, repressor CREB isoforms have been characterized [27,28], indicating the multiple roles played by the AS of *CREB*.

CREB is a bifunctional transcription activator, mediating both the constitutive and kinase-induced transcription activations of numerous genes in a variety of cell types [24,29,30]. It exerts its effects through Q2, a constitutive activation domain, for activating constitutive transcription, and KID for kinase-inducible activities. Although the Q2 domain is crucial for constitutive activation, both Q1 and Q2 are important for basal activation [30]. In contrast to V2, a positive regulator of gene expression, V4 is formed by exon-skipping of exon 4, coding for the Q1 domain in CDS. Therefore, V2 and V4 are produced by exon 4 splicing in and out, respectively, and they are differentiated from one another by Q1 inclusion or deletion. A similar phenomenon was also observed in the cAMP-responsive element modulator (CREM), another member of the CREB family that functions as a transcriptional activator. CREM can produce an isoform that explicitly antagonizes the transcriptional activation. Exon 9 of CREM coding for the Q domain is alternatively spliced. Its inclusion or deletion in the CREM mRNA results in a switch of the protein from functioning as an activator to acting as a repressor during spermatogenesis [31].

Q1 is responsible for transcriptional enhancement. It is unclear why its deletion in V4 results in antagonistic activity. Possibly, the deletion causes conformational alteration of the polypeptide, which renders it unable to form productive interactions with the proteins involved in the basal transcription complex. Additionally, V4 has a bZIP domain responsible for the DNA binding at the C-terminal, and, thus, it might compete with the activator CREB isoforms to bind to the CRE motif and inhibit the cAMP-stimulated gene expression. Nevertheless, V4 can function as a TF, and may account for the tissue-specific repression of cAMP-regulated genes because it is not expressed in all tissues, as revealed by real-time PCR.

Isoforms V1 and V2 have identical domains, similar stability, and tissue expression profiles, except for a higher mRNA level in V2 than in V1 in the same tissues. However, V1 and V2 showed differences in their transcriptional transactivation of genes. To reveal the underlying mechanisms, RNA-seq analysis was performed and the DEGs regulated by V1 and V2 were identified. There are only a few common genes significantly regulated by both V1 and V2 (absolute log2-foldchange > 1 and $p < 0.05$), and ectopic V1 altered the expression of many more genes than V2. These data indicate that there is a great difference between their regulations of gene expression. Specific GO terms and KEGG pathways enriched by DEGs were also identified in each group. There was a small difference between the GO terms enriched by both groups, especially in the CC and BP categories, while in the KEGG pathways, a greater difference was observed. Among the top 10 pathways enriched, only three were common to both groups, which indicates that different pathways might be the main reason for the differential roles of V1 and V2.

4. Materials and Methods

4.1. Animals, Tissues, and cDNA Synthesis

One-month-old and seven-month-old Min pigs were sampled from the Institute of Animal Husbandry, Heilongjiang Academy of Agricultural Sciences, Harbin, China, each with three replicates. Animal handling was carried out in accordance with the protocols approved by the Animal Care and Use Committee at Northeast Agricultural University (Harbin, China). Tissues, including those from the heart, liver, lungs, spleen, kidneys, stomach, large intestine, small intestine, fat, and muscles were collected immediately after the pigs were slaughtered, and these tissues were snap-frozen in liquid nitrogen. RNA was isolated with TRIzol reagent (Invitrogen, Carlsbad, CA, USA), and reverse transcription (RT) was performed, as previously described, to synthesize cDNA [32].

4.2. cDNA Amplification and Sequence Analysis

To characterize the AS events in CDS, a pair of primers were designed according to the porcine *CREB* (pCREB) mRNA deposited in GenBank (NM_001361427). PCR was carried out in a final volume of 25 µL containing 1 U exTaq DNA polymerase (Takara, Dalian, China), 1× PCR buffer, 0.2 µM of each primer, 200 µM of each dNTP (Takara, Dalian, China), and 1 µL of cDNA obtained from muscle tissues. The thermal cycling parameters were as follows: 94 °C for 5 min, followed by 30 cycles at 94 °C for 30 s, 59 °C for 30 s, 72 °C for 1 min, and a final extension at 72 °C for 7 min. All primers used in this study are listed in Table S3.

PCR products were inserted into the pMD18-T vector (Takara) for sequencing by Beijing Genomic Institute (BGI, Beijing, China). The resultant sequences were aligned using SnapGene (v5.2.4). The genomic structure was characterized using the Blat program (http://genome.ucsc.edu/cgi-bin/hgBlat, accessed on 3 March 2022).

4.3. Plasmids

To analyze the transcriptional transactivation potential of the AS variants of pCREB, the CDSs of isoforms encoding polypeptides with functional domains were amplified with primer pair A, containing *Eco*RI and *Xho*I sites in the 5' end, and subcloned into a pCMV-HA vector to construct eukaryotic expression plasmids. Luciferase reporter genes of porcine BCL2L2 and PABPN1 were constructed with pGL3-basic as the backbone at the *Kpn*I and *Hind*III sites. The potential of the upstream sequences of the *BCL2L2* or *PABPN1* genes as promoters and the putative binding sites of CREB were first analyzed with Promoter 2.0, Alibaba 2.1, and Jaspar programs. Then, sequences spanning from −1102 to +54 nucleotides (nt) of *BCL2L2* and −791 to −190 nt of *PABPN1* were amplified from genomic DNA and then inserted into the pGL3-basic. The first nucleotide of the start codon was assigned as +1. Additionally, the predicted CRE elements were deleted from the *BCL2L2* and *PABPN1* reporter genes using an overlapped extension PCR method, as described previously [32].

4.4. Dual-Luciferase Reporter Assay

A dual-luciferase reporter assay was performed, as described previously [32]. Briefly, each reporter gene constructed was transfected alone or together with plasmids over-expressing *pCREB* into PK-15 cells using Lipofectamine 2000 (Invitrogen). The *Renilla* luciferase reporter gene was used as the reference to avoid differences in transfection efficiency among the groups. At 48 h after transfection, the cells were collected and the luciferase activity was measured with the Dual-Glo Luciferase Assay System (Promega, Madison, WI, USA). The relative luciferase activity was calculated as the ratio of firefly activity to that of *Renilla*. A Student's *t*-test was used to compare the differences between the experiment and control groups. Each experiment was independently repeated three times, each with triplicates. Data are presented as mean ± standard deviation.

4.5. Competitive RT-PCR

Isoforms V1 and V2 were amplified simultaneously using the competitive RT-PCR method to compare the expression levels in the same tissue. A pair of primers, complementary to the common sequences of both variants, was used. The PCR volume and reaction protocols were the same as those described above. The products were visualized using 12% polyacrylamide gel electrophoresis.

4.6. Library Preparation and Sequencing

Plasmids overexpressing *pCREB* were transiently transfected into PK-15 cells with Lipofectamine 2000 reagent (Invitrogen) according to the manufacturer's protocol. Cells transfected with an empty vector were used as the control. At 48 h after transfection, cells were collected, and RNA was isolated with TRIzol reagent (Invitrogen) to construct paired-end RNA-seq libraries. Three independent experiments were performed. A total of six libraries were prepared from the experimental and control groups, each with three replicates. The libraries were prepared and sequenced on a NovaSeq 6000 sequence analyzer (Illumina, San Diego, CA, USA) by the Frasergen company (Wuhan, China), as described previously [33].

4.7. RNA-Seq Data Analysis

The raw reads were filtered with SOAPnuke (v2.1.0) to obtain clean reads. The clean reads were aligned to the pig reference genome (Sscrofa 11.1) with the HISAT2 (v2.1.0) program [34]. Bowtie 2 (V2.3.5) [35] was used to map the reads to the ENSEMBLE transcriptome (*Sscrofa 11.1*). The transcript assembly and quantification in fragments per kilobase million (FPKM) were performed with Cufflinks (v2.2.1) [36]. The threshold for expression was set at FPKM > 0.1 in at least one sample, as used previously [33]. The gene expression levels were compared between groups with DESeq2 (v1.22.2) [37], and DEGs were identified according to the criteria of absolute Log2-foldchange > 1 and $p < 0.05$. GO and KEGG enrichment analyses of DEGs were carried out with the hypergeometric distribution method and KOBAS tool (v3.0) [38], respectively. KEGG pathways with a p-value < 0.05 were considered as significantly enriched. Hmmscan (v3.0) [39] was used to predict the TFs.

4.8. Real-Time Quantitative PCR

Real-time quantitative (q) PCR was performed as described previously with β-actin as the reference [32]. Each of the AS transcripts of the *pCREB* gene were amplified specifically to measure the mRNA levels in the tissues and cells. The cells overexpressing *pCREB* were collected at 48 h after transfection for RNA isolation. To measure the mRNA stability of isoforms V1 and V2, PK-15 cells were treated with 5 μg/mL actinomycin D (ActD, Biotopped, Beijing, China) and collected for RNA isolation at each time point described in the Results section. The relative mRNA levels were determined as the ratio of mRNA remaining to that at time zero. For the validation of the RNA-seq data, nine DGEs were selected, and qPCR was performed with cDNA obtained from cells treated identically to those used for RNA-seq analysis. Data are shown as mean ± standard deviation.

5. Conclusions

A total of five AS isoforms of porcine CREB, named V1–5, were cloned. Two of these isoforms, V3 and V5, were identified for the first time. These data indicate that the *CREB* gene is rich in AS. Isoforms V1–3 were ubiquitously expressed in all the tissues sampled, while V4 and V5 were not. V1 and V2 contain all the functional domains of CREB and have similar stability and tissue expression profiles, suggesting that they have the same functional roles. Through dual-luciferase reporter analysis on the promoters of *PABNP1* and *BCL2L2* genes containing the putative CRE motif, we found that V1 and V2 both function as activators, but their roles are gene-specific. To analyze the mechanisms underlying these differences, genes, GO terms, and KEGG pathways specifically regulated by V1 or V2 were identified by RNA-seq combined with bioinformatic methods. We found that V3 had no effect on the promoter activities of *PABNP1* and *BCL2L2*, while V4 acted as a repressor in regulating gene expression. These results demonstrate that CREB isoforms play multiple roles.

Supplementary Materials: The following supporting information can be downloaded at: https: //www.mdpi.com/article/10.3390/genes13081304/s1, Table S1: Differentially expressed genes identified, Table S2: Differentially expressed transcription factors identified, Table S3: Primers used in this study.

Author Contributions: Conceptualization, X.Y. and D.Z.; resources, D.L.; investigation, D.Z., Q.Z., J.L. and W.H.; validation, L.W. and Q.Z.; visualization, Y.S. and J.L.; writing—original draft preparation, D.Z. and X.Y.; writing—reviewing and editing, X.Y. and D.L.; funding acquisition, D.Z. and D.L. All authors have read and agreed to the published version of the manuscript.

Funding: This research was funded by the Heilongjiang Science Fund for Distinguished Young Scholars (JQ2020C005).

Institutional Review Board Statement: The animal study was reviewed and approved by the Animal Care Committee of Northeast Agricultural University.

Informed Consent Statement: Not applicable.

Data Availability Statement: The cDNA sequences of CREB have been deposited in GenBank data and the accepted Nos. are provided in the manuscript. All the relevant data are provided along with the manuscript as Supplementary Files.

Conflicts of Interest: The authors declare no conflict of interest.

References

1. Zhang, X.; Odom, D.T.; Koo, S.H.; Conkright, M.D.; Canettieri, G.; Best, J.; Chen, H.; Jenner, R.; Herbolsheimer, E.; Jacobsen, E.; et al. Genome-wide analysis of cAMP-response element binding protein occupancy, phosphorylation, and target gene activation in human tissues. *Proc. Natl. Acad. Sci. USA* **2005**, *102*, 4459–4464. [CrossRef] [PubMed]
2. Bleckmann, S.C.; Blendy, J.A.; Rudolph, D.; Monaghan, A.P.; Schmid, W.; Schütz, G. Activating transcription factor 1 and CREB are important for cell survival during early mouse development. *Mol. Cell Biol.* **2002**, *22*, 1919–1925. [CrossRef]
3. Mantamadiotis, T.; Lemberger, T.; Bleckmann, S.C.; Kern, H.; Kretz, O.; Martin Villalba, A.; Tronche, F.; Kellendonk, C.; Gau, D.; Kapfhammer, J.; et al. Disruption of CREB function in brain leads to neurodegeneration. *Nat. Genet.* **2002**, *31*, 47–54. [CrossRef] [PubMed]
4. Zhang, L.; Dong, D.L.; Gao, J.H.; Wang, A.K.; Shao, Y.P. β-HB inhibits the apoptosis of high glucose-treated astrocytes via activation of CREB/BDNF axis. *Brain Inj.* **2021**, *35*, 1201–1209. [CrossRef] [PubMed]
5. Xiao, G.; Lian, G.; Wang, T.; Chen, W.; Zhuang, W.; Luo, L.; Wang, H.; Xie, L. Zinc-mediated activation of CREB pathway in proliferation of pulmonary artery smooth muscle cells in pulmonary hypertension. *Cell Commun. Signal.* **2021**, *19*, 103. [CrossRef]
6. Dworkin, S.; Mantamadiotis, T. Targeting CREB signalling in neurogenesis. *Exper. Opin. Ther. Targets.* **2010**, *14*, 869–879. [CrossRef] [PubMed]
7. Zheng, K.B.; Xie, J.; Li, Y.T.; Yuan, Y.; Wang, Y.; Li, C.; Shi, Y.F. Knockdown of CERB expression inhibits proliferation and migration of glioma cells line U251. *Bratisl. Lek. Listy* **2019**, *120*, 309–315. [CrossRef]
8. Montminy, M.R.; Sevarino, K.A.; Wagner, J.A.; Mandel, G.; Goodman, R.H. Identification of a cyclic-AMP-responsive element within the rat somatostatin gene. *Proc. Natl. Acad. Sci. USA* **1986**, *83*, 6682–6686. [CrossRef]
9. Nichols, M.; Weih, F.; Schmid, W.; DeVack, C.; Kowenz-Leutz, E.; Luckow, B.; Boshart, M.; Schütz, G. Phosphorylation of CREB affects its binding to high and low affinity sites: Implications for cAMP induced gene transcription. *EMBO J.* **1992**, *11*, 3337–3346. [CrossRef]
10. Craig, J.C.; Schumacher, M.A.; Mansoor, S.E.; Farrens, D.L.; Brennan, R.G.; Goodman, R.H. Consensus and variant cAMP-regulated enhancers have distinct CREB-binding properties. *J. Bio. Chem.* **2001**, *276*, 11719–11728. [CrossRef]
11. Kovacevic-Grujicic, N.; Mojsin, M.; Popovic, J.; Petrovic, I.; Topalovic, V.; Stevanovic, M. Cyclic AMP response element binding (CREB) protein acts as a positive regulator of SOX3 gene expression in NT2/D1 cells. *BMB Rep.* **2014**, *47*, 197–202. [CrossRef]
12. Hashimoto, T.; Morohashi, K.; Takayama, K.; Honda, S.; Wada, T.; Handa, H.; Omura, T. Cooperative transcription activation between Ad1, a CRE-like element, and other elements in the CYP11B gene promoter. *J. Biochem.* **1992**, *112*, 573–575. [CrossRef] [PubMed]
13. Brenig, B.; Duan, Y.; Xing, Y.; Ding, N.; Huang, L.; Schütz, E. Porcine SOX9 Gene Expression Is Influenced by an 18 bp Indel in the 5′-Untranslated Region. *PLoS ONE* **2015**, *10*, e0139583. [CrossRef] [PubMed]
14. Muchardt, C.; Li, C.; Kornuc, M.; Gaynor, R. CREB regulation of cellular cyclic AMP-responsive and adenovirus early promoters. *J. Virol.* **1990**, *64*, 4296–4305. [CrossRef] [PubMed]
15. Fass, D.M.; Craig, J.C.; Impey, S.; Goodman, R.H. Cooperative mechanism of transcriptional activation by a cyclic AMP-response element modulator alpha mutant containing a motif for constitutive binding to CREB-binding protein. *J. Biol. Chem.* **2001**, *276*, 2992–2997. [CrossRef]
16. Merrett, J.E.; Bo, T.; Psaltis, P.J.; Proud, C.G. Identification of DNA response elements regulating expression of CCAAT/enhancer-binding protein (C/EBP) β and δ and MAP kinase-interacting kinases during early adipogenesis. *Adipocyte* **2020**, *9*, 427–442. [CrossRef]
17. Hai, T.; Curran, T. Cross-family dimerization of transcription factors Fos/Jun and ATF/CREB alters DNA binding specificity. *Proc. Natl. Acad. Sci. USA* **1991**, *88*, 3720–3724. [CrossRef]

18. Glick, Y.; Orenstein, Y.; Chen, D.; Avrahami, D.; Zor, T.; Shamir, R.; Gerber, D. Integrated microfluidic approach for quantitative high-throughput measurements of transcription factor binding affinities. *Nucleic Acids Res.* **2016**, *44*, e51. [CrossRef]
19. Ruppert, S.; Cole, T.J.; Boshart, M.; Schmid, E.; Schütz, G. Multiple mRNA isoforms of the transcription activator protein CREB: Generation by alternative splicing and specific expression in primary spermatocytes. *EMBO J.* **1992**, *11*, 1503–1512. [CrossRef]
20. Daniel, P.B.; Habener, J.F. Cyclical alternative exon splicing of transcription factor cyclic adenosine monophosphate response element-binding protein (CREB) messenger ribonucleic acid during rat spermatogenesis. *Endocrinology* **1998**, *139*, 3721–3729. [CrossRef]
21. Huang, X.; Zhang, J.; Lu, L.; Yin, L.; Xu, M.; Wang, Y.; Zhou, Z.; Sha, J. Cloning and expression of a novel CREB mRNA splice variant in human testis. *Reproduction* **2004**, *128*, 775–782. [CrossRef] [PubMed]
22. Waeber, G.; Habener, J.F. Novel testis germ cell-specific transcript of the CREB gene contains an alternatively spliced exon with multiple in-frame stop codons. *Endocrinology* **1992**, *131*, 2010–2015. [CrossRef] [PubMed]
23. Saura, C.A.; Cardinaux, J.R. Emerging Roles of CREB-Regulated Transcription Coactivators in Brain Physiology and Pathology. *Trends Neurosci.* **2017**, *40*, 720–733. [CrossRef] [PubMed]
24. Amidfar, M.; de Oliveira, J.; Kucharska, E.; Budni, J.; Kim, Y.K. The role of CREB and BDNF in neurobiology and treatment of Alzheimer's disease. *Life Sci.* **2020**, *257*, 118020. [CrossRef]
25. Berkowitz, L.A.; Gilman, M.Z. Two distinct forms of active transcription factor CREB (cAMP response element binding protein). *Proc. Natl. Acad. Sci. USA* **1990**, *87*, 5258–5262. [CrossRef]
26. Steven, A.; Friedrich, M.; Jank, P.; Heimer, N.; Budczies, J.; Denkert, C.; Seliger, B. What turns CREB on? And off? And why does it matter? *Cell Mol. Life Sci.* **2020**, *77*, 4049–4067. [CrossRef]
27. Walker, W.H.; Girardet, C.; Habener, J.F. Alternative exon splicing controls a translational switch from activator to repressor isoforms of transcription factor CREB during spermatogenesis. *J. Biol. Chem.* **1996**, *271*, 20145–21050. [CrossRef]
28. Walker, W.H.; Daniel, P.B.; Habener, J.F. Inducible cAMP early repressor ICER down-regulation of CREB gene expression in Sertoli cells. *Mol. Cell. Endocrinol.* **1998**, *143*, 167–178. [CrossRef]
29. Zhang, H.; Kong, Q.; Wang, J.; Jiang, Y.; Hua, H. Complex roles of cAMP-PKA-CREB signaling in cancer. *Exp. Hematol. Oncol.* **2020**, *9*, 32. [CrossRef]
30. Quinn, P.G. Distinct activation domains within cAMP response element-binding protein (CREB) mediate basal and cAMP-stimulated transcription. *J. Biol. Chem.* **1993**, *268*, 16999–17009. [CrossRef]
31. Foulkes, N.S.; Sassone-Corsi, P. More is better: Activators and repressors from the same gene. *Cell* **1992**, *68*, 411–414. [CrossRef]
32. Yang, X.Q.; Zhang, C.X.; Wang, J.K.; Wang, L.; Du, X.; Song, Y.F.; Liu, D. Transcriptional regulation of the porcine miR-17-92 cluster. *Mol. Genet. Genom.* **2019**, *294*, 1023–1036. [CrossRef] [PubMed]
33. Hao, W.; Yang, Z.; Sun, Y.; Li, J.; Zhang, D.; Liu, D.; Yang, X. Characterization of Alternative Splicing Events in Porcine Skeletal Muscles with Different Intramuscular Fat Contents. *Biomolecules* **2022**, *12*, 154. [CrossRef] [PubMed]
34. Kim, D.; Langmead, B.; Salzberg, S.L. HISAT: A fast spliced aligner with low memory requirements. *Nat. Methods* **2015**, *12*, 357–360. [CrossRef]
35. Langmead, B. Aligning short sequencing reads with Bowtie. *Curr. Protoc. Bioinform.* **2010**, *32*, 11.7.1–11.7.14. [CrossRef] [PubMed]
36. Trapnell, C.; Williams, B.A.; Pertea, G.; Mortazavi, A.; Kwan, G.; van Baren, M.J.; Salzberg, S.L.; Wold, B.J.; Pachter, L. Transcript assembly and quantification by RNA-Seq reveals unannotated transcripts and isoform switching during cell differentiation. *Nat. Biotechnol.* **2010**, *28*, 511–515. [CrossRef] [PubMed]
37. Love, M.I.; Huber, W.; Anders, S. Moderated estimation of fold change and dispersion for RNA-seq data with DESeq2. *Genome. Biol.* **2014**, *15*, 550. [CrossRef]
38. Xie, C.; Mao, X.; Huang, J.; Ding, Y.; Wu, J.; Dong, S.; Kong, L.; Gao, G.; Li, C.Y.; Wei, L. KOBAS 2.0: A web server for annotation and identification of enriched pathways and diseases. *Nucleic Acids Res.* **2011**, *39*, W316–W322. [CrossRef]
39. Potter, S.C.; Luciani, A.; Eddy, S.R.; Park, Y.; Lopez, R.; Finn, R.D. HMMER web server: 2018 update. *Nucleic Acids Res.* **2018**, *46*, W200–W204. [CrossRef]

genes

Article

Transcriptomic Profile of Genes Regulating the Structural Organization of Porcine Atrial Cardiomyocytes during Primary In Vitro Culture

Mariusz J. Nawrocki [1], Karol Jopek [2], Mariusz Kaczmarek [3,4], Maciej Zdun [5], Paul Mozdziak [6,7], Marek Jemielity [8], Bartłomiej Perek [8], Dorota Bukowska [9] and Bartosz Kempisty [1,2,7,10,11,*]

1 Department of Anatomy, Poznan University of Medical Sciences, 60-781 Poznan, Poland; mjnawrocki@ump.edu.pl
2 Department of Histology and Embryology, Poznan University of Medical Sciences, 60-781 Poznan, Poland; karoljopek@ump.edu.pl
3 Department of Cancer Immunology, Chair of Medical Biotechnology, Poznan University of Medical Sciences, 61-866 Poznan, Poland; markacz@ump.edu.pl
4 Gene Therapy Laboratory, Department of Cancer Diagnostics and Immunology, Greater Poland Cancer Centre, 61-866 Poznan, Poland
5 Department of Basic and Preclinical Sciences, Institute of Veterinary Medicine, Nicolaus Copernicus University in Toruń, 87-100 Toruń, Poland; maciejzdun@umk.pl
6 Physiology Graduate Program, North Carolina State University, Raleigh, NC 27695, USA; pemozdzi@ncsu.edu
7 Prestage Department of Poultry Science, North Carolina State University, Raleigh, NC 27695, USA
8 Department of Cardiac Surgery and Transplantology, Poznan University of Medical Sciences, 61-848 Poznan, Poland; kardiock@ump.edu.pl (M.J.); bperek@ump.edu.pl (B.P.)
9 Department of Diagnostics and Clinical Sciences, Institute of Veterinary Medicine, Nicolaus Copernicus University in Toruń, 87-100 Toruń, Poland; dbukowska@umk.pl
10 Department of Veterinary Surgery, Institute of Veterinary Medicine, Nicolaus Copernicus University in Toruń, 87-100 Toruń, Poland
11 Bartosz Kempisty, Department of Histology and Embryology, Department of Anatomy, Poznań University of Medical Sciences, 6 Święcickiego St., 60-781 Poznań, Poland
* Correspondence: bkempisty@ump.edu.pl; Tel.: +48-618-546-054; Fax: +48-618-546-107

check for **updates**

Citation: Nawrocki, M.J.; Jopek, K.; Kaczmarek, M.; Zdun, M.; Mozdziak, P.; Jemielity, M.; Perek, B.; Bukowska, D.; Kempisty, B. Transcriptomic Profile of Genes Regulating the Structural Organization of Porcine Atrial Cardiomyocytes during Primary In Vitro Culture. *Genes* 2022, 13, 1205. https://doi.org/10.3390/genes13071205

Academic Editor: Katarzyna Piórkowska

Received: 15 June 2022
Accepted: 4 July 2022
Published: 5 July 2022

Publisher's Note: MDPI stays neutral with regard to jurisdictional claims in published maps and institutional affiliations.

Abstract: Numerous cardiovascular diseases (CVD) eventually lead to severe myocardial dysfunction, which is the most common cause of death worldwide. A better understanding of underlying molecular mechanisms of cardiovascular pathologies seems to be crucial to develop effective therapeutic options. Therefore, a worthwhile endeavor is a detailed molecular characterization of cells extracted from the myocardium. A transcriptomic profile of atrial cardiomyocytes during long-term primary cell culture revealed the expression patterns depending on the duration of the culture and the heart segment of origin (right atrial appendage and right atrium). Differentially expressed genes (DEGs) were classified as involved in ontological groups such as: "cellular component assembly", "cellular component organization", "cellular component biogenesis", and "cytoskeleton organization". Transcriptomic profiling allowed us to indicate the increased expression of *COL5A2*, *COL8A1*, and *COL12A1*, encoding different collagen subunits, pivotal in cardiac extracellular matrix (ECM) structure. Conversely, genes important for cellular architecture, such as *ABLIM1*, *TMOD1*, *XIRP1*, and *PHACTR1*, were downregulated during in vitro culture. The culture conditions may create a favorable environment for reconstruction of the ECM structures, whereas they may be suboptimal for expression of some pivotal transcripts responsible for the formation of intracellular structures.

Keywords: cardiomyocyte structure; cytoskeleton organization; extracellular matrix; cell culture; transcriptomic analysis

1. Introduction

Many cardiovascular diseases (CVD) are associated with myocardial damage, and when untreated they eventually lead to severe myocardial dysfunction within a relatively short time. Clinically evident impairment of systolic myocardial performance predominantly results from the insufficient regeneration of cardiomyocytes (CMs) [1]. Currently available therapies are unable to replace lost CMs and largely irreversible cardiac dysfunction ensues. Therapeutic strategies aimed at stimulating the regenerative potential of myocardium may improve patient outcomes [2].

Strategies involving the transplantation of multipotent cardiac progenitor cells (CPCs) into the area of the injured myocardium, to promote regeneration of new functioning myocytes, are promising treatments. However, the effectiveness of the implantation strategy remains to be fully elucidated. Recent studies suggest that the adult heart is capable of limited cardiomyocyte turnover via the CPCs population [3–6]. Nevertheless, culture conditions and propagation of candidate cells need to be refined and standardized. A key developmental step will be to ensure in vitro culture conditions that allow cells extracted from the myocardium to expand. The description of the structure of the cultured cells will undoubtedly be an important element enabling the assessment of the condition of the cultured cardiomyocytes.

Understanding the structural property changes during myocardial damage may lead to the development of novel prevention therapies. Moreover, knowledge at the molecular level about the changes occurring in the participation of individual protein isoforms during either pathological processes or during in vitro culture will be crucial for determining the properties and usefulness of myocardial cells [7]. Therefore, novel measurement systems may lead to greater insight into cardiac tissue structure, properties, and performance [8]. Organization of the intracellular network cytoskeleton, or composition of the extracellular matrix (ECM), require deepening the knowledge at the molecular level to better understand the molecular changes in cells. The cytoskeleton tightly regulates myofibrillar activity and maintains muscle contraction/relaxation [9], while the ECM plays essential structural and regulatory roles in establishing and maintaining tissue architecture and cellular function [10].

The aim of the present study was to analyze the transcriptomic profile of genes involved in organization of the cardiomyocyte structure. Moreover, structural changes of cardiomyocytes isolated form the porcine right atrial appendage (RAA) and right atrium (RA) during long-term primary cell culture were evaluated to understand the CMs transcript expression patterns over the duration of the culture.

2. Materials and Methods

2.1. Animal Tissues

Pubertal crossbred Polish Landrace (PBZ × WBP) gilts (*Sus scrofa f. domestica*), bred with a mean age of 155 days (range 140–170 days) and a mean weight of 100 kg (95–120 kg), were used as a source of tissue. All animals were housed under identical conditions and fed the same forage. Porcine hearts were excised within 25 min of slaughter. After cutting the sternum and the diaphragm, the heart was removed. The hearts in intact pericardium were then preserved and transported to the laboratory on ice as soon as possible from a local slaughterhouse. After removing hearts from the pericardial sacs, the right atrial appendage and right atrial free wall were dissected and manually prepared with sterile surgical instruments to remove the visceral layer of the serous pericardium. Fat was also removed, and the tissue was the source of cells for in vitro culture. Each time, the delivered hearts were assessed for their suitability for downstream analyses. The following hearts were disqualified from further study: those with macroscopic injury after slaughter, those contaminated during transport or preliminary preparation, and those with connective tissue adhesions (signs of inflammation) or ischemia. As the research material is usually disposed of after slaughter, being a remnant by-product, no ethical committee approval was needed for the project.

2.2. Enzymatic Dissociation and Primary Cell Culture

The right atrial appendage (right auricle) and right atrium were extracted from the delivered hearts and washed in ice-cold PBS solution to remove the blood. After the two-step mincing by sterilized tools, the tissue underwent enzymatic digestion in DMEM + collagenase type II (2 mg/mL) solution at 37 °C for 40 min with gentle mixing. After the end of digestion, using nylon strainers of 70 μm pore size, the remaining tissue was separated from cell debris. The filtrate containing cells of interest was subject to centrifugation (5 min, 200× g, RT) to pellet the cells. The cells were washed with the PBS solution and then pre-plated on 25 mL culture bottles with culture medium (DMEM/F12, Sigma-Aldrich, Saint Louis, MO, USA), 20% FBS (Fetal Bovine Serum, Gibco, Thermo-Fischer Scientific, Waltham, MA, USA), 10% HS (Horse Serum, Gibco, Thermo-Fischer Scientific, Waltham, MA, USA), EGF (20 ng/mL; Sigma-Aldrich, Saint Louis, MO, USA), bFGF (10 ng/mL; Sigma-Aldrich, Saint Louis, MO, USA), and 1% P/S, and initially incubated for 4 h at 37 °C, 5% CO_2. This stage aims to deplete the fibroblasts, which show much higher adhesion affinity [11,12]. After this stage, the supernatant (including nonadherent cardiomyocytes) was pelleted by centrifugation (5 min, 200× g, RT) and placed into the new 25 mL culture bottle, previously coated with 0.1% gelatin solution. The cells were cultured in DMEM/F12 medium complemented with 20% FBS, 10% HS, EGF (20 ng/mL), LIF (10 ng/mL), and 1% P/S. The procedure of cell isolation and culture is based on the current literature sources [11,12] modified to fit the aims of the experiments. The culture medium was changed every three days. Cultures selected for downstream analysis exhibited a cell viability of 90% and more, measured by the ADAM-MC Automated Cell Counter (Bulldog Bio Inc., Rochester, NY, USA) at all time intervals (7D, 15D, and 30D of culture).

2.3. Morphological Observation of Cells during Long-Term Primary In Vitro Culture

Using an inverted light microscope with relief contrast (IX73, Olympus, Tokyo, Japan), daily observation of cultured cells was performed. Images obtained during the culture period correspond to the time intervals used in the molecular analysis (7 days (D), 15D, and 30D).

2.4. Flow Cytometry Analysis

The cardiomyocytes levels were measured by flow cytometry. Cells were stained with combinations of the following antibodies: anti-α-Actinin (Sarcomeric)-FITC (cat.: 130-106-997), anti-Myosin Heavy Chain-APC (cat.: 130-106-253, Miltenyi Biotec, San Diego, CA, USA), and anti-GATA4 Alexa Fluor 488 (cat.: 560330, BD Biosciences, San Jose, CA, USA). Samples were fixed and permeabilized with ice-cold BD Perm/Wash™ buffer. For intracellular staining, to 100 μL of cell sediments, 300 μL of BD Perm/Wash™ buffer was added to each tube, mixed, and incubated on ice for 10 min. Following incubation, cells were centrifuged (1500 rpm, 5 min), supernatant was removed, and cell sediments were incubated on ice for 30 min with antibodies. Finally, cells were washed with BD Perm/Wash™ buffer. Stained samples were evaluated by flow cytometry using the BD FACSAria™ equipment (Becton Dickinson, Franklin Lakes, NJ, USA).

2.5. RNA Extraction and Reverse Transcription

RNA samples (both before and after in vitro culture) were isolated according to the Chomczyński and Sacchi [13] method, employing TRI reagent (Sigma-Aldrich; Merck KGaA, Saint Louis, MO, USA). Total RNA isolation was conducted using, successively, chloroform, 2-propanol, and 75% ethanol solution during the procedure. Next, the obtained RNA samples were re-suspended in 20–40 μL of RNase-free water and frozen at −80 °C. RNA integrity was determined by denaturing agarose gel (2%) electrophoresis. Subsequently, the total amount of the collected RNA was evaluated by measuring the optical density (OD) at 260 nm (NanoDrop spectrophotometer; Thermo Scientific, Inc., Waltham, MA, USA). RNA samples were subjected to reverse transcription using the RT2

First Strand kit (Qiagen, Hilden, Germany), according to the manufacturer's protocol. To reverse-transcribe RNA into cDNA, 500 ng of an RNA sample was used [14].

2.6. Microarray Expression Study and Data Analysis

The microarray study was carried out as previously described [15]. Previously isolated total RNA (50 ng) from each pooled sample (during analyses, for each time interval we pooled RNA from 4 different cultures obtained from other hearts) was used in two rounds of sense cDNA amplification (Ambion® WT Expression Kit). In the first step, synthesis of cRNA was performed by in vitro transcription (16 h, 40 °C). Subsequently, cRNA after purification was re-transcribed into cDNA. Then, cDNA samples obtained via the Affymetrix GeneChip® WT Terminal Labeling and Hybridization kit (Affymetrix, Santa Clara, CA, USA) have been subjected to biotin labeling and fragmentation processes. Hybridization to the Affymetrix® Porcine Gene 1.1 ST Array Strip was conducted at 48 °C for 20 h, in an AccuBlock™ Digital Dry Bath (Labnet International, Inc., NY, USA) hybridization oven. Our experiment employed 4 array strips. Two biological samples (each obtained by pooling RNA from four different animals to mitigate the effect of interindividual variation) were used for the analysis of each individual time interval. After hybridization, using an Affymetrix GeneAtlas™ Fluidics Station (Affymetrix, Santa Clara, CA, USA), all array strips were washed and stained, according to the technical protocol. The array strips were scanned using an Affymetrix GeneAtlas™ Imaging Station (Affymetrix, Santa Clara, CA, USA). The preliminary analysis of the scanned chips was performed using Affymetrix GeneAtlas TM Operating Software (Affymetrix, Santa Clara, CA, USA). The quality of gene expression data was checked according to quality control (QC) criteria following the manufacturer's standards. The scans of the microarrays were saved as *.CEL files for downstream data analysis.

The created *.CEL files were subjected to further analysis performed using Bioconductor software, based on the R statistical language with the relevant Bioconductor libraries, as described previously [16,17]. To conduct background correction, normalization, and summarization of the results, we used the Robust Multiarray Averaging (RMA) algorithm. Assigned biological annotations were obtained from the "pd.porgene.1.1.st" library, employed for the mapping of normalized gene expression values with their symbols, gene names, and Entrez IDs, allowing us to generate a complex gene data table. To determine the statistical significance of the analyzed genes, moderated t-statistics from the empirical Bayes method were performed. The obtained p-values were corrected for multiple comparisons using Benjamini and Hochberg's false discovery rate and described as adjusted p-values. The selection criteria of a significantly changed gene expression were based on an expression fold difference higher than abs. 2 and an adjusted p-value < 0.05. The list of differentially expressed genes (DEGs) was uploaded to the DAVID software (Database for Annotation, Visualization, and Integrated Discovery). Ontology groups that contained at least 5 genes and expressed a p-value (Benajamini) < 0.05 were selected for further analysis. Particularly, the "cellular component assembly", "cellular component organization", "cellular component biogenesis", and "cytoskeleton organization" were selected as GO BPs of interest.

It is important to compare the expression profile in RA and RAA because the objective of the research was to characterize the molecular basis for structural changes. A Venn diagram was employed to detect relations between lists of DEGs in used hearts' compartments and to explore the intersection of genes of analyzed terms from the functional analysis. Fifty of the most altered genes for both heart compartments were analyzed.

2.7. Real-Time Quantitative Polymerase Chain Reaction (RT-qPCR) Analysis

The RT-qPCR validation was performed using a Light Cycler® 96 Real-Time PCR System, Roche Diagnostics GmbH (Mannheim, Germany), with SYBR Green as a detection dye. To the RNA material remaining after pooling of samples used in the microarray analysis (from 8 animals), 4 new biological replicates were added; as a result, the quantitative validation was performed for 12 samples for each time interval of culture. Levels of analyzed transcripts were standardized in each sample, in reference to hypoxanthine 1 phosphoribosyltransferase (*HPRT1*) and β-actin (*ACTB*) as an internal control. The final reaction mix consisted of 1 μL of cDNA, 5 μL of mastermix (RT^2 SYBR Green FAST Mastermix, Qiagen), 1 μL of forward + reverse primer mix (10 μM), and finally, 3 μL of PCR-grade water. We have used the Primer3 software for primer design (Table 1), based on Ensembl database transcript sequences. The exon–exon design method was used as an additional method to avoid potential remnant genomic DNA fragments' amplification. For target cDNA quantification, we have performed relative quantification with the $2^{-\Delta\Delta Cq}$ method.

Table 1. Primers. Oligonucleotide sequences of primers used for RT-qPCR analysis.

Gene		Primer Sequence (5′-3′)	Product Size (bp)
ITGA8	F	CCAGCAGACCAAAACCCTTC	164
	R	AAGAAGTTGTGCAGCTGTGG	
TNC	F	TTTCAGATGCCACCCCAGAT	169
	R	GTGGCTTCTCTGAGACCTGT	
FGF7	F	TGGAAATCAGGACAGTGGCT	192
	R	CTCCTCCACTGTGTGTCCAT	
COL12A1	F	TCCACAGGTTCAAGAGGTCC	150
	R	TTGTTAGCCGGAACCTGGAT	
KIF23	F	TGCAACAGGAGCTTGAAACC	243
	R	AGGGTCTCTCTGGCTTTTCA	
FLNB	F	AACATCCCGAACAGCCCTTA	159
	R	ACTGACATCACCTTCCCCAG	
COL5A2	F	TGGTGAAAATGGCCCAACTG	193
	R	TCCTCGACCACCTTTCAGTC	
COL8A1	F	GGAGAGAAGGGCTTTGGGAT	249
	R	GATCCCATCCTGACCTGGTT	
MYO5A	F	TGAGAAGAAGGTGCCTCTGG	199
	R	TTCCTGACGCTTGAGTGACT	
TRAM1	F	CCTCGTCAGCTCGTCTACAT	241
	R	AGCCAACAGTGAGTACCGAA	
TRAM2	F	ACATCTGCCTGTACCTGGTC	209
	R	GGCGAGAGTGAGGATGAAGA	
DMD	F	TCCACTTCTGTCCAAGGTCC	187
	R	GCAGTCTTCGGAGCTTCATG	
ABLIM1	F	ATGAAGCTCAACTCAGGCCT	161
	R	TAGCCTGGGAGAGATGAGGT	
CASQ2	F	TCCTTGTCTATGCAACGGGT	187
	R	GCTTTTCCCAGGTGTTGAGG	
TMOD1	F	ACAGCCGGGTCATAGATCAG	159
	R	GTCAGGGTCCAACTCATCCA	
XIRP1	F	AGAGCAATGCAGTGAGGACT	201
	R	AGTCCTTCTCGTCCACCAAG	

Table 1. *Cont.*

Gene		Primer Sequence (5′-3′)	Product Size (bp)
PHACTR1	F	TTAACTCGGAAGCTCAGCCA	174
	R	GAGCTCCTTTCGAATGGCAG	
MATN2	F	ACGACTTGCAGAATCCAGGA	156
	R	TGAGGCACAGTAGTCCACAG	
VWF	F	TGCAACACTTGTGTCTGTCG	229
	R	TGCATTTCAGGGAGGGGTAG	
ITGB6	F	TGACGACCTCAACACGATCA	190
	R	TCCAAAGGTAGGCAAGCAGA	
ACTB	F	CCCTGGAGAAGAGCTACGAG	156
	R	CGTCGCACTTCATGATGGAG	
HPRT1	F	CCATCACATCGTAGCCCTCT	166
	R	TATATCGCCCGTTGACTGGT	

3. Results

Daily observation of cell morphology using an inverted microscope employing relief contrast (IX73, Olympus, Tokyo, Japan) was performed and documented (Figures 1 and 2). Importantly, the morphology of cells obtained from different segments of the heart was similar. However, to obtain a complete view of the morphology of RAA and RA cells cultured in vitro, the information from the micrographs presented in two figures (Figures 1 and 2) should be supplemented with previously published data [15]. The initially seeded cells exhibited an irregular, slightly elongated shape. After several days in culture (15D), the cells presented a clearly elongated, spindle-like morphology (Figure 2), resulting from the increase in confluency, until almost the entire available surface of the bottom of the culture bottle was covered at the end of the culture period (30D). Characteristically, cultured cells often tended to overlap and form three-dimensional structures, even in incomplete confluence (Figure 1, at 15D and 30D of culture). A similar observation was made regardless of the source of the cells (both RA and RAA cells showed a similar trend).

Flow cytometric analysis (Figure 3) revealed that both cells isolated from RAA and RA expressed cytoplasmic markers specific for CMs at all analyzed culture periods (at 7D, 15D, and 30D of culture). The expression of α-Actinin (Sarcomeric), Myosin Heavy Chain, and GATA4 was observed, and thus the obtained results confirm the effectiveness of a pre-plating step to select the CMs population [18–20]. The flow cytometry gating strategy is presented in Supplementary Figure S1.

The dynamic changes in the transcriptome profile of cultured cardiomyocytes were observed at individual time intervals. Whole transcriptome profiling by Affymetrix microarray revealed gene expression after enzymatic dissociation and 7D, 15D, and 30D of cardiomyocytes long-term in vitro primary culture. Employing the Porcine Gene 1.1 ST Array Strip, microarray transcriptome screening was performed. The cells obtained from both RAA and RA at individual time intervals resulted in changes in gene expression levels. Genes with fold change >|2| and an adjusted *p*-value of <0.05 were considered as differentially expressed. Global expression analysis yielded 4239 DEGs for RA and 4662 DEGs for RAA. Principal component analysis (PCA) of overall DEGs allowed to examine the variance between the analyzed sample groups [15].

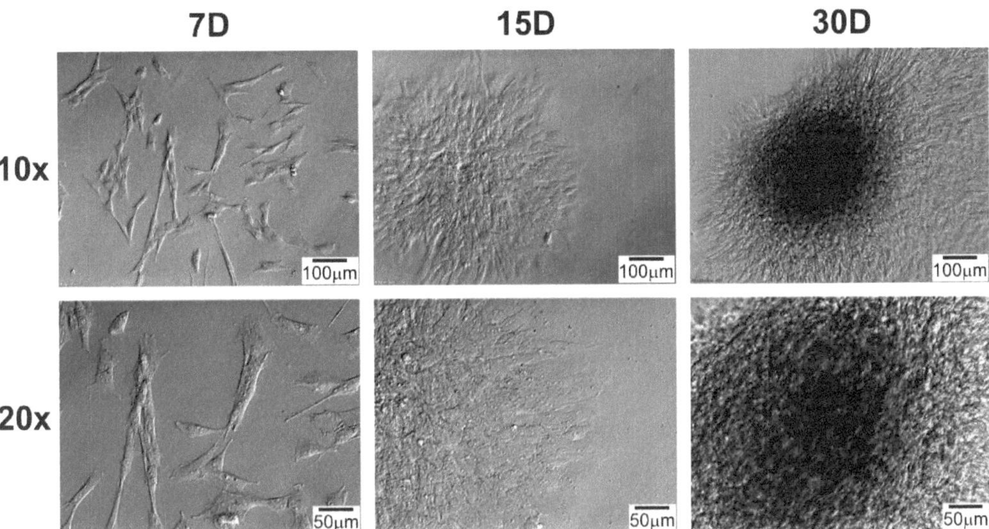

Figure 1. Changes in right atrial appendage (RAA) cells' morphology during long-term in vitro primary culture at individual time intervals. D: day of culture; $10\times$, $20\times$: magnification.

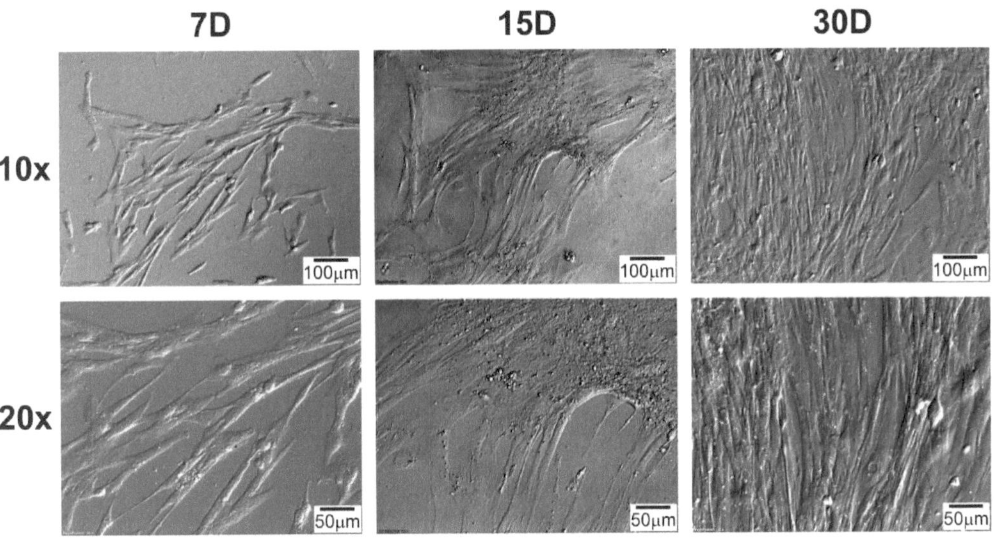

Figure 2. Changes in right atrium (RA) cells' morphology during long-term in vitro primary culture at individual time intervals. D: day of culture; $10\times$, $20\times$: magnification.

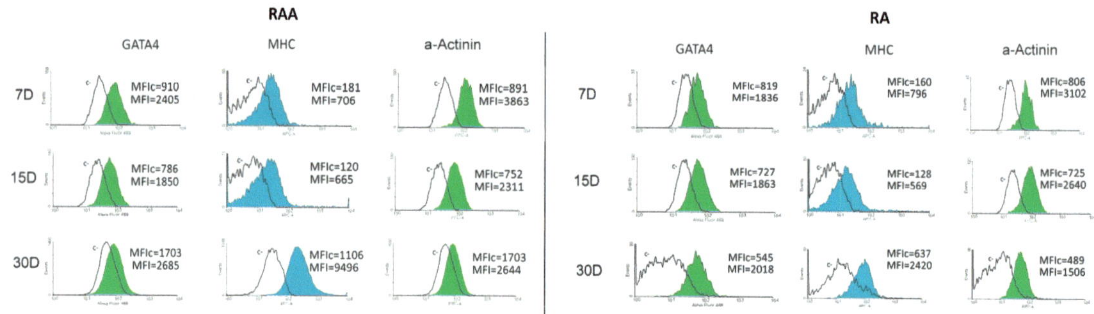

Figure 3. The results of flow cytometry analysis of selected cardiomyocytes markers (α-Actinin (Sarcomeric), Myosin Heavy Chainm and GATA4) in the cell samples at different time intervals subjected to in vitro culture. D: day of culture; MFI: mean fluorescent intensity; MHC: Myosin Heavy Chain.

Genes related to the cellular structure organization during development and maturation were analyzed, revealing 237 different transcripts for RA and 276 for RAA. In the next step, using DAVID software, selected DEGs were classify as involved in ontological groups, such as: "cellular component assembly", "cellular component organization", "cellular component biogenesis", and "cytoskeleton organization". Genes selected by DAVID software analysis were subjected to hierarchical clustering and are presented as heatmaps (Figure 4). Additionally, comprehensive DEGs enrichment analysis was performed, and obtained results are presented in the form of a clustergram (Supplementary Figure S2). We used the *Enrichr* database [21–23] for the determination of ontological groups. Selected DEGs were used as input genes for the analysis. The obtained enriched processes belonging to the GO Biological Process 2021 group were ranked according to *p*-value. The names and GO signatures of the ten most enriched processes are listed in a bar graph. The same ten processes were used to prepare the clustergram. The most enriched processes are related to the ECM (extracellular matrix organization GO:0030198, extracellular structure organization GO:0043062, external encapsulating structure organization GO:0045229, and supramolecular fiber organization GO:0097435).

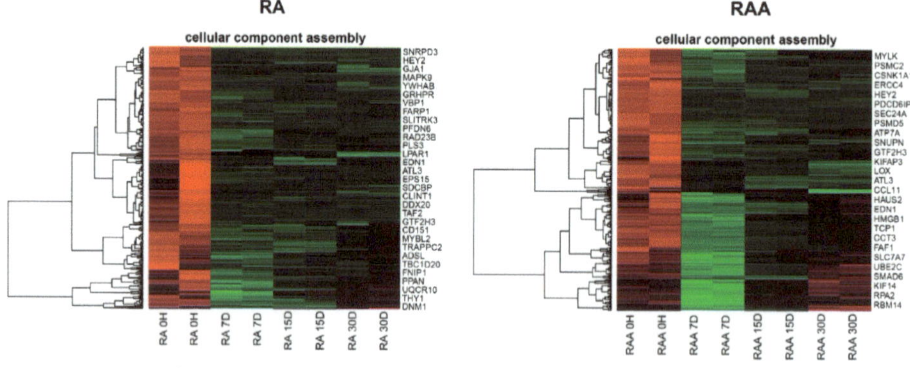

Figure 4. *Cont.*

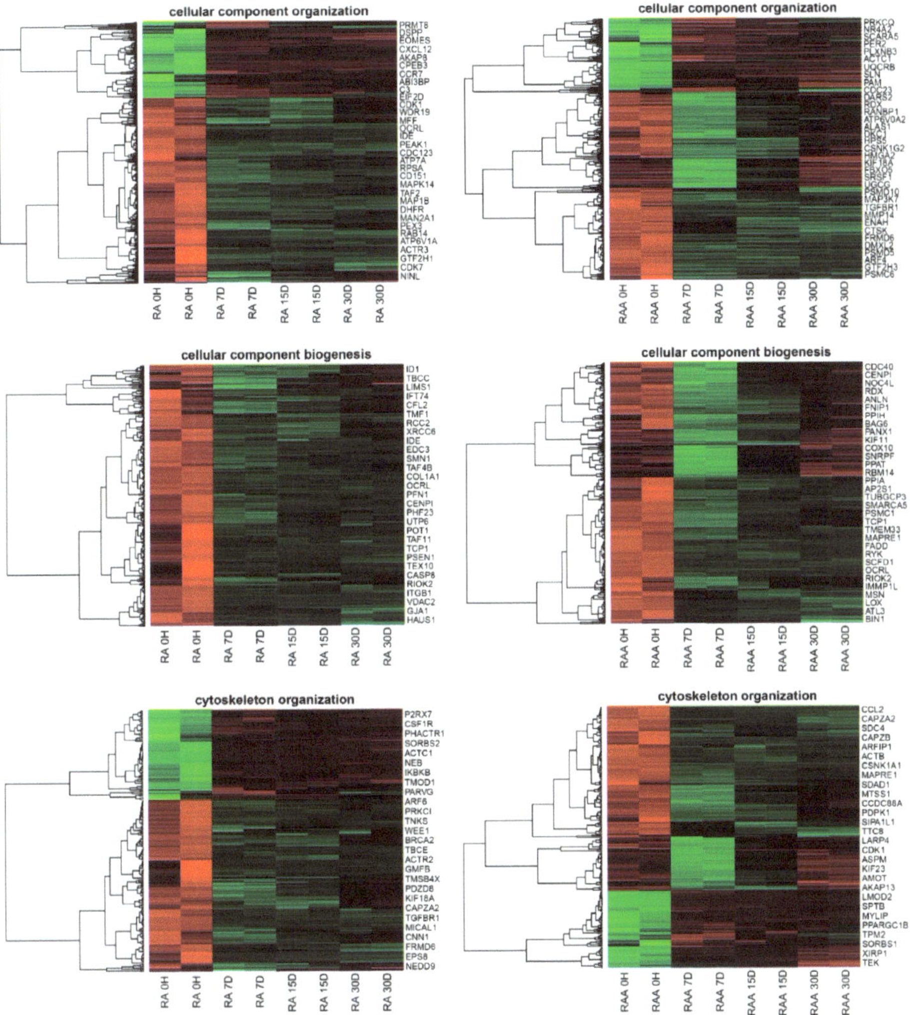

Figure 4. Heatmaps with hierarchical clustering of the differentially expressed genes, both in the right atrial appendage (RAA) and right atrium (RA), involved in "cellular component assembly", "cellular component organization", "cellular component biogenesis", and "cytoskeleton organization", based on GO BP terms. Each separate row on the y-axis represents a single transcript. The red color indicates downregulated, and the green indicates upregulated genes at subsequent time intervals of the analysis.

Due to the structure of the GO database and since any given gene can belong to multiple GO groups, gene intersections between selected ontological terms were examined. The relationships between genes and GO terms were mapped with circle plots, with visualization of logFC values and gene symbols (Figure 5). All the genes were either upregulated or downregulated in the cell culture intervals compared to the inoculation.

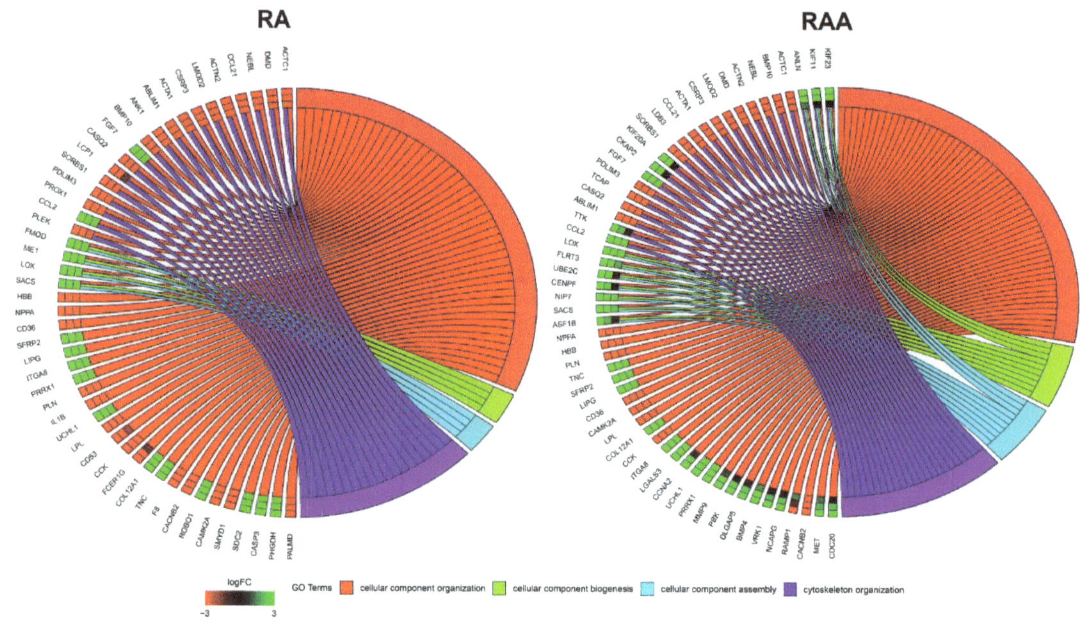

Figure 5. Analysis of enriched Gene Ontology groups involved in general cellular structure organization in cultured porcine cardiac muscle cells. The ribbons show the genes belonging to the given categories. The color bars near each gene correspond to logFC between culture intervals.

The multitude of results obtained for cells from two separate cardiac segments contributed to the creation of a Venn diagram visualizing the gene expression patterns' comparison (Figure 6). Out of differentially expressed genes (DEGs) sets for RA and RAA containing genes belonging to the GO BP terms of interest, the 50 most altered genes for both segments were selected and presented on the diagram. Commonly upregulated and commonly downregulated genes in RA and RAA are visible in the middle part of the diagram (Figure 6). Moreover, the same direction of changes in the transcript levels can be observed for RA and RAA. In the lateral parts of the Venn diagram, the gene names listed indicate the DEGs specific only to RAA (right side) and RA (left side of the diagram).

High-throughput global transcript analysis employing the microarray approach is largely qualitative. Therefore, the RT-qPCR method was applied. RT-qPCR analysis together with microarray results are illustrated as a heatmap (Figure 7). As can be seen, the RT-qPCR analysis confirmed the direction of expression change for all the selected DEGs. Importantly, none of the analyzed genes showed any other direction of expression changes than those indicated by the results of the expression microarrays. However, the scale of differences in transcript levels differed between both methods analyzed. The most upregulated genes, in RT-qPCR, from the examined DEGs included, among others, *TNC* (tenascin-C), *ITGA8* (integrin, α 8), and *COL12A1* (collagen type XII α 1 chain). The strongest downregulated genes were *ABLIM1* (actin-binding LIM protein 1), *DMD* (dystrophin), and *MATN2* (matrilin 2).

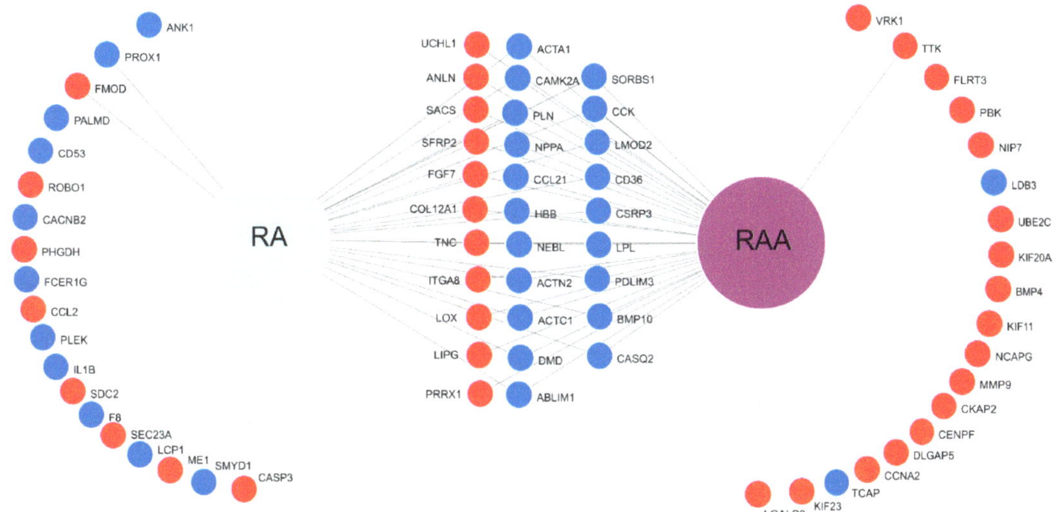

Figure 6. Venn diagram. The 50 most altered genes from analyzed GO BP terms for RAA and RA were used in the comparison. Blue circles show downregulated genes, whereas red circles show upregulated.

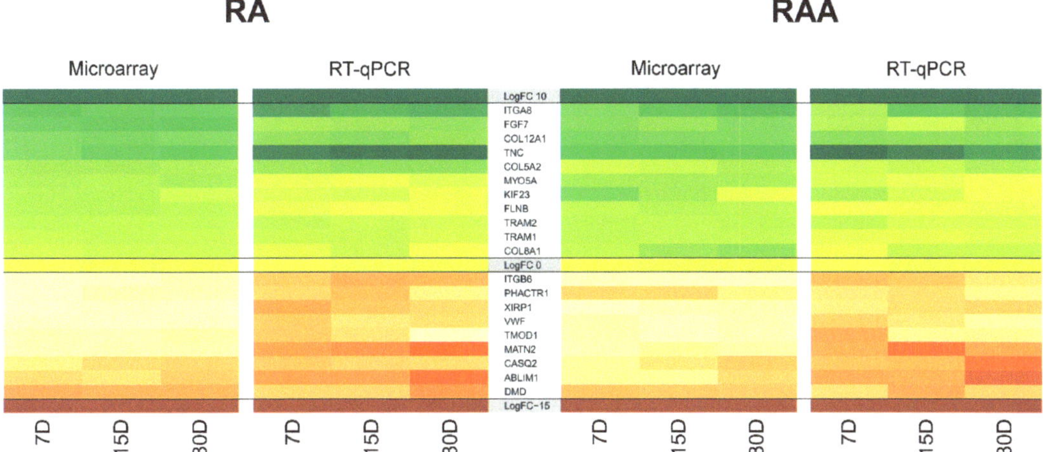

Figure 7. Heatmap representation of selected differentially expressed genes. Microarray results are presented together with RT-qPCR quantitative validation to indicate common patterns of transcript expression. Quantitative data (LogFC), used to compile the figure, are available in Supplementary Table S1. All the presented sample means were deemed to be statistically significant ($p < 0.05$). A scale indicating the color and its intensity depending on the direction and size of the changes has been integrated into the figure.

4. Discussion

It is necessary to understand the mechanisms regulating cardiac muscle cells that may enable application in cell-based therapy. The myocardium could serve as a source of cells that could be expanded as a treatment platform [5,24]. Therefore, the detailed characterization of cells obtained from the myocardium is important. Cardiomyocytes have been demonstrated to exhibit the ability to develop new vessels [15]. Furthermore, epigenetic alterations by measurement of mRNA expression levels of the DNA methyltransferases during CMs in vitro culture were evaluated in the current study [25]. The reported culture conditions may create a favorable environment for neovascularization; nevertheless, expression levels of genes encoding structural proteins, such as myosin (*MYH7*, *MYL3*) or actin (*ACTC1*), were significantly lower during culture. The previous results and the fact that the correct structure and properties of cardiomyocytes are necessary for the overall systolic/diastolic activity of this essential organ directed us to focus on issues related to the analysis of the molecular background of key factors for the correct structure of the growing cultured cells. Porcine hearts were chosen for this study because they are important due to their many similarities to human beings. Moreover, large animal models are essential to develop the discoveries from murine models into clinical therapies and interventions.

The extracellular matrix (ECM) is the non-cellular, highly dynamic structural network composed of numerous glycoproteins, such as collagens, elastin, fibronectin, and laminins [26]. ECM remodeling is crucial for regulating the cellular behavior, development, and maturation [27]. The ECM provides structural support for the developing myocardium and plays a crucial role in cardiac homeostasis by transducing key signals to cardiomyocytes or vascular cells [28]. Various natural components (such as collagen [29] or Matrigel [30]) and synthetic [31] biomaterials have been used to increase the efficiency of cardiomyocyte culture and influence the degree of maturation obtained. However, none of these biomaterials can fully recapitulate the architecture and functional composition of the cardiac ECM. It has been reported that human pluripotent stem cell-derived cardiomyocytes (hPSC-CMs) obtain the most favorable conditions for growth and maturation when used in artificial ECM [32] or decellularized myocardial ECM [33–35]. Therefore, any changes in the structure and composition of the ECM require analysis as it is crucial for the proper functioning and maturation of cultured cells.

The cardiac ECM is primarily composed of fibrillar collagen type I, which is commonly found in tissues' ECM together with type V collagen, which helps to organize type I collagen and makes contact with basement membranes [36], and the less abundant collagen type XII [37]. During long-term in vitro culture, cells isolated from both the right atrial appendage (RAA) and right atrial (RA) wall exhibited increased expression of three genes, *COL5A2*, *COL8A1*, and *COL12A1*, encoding different collagen subunits. Interestingly, increased transcript levels of these genes were observed as the culture continued in RAA cells. The obtained results on the mRNA level may suggest the readiness of cells maintained in the conditions of long-term in vitro culture to reconstruct the ECM structures precisely by the increased production of the analyzed transcripts. The transcript level that is translated into protein under 2D culture conditions would indicate the extent that this structure can be reconstructed by collagen overproduction. Azuaje et al. showed that *Col5a2* transcript is highly expressed in the left ventricle after myocardial infarction (MI), suggesting a role of collagen V α 2 chain in the post-MI response via ventricular remodeling by recruiting and deposition of sufficient amounts of collagen type I, thereby increasing myocardium healing [38]. In contrast, transgenic mice expressing a non-functional form of *Col5a2* do not present ventricular defects [39]. Moreover, patients suffering from classic Ehlers–Danlos syndrome, a rare connective tissue disorder mainly caused by mutations in *COL5A1* or *COL5A2*, do not appear to show ventricular malformations [40]. During transcriptomic analyses on a mouse model of cardiac remodeling, Wang et al. reported upregulation of transcript levels of the gene encoding collagen type VIII α 1 chain [41]. The authors suggest that the *Col8a1* gene may be a candidate for one of the biomarkers for cardiac remodeling. Interesting results presented in studies comparing mRNA expression levels of extracellular matrix protein genes with functional changes in the heart that occur with reproductive status indicated that *Col8a1* mRNA levels increased in the early postpartum period [42]. Other investigators showed the effect of deletion of the *Col8a1/2* genes in mice on induced pressure overload of the heart and reduced fibrosis, but increased dilatation and mortality compared to wildtype controls [43]. Therefore, different collagens show a significant role in the process of restoring matrix structures.

Among the differentially expressed genes identified in the present study, two genes encoding integrins (*ITGA8* and *ITGB6*) indicated significant changes in the mRNA levels during culture in relation to the starting point of in vitro culture. Integrins are heterodimeric, transmembrane glycoprotein receptors important for the ECM, and thus the cells, structure, and signaling. Members of this family are capable of bidirectionally transmitting signals between the ECM and the intracellular environment [44]. The obtained results are not as unequivocal as in the case of collagen-encoding genes, because *ITGA8* showed increased mRNA expression levels, while *ITGB6* showed downregulation. Liu et al. suggested an important role of *Itga8* expression in the molecular mechanism of hypoplastic left heart syndrome (HLHS), based on the results of a mouse model [45]. Employing the next-generation high-throughput RNA sequencing, Chen et al. showed, in the neonatal mouse heart, that Itgb6 belongs to targets of highly dysregulated lncRNAs during the first seven days of cardiac development [46]. In cells isolated from both the right atrial appendage (RAA) and the right atrial (RA) wall during long-term culture, the highest level of increased transcript expression was observed for genes encoding another extracellular matrix protein: tenascin-C (TNC). It is important to note that tenascin-C is abundantly expressed in the heart during embryonic development but is rarely detected in healthy subjects [47]. Nevertheless, TNC is a very important biomarker for many pathological conditions associated with inflammation, such as myocardial infarction, hypertensive cardiac fibrosis, or heart failure with preserved ejection fraction [47–49]. These observations suggest that tenascin-C is re-expressed in the affected heart. The in vitro culture conditions shock the cultured heart muscle cells by radically changing the environment in which they must exist. Therefore, another of the mechanisms of cell adaptation, aimed at restoring optimal conditions for development, is a significant overproduction of the *TNC* transcript.

The pivotal role for ECM homeostasis and its composition dynamic remodeling has been confirmed for matrix metalloproteinases (MMPs), which are an important environmental mediator of cardiac diastolic or systolic dysfunction [50]. Members of the matrix metalloproteinase (*MMP*) gene family, and encoded by these genes' proteins, that are zinc-dependent proteases, are involved in the breakdown of ECM [51]. MMP2 and MMP9 belong to the most frequently analyzed MMPs in relation to HF syndrome and other cardiovascular diseases [52,53]. Interestingly, the screening analyses of the transcriptome showed overexpression for both *MMP2* and *MMP9* at all time intervals of the culture. It should be emphasized that the increased levels of *MMP2* mRNA were only observed in RA, while the overexpression of the *MMP9* gene was demonstrated in cells obtained from RAA. Hayashidani et al. reported the response of MMP2 KO mice to in vivo myocardial infraction (MI) and described decreased mortality rates in the week following MI [54]. Moreover, the authors found protective effects in terms of hemodynamic performance or remodeling [54]. Other studies [55] employing a swine model, which as a large animal model may be more clinically representative than the most commonly used small rodent models, also confirmed the role of MMP in heart disease progression. Apple et al., using the MMP inhibitor (1 mg/kg PGE-530742) and starting animal treatment 3 days prior to MI, showed decreased left ventricular (LV) dilation post-myocardial infarction and improved LV contractile performance [55]. Numerous scientific reports confirm that MMP2 and MMP9 have been recognized to play an important role in matrix remodeling in these cardiac disease states [52]. However, results presented by Kiczak et al. indicated no differences in the expression of transcripts and proteins of MMP9 and MMP2 in either LV myocardium or skeletal muscles between diseased and healthy animals. These data may suggest that the activity of these enzymes may be altered post-translationally in HF and is not dependent on the expression of the mRNA or protein [53]. It should be emphasized that the MMPs are associated with the state of active myocardial remodeling and could be a potentially useful marker for the identification of patients at risk for heart failure development and poor outcome.

Variable expression levels of transcripts for genes encoding key structural proteins were also observed. Myosin, the major component of the thick filaments, is an essential motor protein of the myocardium. Lower mRNA levels of *MYH7* and *MYL3*, genes encoding two isoforms of myosin heavy and light chains, were previously observed [15]. In all culture intervals, both in RA and RAA cells, upregulation of *MYO5A* transcript levels was observed. The myosin Va class is involved in the cytoplasmic transport of vesicles along actin filaments to membrane docking sites [56]. MYO5A is a member of the myosin V heavy-chain class of actin-based motor proteins and may play an important role in potassium channel functioning via modulation of trafficking pathways and cell surface density of Kv1.5 in the myocardium [57]. A change in the expression levels of several genes encoding proteins closely related to the actin structure was observed. In contrast to an upregulation of *MYO5A* mRNA levels, *ABLIM1*, *TMOD1*, *XIRP1*, and *PHACTR1* were downregulated during in vitro culture. Regulation of actin filament organization and dynamics is important for cellular architecture and numerous biological functions, including muscle contraction. Actin-binding LIM 1 (abLIM1, gene: *ABLIM1*) is a cytoskeletal protein that has been implicated in interactions between actin filaments and cytoplasmic targets [58]. Additionally, this actin-binding intracellular protein is localized to the sarcomeric Z-disk of the myocardium, playing a role in force transmission and muscle integrity [59]. Tropomodulin1 (Tmod1, gene: *TMOD1*) is an actin-capping protein necessary in actin thin-filament pointed-end dynamics and length in striated muscle, including cardiac muscle [60]. Bliss et al. demonstrated that in cardiac myocytes, the length regulatory function is modulated by the Tmod1 phosphorylation state [61]. Filamins (FLNs) are large dimeric actin-binding proteins, and as scaffolding proteins, FLNs regulate actin cytoskeleton remodeling [62]. Employing transcriptomic analyses, we indicated increasing levels of *FLNB*, encoding filamin B isoform. Interestingly, Xin actin-binding repeat-containing protein (gene: *XIRP1*), an important filamins ligand affecting the behavior of FLNs family members in Z-discs and myotendinous junctions of striated muscle cells [63], showed significantly decreased expression of the transcripts under the provided culture conditions. Leber et al. emphasized the role and correlation between Xin actin-binding repeat-containing protein and filamin family members in myofibrillar myopathy and skeletal and cardiac diseases [64]. Finally, it was demonstrated that lower mRNA levels of the phosphatase and actin regulating protein 1 were encoded by the *PHACTR1* gene during cell culture. *PHACTR1* transcript expression was originally observed in brain tissue [65], whereas subsequent studies indicated the presence of different variants of transcripts also in other types of tissues, including heart and coronary vessels [66]. The PHACTR protein family interacts directly with actin and protein phosphatase 1 (PP1) [65]. By binding to PP1, PHACTR1 acts as an inhibitor of its activity. Other results indicated a role of PHACTR1 in altering the structure of actin in HUVECs cell lines, where a reduction of F-actin filament numbers and repartitioning as well as increasing cell protrusion dynamics were observed [67]. The authors suggest that through its correlation with VEGF, PHACTR1 is a key component in the angiogenic process [67]. PHACTR1 function is linked with atherosclerosis-relevant phenotypes, such as angiogenesis [68], extracellular matrix protein production [69], and inflammation [70] in vitro. Numerous scientific reports emphasize the connection of *PHACTR1* transcript expression with coronary artery disease (CAD) initiation and progression [66,71–73].

Other proteins crucial for the structure of the heart tissue, whose transcript levels decreased significantly under in vitro culture conditions, are dystrophin (gene: *DMD*) and calsequestrin 2 (gene: *CASQ2*). The protein encoded by the *DMD* gene forms a component of the dystrophin-associated glycoprotein complex (DGC), which bridges the inner cytoskeleton and the ECM [74]. Heart muscles lacking functional dystrophin are mechanically weak, and contraction of cardiac myocytes leads to membrane damage, which affects the calcium channels, resulting in increased levels of intracellular calcium and activating proteases that consequently degrade contractile proteins, promoting cellular death and fibrosis [75]. Calcium handling in heart tissue belongs to the roles of calsequestrin family members. CASQ2 is localized to the sarcoplasmic reticulum (SR) in myocardium and functions as a calcium storage protein [76]. Refaat et al. showed a correlation between

CASQ2 gene polymorphism and sudden cardiac arrest (SCA) due to ventricular arrhythmias (VA) in patients with CAD [77].

The final aim of our project was to develop a strong background to understand the in vitro situation of porcine cardiomyocytes as a mode that will eventually be applicable in clinical practice. Atrial appendices, either left or right, may be harvested during minimally invasive procedures without any consequences. Moreover, during routine cardiac procedures employed in cardio-pulmonary bypass (all standard cardiac surgical procedures on the cardiac valves, aorta, heart transplantations, and many congenital malformations), the right atrial appendage is usually removed (the site of choice to introduce venous canula). The volume of the removed atrium can reach even a few square centimeters. Contrary to the atrium, in daily cardiac surgical practice, it is not possible to harvest such an enormously large volume from the left or right ventricular myocardium. The biggest biopsies are taken routinely from the right aspect of the ventricular septum in patients after heart transplantation to check for rejection (endomyocardial biopsies). However, the volume of the largest of them is approximately 1 mm^3. Therefore, the best candidate source of cells is the atrial myocardium.

The transcriptomic profile of genes encoding important structural proteins clearly demonstrates that long-term 2D culture conditions do not sufficiently allow for the reconstruction of important elements of the cytoskeleton. Moreover, further analysis of the observed changes at the protein level will provide more complete information about changes occurring during the conducted long-term primary cell culture. The present report is an initial step in developing a novel strategy to understand the in vitro situation of porcine cardiomyocytes. Our goal was to better understand the molecular background of the physiological potential of myocardial regeneration in an animal model of healthy hearts. The initiated research is required to be continued through, among others, determination of proteins for which changes at the transcript level have been considered essential for structural changes in cells. Similarly, the implementation of analyses using 3D cultures as well as functional analyses should be the next step to characterize CMs under long-term in vitro culture conditions. Significantly decreased mRNA expression levels indicate a potential problem with the lack of specific "building blocks" for intracellular structures, and thus cardiomyocytes' development and final maturation may be limited.

5. Conclusions

The effect of long-term primary cell culture conditions on the expression levels of gene transcripts encoding important structural proteins demonstrates increasing mRNA expression levels of genes encoding collagen subunits (*COL5A2*, *COL8A1*, and *COL12A1*) or tenascin-C (*TNC*), which may create a favorable environment for reconstruction of the ECM structures. A significant limit of the applied 2D in vitro culture conditions is the significantly lower expression of some pivotal transcripts responsible for the formation of intracellular structures. Decreased levels of genes such as *DMD*, *CASQ2*, *ABLIM1*, *TMOD1*, *XIRP1*, and *PHACTR1* confirm the necessity for further development and optimization of cardiomyocyte culture conditions. The conclusions drawn from the results are an important cognitive step in revealing the therapeutic potential of the cells.

Supplementary Materials: The following supporting information can be downloaded at: https://www.mdpi.com/article/10.3390/genes13071205/s1, Table S1: The list of differentially expressed genes of interest analyzed in this study. Quantitative data are presented as LogFC. All the presented sample means were deemed to be statistically significant ($p < 0.05$). Figure S1: The flow cytometry gating strategy. Figure S2: Enrichment analysis. In the clustergram, enriched GO terms are the columns, input genes are the rows, and cells in the matrix indicate if a gene is associated with a term.

Author Contributions: Conceptualization, M.J.N., D.B. and B.K.; methodology, M.J.N., M.Z., B.P., M.K. and B.K.; software, K.J. and M.J.N.; validation, M.J.N. and P.M.; formal analysis, M.J., D.B. and B.K.; investigation, M.J.N., M.K. and K.J.; resources, M.J. and B.K.; data curation, D.B. and B.K.; writing—original draft preparation, M.J.N.; writing—review and editing, K.J., B.P., P.M., M.K. and B.K.; visualization, M.J.N., M.Z. and K.J.; supervision, M.J., D.B. and B.K.; project administration, B.K.; funding acquisition, M.J. and P.M. All authors have read and agreed to the published version of the manuscript.

Funding: This research was funded in part by USDA/NIFA multistate project NC 1184.

Institutional Review Board Statement: Not applicable.

Informed Consent Statement: Not applicable.

Data Availability Statement: Not applicable.

Conflicts of Interest: The authors declare no conflict of interest.

References

1. Gummert, J.; Barten, M.; Rahmel, A.; Doll, S.; Krakor, R.; Garbade, J.; Autschbach, R. Therapy for Heart Failure: The Leipzig Approach. *Thorac. Cardiovasc. Surg.* **2017**, *65*, S205–S208. [CrossRef] [PubMed]
2. Ratajczak, M.Z.; Zuba-Surma, E.K.; Wysoczynski, M.; Wan, W.; Ratajczak, J.; Wojakowski, W.; Kucia, M. Hunt for pluripotent stem cell—Regenerative medicine search for almighty cell. *J. Autoimmun.* **2008**, *30*, 151–162. [CrossRef] [PubMed]
3. Bergmann, O.; Bhardwaj, R.D.; Bernard, S.; Zdunek, S.; Barnabé-Heider, F.; Walsh, S.; Zupicich, J.; Alkass, K.; Buchholz, B.A.; Druid, H.; et al. Evidence for cardiomyocyte renewal in humans. *Science* **2009**, *324*, 98–102. [CrossRef] [PubMed]
4. Rasmussen, T.L.; Raveendran, G.; Zhang, J.; Garry, D.J. Getting to the heart of myocardial stem cells and cell therapy. *Circulation* **2011**, *123*, 1771–1779. [CrossRef] [PubMed]
5. Senyo, S.E.; Steinhauser, M.L.; Pizzimenti, C.L.; Yang, V.K.; Cai, L.; Wang, M.; Wu, T.D.; Guerquin-Kern, J.L.; Lechene, C.P.; Lee, R.T. Mammalian heart renewal by pre-existing cardiomyocytes. *Nature* **2013**, *493*, 433–436. [CrossRef]
6. Ellison, G.M.; Vicinanza, C.; Smith, A.J.; Aquila, I.; Leone, A.; Waring, C.D.; Henning, B.J.; Stirparo, G.G.; Papait, R.; Scarfò, M.; et al. Adult c-kitpos Cardiac Stem Cells Are Necessary and Sufficient for Functional Cardiac Regeneration and Repair. *Cell* **2013**, *154*, 827–842. [CrossRef]
7. Hamdani, N.; Kooij, V.; Van Dijk, S.; Merkus, D.; Paulus, W.J.; Dos Remedios, C.; Duncker, D.J.; Stienen, G.J.M.; Van Der Velden, J. Sarcomeric dysfunction in heart failure. *Cardiovasc. Res.* **2008**, *77*, 649–658. [CrossRef]
8. Golob, M.; Moss, R.L.; Chesler, N.C. Cardiac Tissue Structure, Properties, and Performance: A Materials Science Perspective. *Ann. Biomed. Eng.* **2014**, *42*, 2003–2013. [CrossRef]
9. Kuznetsov, A.V.; Javadov, S.; Grimm, M.; Margreiter, R.; Ausserlechner, M.J.; Hagenbuchner, J. Crosstalk between Mitochondria and Cytoskeleton in Cardiac Cells. *Cells* **2020**, *9*, 222. [CrossRef]
10. Song, R.; Zhang, L. Cardiac ECM: Its epigenetic regulation and role in heart development and repair. *Int. J. Mol. Sci.* **2020**, *21*, 8610. [CrossRef]
11. Sreejit, P.; Kumar, S.; Verma, R.S. An improved protocol for primary culture of cardiomyocyte from neonatal mice. *Vitr. Cell. Dev. Biol.—Anim.* **2008**, *44*, 45–50. [CrossRef] [PubMed]
12. Wysoczynski, M.; Guo, Y.; Moore, J.B.; Muthusamy, S.; Li, Q.; Nasr, M.; Li, H.; Nong, Y.; Wu, W.; Tomlin, A.A.; et al. Myocardial Reparative Properties of Cardiac Mesenchymal Cells Isolated on the Basis of Adherence. *J. Am. Coll. Cardiol.* **2017**, *69*, 1824–1838. [CrossRef] [PubMed]
13. Chomczynski, P.; Sacchi, N. Single-step method of RNA isolation by acid guanidinium thiocyanate-phenol-chloroform extraction. *Anal. Biochem.* **1987**, *162*, 156–159. [CrossRef]
14. Kałuzna, S.; Nawrocki, M.J.; Bryl, R.; Stefańska, K.; Jemielity, M.; Mozdziak, P.; Nowicki, M.; Perek, B. Study of the expression of genes associated with post-translational changes in histones in the internal thoracic artery and the saphenous vein grafts used in coronary artery bypass grafting procedure. *Med. J. Cell Biol.* **2020**, *8*, 183–189. [CrossRef]
15. Nawrocki, M.J.; Jopek, K.; Zdun, M.; Mozdziak, P.; Jemielity, M.; Perek, B.; Bukowska, D.; Kempisty, B. Expression profile of genes encoding proteins involved in regulation of vasculature development and heart muscle morphogenesis—A transcriptomic approach based on a porcine model. *Int. J. Mol. Sci.* **2021**, *22*, 8794. [CrossRef]
16. Kałuzna, S.; Nawrocki, M.J.; Jopek, K.; Hutchings, G.; Perek, B.; Jemielity, M.; Kempisty, B.; Malińska, A.; Mozdziak, P.; Nowicki, M. In search of markers useful for evaluation of graft patency-molecular analysis of "muscle system process" for internal thoracic artery and saphenous vein conduits. *Med. J. Cell Biol.* **2020**, *8*, 12–23. [CrossRef]
17. Nawrocki, M.J.; Kałuzna, S.; Jopek, K.; Hutchings, G.; Perek, B.; Jemielity, M.; Malińska, A.; Kempisty, B.; Mozdziak, P.; Nowicki, M. Aortocoronary conduits may show a different inflammatory response-comparative study at transcript level. *Med. J. Cell Biol.* **2020**, *8*, 24–34. [CrossRef]

18. Ichimura, H.; Kadota, S.; Kashihara, T.; Yamada, M.; Ito, K.; Kobayashi, H.; Tanaka, Y.; Shiba, N.; Chuma, S.; Tohyama, S.; et al. Increased predominance of the matured ventricular subtype in embryonic stem cell-derived cardiomyocytes in vivo. *Sci. Rep.* **2020**, *10*, 11883. [CrossRef]
19. Qian, L.; Huang, Y.; Spencer, C.I.; Foley, A.; Vedantham, V.; Liu, L.; Conway, S.J.; Fu, J.; Srivastava, D. In vivo reprogramming of murine cardiac fibroblasts into induced cardiomyocytes. *Nature* **2012**, *485*, 593–598. [CrossRef]
20. Hassan, N.; Tchao, J.; Tobita, K. Concise Review: Skeletal Muscle Stem Cells and Cardiac Lineage: Potential for Heart Repair. *Stem Cells Transl. Med.* **2014**, *3*, 183. [CrossRef]
21. Chen, E.Y.; Tan, C.M.; Kou, Y.; Duan, Q.; Wang, Z.; Meirelles, G.V.; Clark, N.R.; Ma'ayan, A. Enrichr: Interactive and collaborative HTML5 gene list enrichment analysis tool. *BMC Bioinform.* **2013**, *14*, 128. [CrossRef] [PubMed]
22. Kuleshov, M.V.; Jones, M.R.; Rouillard, A.D.; Fernandez, N.F.; Duan, Q.; Wang, Z.; Koplev, S.; Jenkins, S.L.; Jagodnik, K.M.; Lachmann, A.; et al. Enrichr: A comprehensive gene set enrichment analysis web server 2016 update. *Nucleic Acids Res.* **2016**, *44*, W90–W97. [CrossRef] [PubMed]
23. Xie, Z.; Bailey, A.; Kuleshov, M.V.; Clarke, D.J.B.; Evangelista, J.E.; Jenkins, S.L.; Lachmann, A.; Wojciechowicz, M.L.; Kropiwnicki, E.; Jagodnik, K.M.; et al. Gene Set Knowledge Discovery with Enrichr. *Curr. Protoc.* **2021**, *1*, e90. [CrossRef] [PubMed]
24. Hutchings, G.; Nawrocki, M.J.; Mozdziak, P.; Kempisty, B. Cardiac Stem Cell Therapy, Resident Progenitor Cells and the role of Cellular Signalling; a Review. *Med. J. Cell Biol.* **2019**, *7*, 112–118. [CrossRef]
25. Nawrocki, M.J.; Bryl, R.; Kałużna, S.; Stefańska, K.; Stelmach, B.; Jemielity, M.; Perek, B.; Bukowska, D.; Mozdziak, P.; Petitte, J.N.; et al. Increased transcript expression levels of DNA methyltransferases type 1 and 3A during cardiac muscle long-term cell culture. *Med. J. Cell Biol.* **2021**, *9*, 27–32. [CrossRef]
26. Bonnans, C.; Chou, J.; Werb, Z. Remodelling the extracellular matrix in development and disease. *Nat. Rev. Mol. Cell Biol.* **2014**, *15*, 786–801. [CrossRef]
27. Li, L.; Zhao, Q.; Kong, W. Extracellular matrix remodeling and cardiac fibrosis. *Matrix Biol.* **2018**, *68*, 490–506. [CrossRef]
28. Frangogiannis, N.G. The extracellular matrix in ischemic and nonischemic heart failure. *Circ. Res.* **2019**, *125*, 117–146. [CrossRef]
29. Nunes, S.S.; Miklas, J.W.; Liu, J.; Aschar-Sobbi, R.; Xiao, Y.; Zhang, B.; Jiang, J.; Massé, S.; Gagliardi, M.; Hsieh, A.; et al. Biowire: A platform for maturation of human pluripotent stem cell-derived cardiomyocytes. *Nat. Methods* **2013**, *10*, 781–787. [CrossRef]
30. Feaster, T.K.; Cadar, A.G.; Wang, L.; Williams, C.H.; Chun, Y.W.; Hempel, J.E.; Bloodworth, N.; Merryman, W.D.; Lim, C.C.; Wu, J.C.; et al. A method for the generation of single contracting human-induced pluripotent stem cell-derived cardiomyocytes. *Circ. Res.* **2015**, *117*, 995–1000. [CrossRef]
31. Chen, Y.; Wang, J.; Shen, B.; Chan, C.W.Y.; Wang, C.; Zhao, Y.; Chan, H.N.; Tian, Q.; Chen, Y.; Yao, C.; et al. Engineering a freestanding biomimetic cardiac patch using biodegradable poly(lactic-co-glycolic acid) (PLGA) and human embryonic stem cell-derived ventricular cardiomyocytes (hESC-VCMs). *Macromol. Biosci.* **2015**, *15*, 426–436. [CrossRef] [PubMed]
32. Herron, T.J.; Da Rocha, A.M.; Campbell, K.F.; Ponce-Balbuena, D.; Willis, B.C.; Guerrero-Serna, G.; Liu, Q.; Klos, M.; Musa, H.; Zarzoso, M.; et al. Extracellular matrix-mediated maturation of human pluripotent stem cell-derived cardiac monolayer structure and electrophysiological function. *Circ. Arrhythmia Electrophysiol.* **2016**, *9*, e003638. [CrossRef] [PubMed]
33. Fong, A.H.; Romero-López, M.; Heylman, C.M.; Keating, M.; Tran, D.; Sobrino, A.; Tran, A.Q.; Pham, H.H.; Fimbres, C.; Gershon, P.D.; et al. Three-Dimensional Adult Cardiac Extracellular Matrix Promotes Maturation of Human Induced Pluripotent Stem Cell-Derived Cardiomyocytes. *Tissue Eng.* **2016**, *22*, 1016–1025. [CrossRef] [PubMed]
34. Garreta, E.; de Oñate, L.; Fernández-Santos, M.E.; Oria, R.; Tarantino, C.; Climent, A.M.; Marco, A.; Samitier, M.; Martínez, E.; Valls-Margarit, M.; et al. Myocardial commitment from human pluripotent stem cells: Rapid production of human heart grafts. *Biomaterials* **2016**, *98*, 64–78. [CrossRef]
35. Guyette, J.P.; Charest, J.M.; Mills, R.W.; Jank, B.J.; Moser, P.T.; Gilpin, S.E.; Gershlak, J.R.; Okamoto, T.; Gonzalez, G.; Milan, D.J.; et al. Bioengineering Human Myocardium on Native Extracellular Matrix. *Circ. Res.* **2016**, *118*, 56–72. [CrossRef] [PubMed]
36. Mak, K.M.; Png, C.Y.M.; Lee, D.J. Type V Collagen in Health, Disease, and Fibrosis. *Anat. Rec.* **2016**, *299*, 613–629. [CrossRef]
37. Fan, D.; Takawale, A.; Lee, J.; Kassiri, Z. Cardiac fibroblasts, fibrosis and extracellular matrix remodeling in heart disease. *Fibrogenes. Tissue Repair* **2012**, *5*, 1–13. [CrossRef]
38. Azuaje, F.; Zhang, L.; Jeanty, C.; Puhl, S.L.; Rodius, S.; Wagner, D.R. Analysis of a gene co-expression network establishes robust association between Col5a2 and ischemic heart disease. *BMC Med. Genom.* **2013**, *6*, 13. [CrossRef]
39. Liu, X.; Keene, D.R.; Jaenisch, R.; Ramirez, F. Targeted mutation in the col5a2 gene reveals a regulatory role for type V collagen during matrix assembly. *Nat. Genet.* **1995**, *9*, 31–36. [CrossRef]
40. Malfait, F.; Wenstrup, R.J.; De Paepe, A. Clinical and genetic aspects of Ehlers-Danlos syndrome, classic type. *Genet. Med.* **2010**, *12*, 597–605. [CrossRef]
41. Wang, H.B.; Huang, R.; Yang, K.; Xu, M.; Fan, D.; Liu, M.X.; Huang, S.H.; Liu, L.B.; Wu, H.M.; Tang, Q.Z. Identification of differentially expressed genes and preliminary validations in cardiac pathological remodeling induced by transverse aortic constriction. *Int. J. Mol. Med.* **2019**, *44*, 1447–1461. [CrossRef] [PubMed]
42. Parrott, M.E.; Aljrbi, E.; Biederman, D.L.; Montalvo, R.N.; Barth, J.L.; LaVoie, H.A. Feature Article: Maternal cardiac messenger RNA expression of extracellular matrix proteins in mice during pregnancy and the postpartum period. *Exp. Biol. Med.* **2018**, *243*, 1220–1232. [CrossRef] [PubMed]

43. Skrbic, B.; Engebretsen, K.V.T.; Strand, M.E.; Lunde, I.G.; Herum, K.M.; Marstein, H.S.; Sjaastad, I.; Lunde, P.K.; Carlson, C.R.; Christensen, G.; et al. Lack of collagen VIII reduces fibrosis and promotes early mortality and cardiac dilatation in pressure overload in mice. *Cardiovasc. Res.* **2015**, *106*, 32–42. [CrossRef] [PubMed]

44. Chen, C.; Manso, A.M.; Ross, R.S. Talin and Kindlin as Integrin-Activating Proteins: Focus on the Heart. *Pediatr. Cardiol.* **2019**, *40*, 1401–1409. [CrossRef] [PubMed]

45. Liu, X.; Shang, H.; Li, B.; Zhao, L.; Hua, Y.; Wu, K.; Hu, M.; Fan, T. Exploration and validation of hub genes and pathways in the progression of hypoplastic left heart syndrome via weighted gene co-expression network analysis. *BMC Cardiovasc. Disord.* **2021**, *21*, 300. [CrossRef]

46. Chen, Y.M.; Li, H.; Fan, Y.; Zhang, Q.J.; Li, X.; Wu, L.J.; Chen, Z.J.; Zhu, C.; Qian, L.M. Identification of differentially expressed lncRNAs involved in transient regeneration of the neonatal C57BL/6J mouse heart by next-generation high-throughput RNA sequencing. *Oncotarget* **2017**, *8*, 28052–28062. [CrossRef]

47. Imanaka-Yoshida, K. Tenascin-c in heart diseases—The role of inflammation. *Int. J. Mol. Sci.* **2021**, *22*, 5828. [CrossRef]

48. Kanagala, P.; Arnold, J.R.; Khan, J.N.; Singh, A.; Gulsin, G.S.; Chan, D.C.S.; Cheng, A.S.H.; Yang, J.; Li, Z.; Gupta, P.; et al. Plasma Tenascin-C: A prognostic biomarker in heart failure with preserved ejection fraction. *Biomarkers* **2020**, *25*, 556–565. [CrossRef]

49. Imanaka-Yoshida, K.; Tawara, I.; Yoshida, T. Tenascin-C in cardiac disease: A sophisticated controller of inflammation, repair, and fibrosis. *Am. J. Physiol.—Cell Physiol.* **2020**, *319*, C781–C796. [CrossRef]

50. Radosinska, J.; Barancik, M.; Vrbjar, N. Heart failure and role of circulating MMP-2 and MMP-9. *Panminerva Med.* **2017**, *59*, 241–253. [CrossRef]

51. Hughes, B.G.; Schulz, R. Targeting MMP-2 to treat ischemic heart injury. *Basic Res. Cardiol.* **2014**, *109*, 1–19. [CrossRef] [PubMed]

52. Spinale, F.G. Myocardial matrix remodeling and the matrix metalloproteinases: Influence on cardiac form and function. *Physiol. Rev.* **2007**, *87*, 1285–1342. [CrossRef] [PubMed]

53. Kiczak, L.; Tomaszek, A.; Bania, J.; Paslawska, U.; Zacharski, M.; Noszczyk-Nowak, A.; Janiszewski, A.; Skrzypczak, P.; Ardehali, H.; Jankowska, E.A.; et al. Expression and complex formation of MMP9, MMP2, NGAL, and TIMP1 in porcine myocardium but not in skeletal muscles in male pigs with tachycardia-induced systolic heart failure. *Biomed. Res. Int.* **2013**, *2013*, 283856. [CrossRef] [PubMed]

54. Hayashidani, S.; Tsutsui, H.; Ikeuchi, M.; Shiomi, T.; Matsusaka, H.; Kubota, T.; Imanaka-Yoshida, K.; Itoh, T.; Takeshita, A. Targeted deletion of MMP-2 attenuates early LV rupture and late remodeling after experimental myocardial infarction. *Am. J. Physiol. Heart Circ. Physiol.* **2003**, *285*, H1229–H1235. [CrossRef]

55. Apple, K.A.; Yarbrough, W.M.; Mukherjee, R.; Deschamps, A.M.; Escobar, P.G.; Mingoia, J.T.; Sample, J.A.; Hendrick, J.W.; Dowdy, K.B.; McLean, J.E.; et al. Selective targeting of matrix metalloproteinase inhibition in post-infarction myocardial remodeling. *J. Cardiovasc. Pharmacol.* **2006**, *47*, 228–235. [CrossRef]

56. Langford, G.M. Myosin-V, a versatile motor for short-range vesicle transport. *Traffic* **2002**, *3*, 859–865. [CrossRef]

57. Schumacher-Bass, S.M.; Vesely, E.D.; Zhang, L.; Ryland, K.E.; McEwen, D.P.; Chan, P.J.; Frasier, C.R.; McIntyre, J.C.; Shaw, R.M.; Martens, J.R. Role for myosin-V motor proteins in the selective delivery of Kv channel isoforms to the membrane surface of cardiac myocytes. *Circ. Res.* **2014**, *114*, 982–992. [CrossRef]

58. Krupp, M.; Weinmann, A.; Galle, P.R.; Teufel, A. Actin binding LIM protein 3 (abLIM3). *Int. J. Mol. Med.* **2006**, *17*, 129–133. [CrossRef]

59. Roof, D.J.; Hayes, A.; Adamian, M.; Chishti, A.H.; Li, T. Molecular characterization of abLIM, a novel actin-binding and double zinc finger protein. *J. Cell Biol.* **1997**, *138*, 575–588. [CrossRef]

60. Kostyukova, A.S.; Rapp, B.A.; Choy, A.; Greenfield, N.J.; Hitchcock-DeGregori, S.E. Structural requirements of tropomodulin for tropomyosin binding and actin filament capping. *Biochemistry* **2005**, *44*, 4905–4910. [CrossRef]

61. Bliss, K.T.; Tsukada, T.; Novak, S.M.; Dorovkov, M.V.; Shah, S.P.; Nworu, C.; Kostyukova, A.S.; Gregorio, C.C. Phosphorylation of tropomodulin1 contributes to the regulation of actin filament architecture in cardiac muscle. *FASEB J.* **2014**, *28*, 3987–3995. [CrossRef] [PubMed]

62. Rosa, J.P.; Raslova, H.; Bryckaert, M. Filamin A: Key actor in platelet biology. *Blood* **2019**, *134*, 1279–1288. [CrossRef] [PubMed]

63. Kley, R.A.; Maerkens, A.; Leber, Y.; Theis, V.; Schreiner, A.; Van Der Ven, P.F.M.; Uszkoreit, J.; Stephan, C.; Eulitz, S.; Euler, N.; et al. A combined laser microdissection and mass spectrometry approach reveals new disease relevant proteins accumulating in aggregates of filaminopathy patients. *Mol. Cell. Proteom.* **2013**, *12*, 215–227. [CrossRef]

64. Leber, Y.; Ruparelia, A.A.; Kirfel, G.; van der Ven, P.F.M.; Hoffmann, B.; Merkel, R.; Bryson-Richardson, R.J.; Fürst, D.O. Filamin C is a highly dynamic protein associated with fast repair of myofibrillar microdamage. *Hum. Mol. Genet.* **2016**, *25*, 2776–2788. [CrossRef]

65. Allen, P.B.; Greenfield, A.T.; Svenningsson, P.; Haspeslagh, D.C.; Greengard, P. Phactrs 1–4: A family of protein phosphatase 1 and actin regulatory proteins. *Proc. Natl. Acad. Sci. USA* **2004**, *101*, 7187–7192. [CrossRef] [PubMed]

66. Codina-Fauteux, V.A.; Beaudoin, M.; Lalonde, S.; Lo, K.S.; Lettre, G. PHACTR1 splicing isoforms and eQTLs in atherosclerosis-relevant human cells. *BMC Med. Genet.* **2018**, *19*, 97. [CrossRef]

67. Allain, B.; Jarray, R.; Borriello, L.; Leforban, B.; Dufour, S.; Liu, W.Q.; Pamonsinlapatham, P.; Bianco, S.; Larghero, J.Ô.; Hadj-Slimane, R.; et al. Neuropilin-1 regulates a new VEGF-induced gene, Phactr-1, which controls tubulogenesis and modulates lamellipodial dynamics in human endothelial cells. *Cell. Signal.* **2012**, *24*, 214–223. [CrossRef]

68. Jarray, R.; Allain, B.; Borriello, L.; Biard, D.; Loukaci, A.; Larghero, J.; Hadj-Slimane, R.; Garbay, C.; Lepelletier, Y.; Raynaud, F. Depletion of the novel protein PHACTR-1 from human endothelial cells abolishes tube formation and induces cell death receptor apoptosis. *Biochimie* **2011**, *93*, 1668–1675. [CrossRef]
69. Jarray, R.; Pavoni, S.; Borriello, L.; Allain, B.; Lopez, N.; Bianco, S.; Liu, W.Q.; Biard, D.; Demange, L.; Hermine, O.; et al. Disruption of phactr-1 pathway triggers pro-inflammatory and pro-atherogenic factors: New insights in atherosclerosis development. *Biochimie* **2015**, *118*, 151–161. [CrossRef]
70. Reschen, M.E.; Lin, D.; Chalisey, A.; Soilleux, E.J.; O'Callaghan, C.A. Genetic and environmental risk factors for atherosclerosis regulate transcription of phosphatase and actin regulating gene PHACTR1. *Atherosclerosis* **2016**, *250*, 95–105. [CrossRef]
71. Hao, K.; Ermel, R.; Li, L.; Amadori, L.; Koplev, S.; Franzén, O.; D'Escamard, V.; Chandel, N.; Wolhuter, K.; Bryce, N.S.; et al. Integrative Prioritization of Causal Genes for Coronary Artery Disease. *Circ. Genom. Precis. Med.* **2022**, *15*, E003365. [CrossRef] [PubMed]
72. Georges, A.; Yang, M.L.; Berrandou, T.E.; Bakker, M.K.; Dikilitas, O.; Kiando, S.R.; Ma, L.; Satterfield, B.A.; Sengupta, S.; Yu, M.; et al. Genetic investigation of fibromuscular dysplasia identifies risk loci and shared genetics with common cardiovascular diseases. *Nat. Commun.* **2021**, *12*, 1–46. [CrossRef] [PubMed]
73. Saw, J.; Yang, M.L.; Trinder, M.; Tcheandjieu, C.; Xu, C.; Starovoytov, A.; Birt, I.; Mathis, M.R.; Hunker, K.L.; Schmidt, E.M.; et al. Chromosome 1q21.2 and additional loci influence risk of spontaneous coronary artery dissection and myocardial infarction. *Nat. Commun.* **2020**, *11*, 1–14. [CrossRef] [PubMed]
74. D'Amario, D.; Amodeo, A.; Adorisio, R.; Tiziano, F.D.; Leone, A.M.; Perri, G.; Bruno, P.; Massetti, M.; Ferlini, A.; Pane, M.; et al. A current approach to heart failure in Duchenne muscular dystrophy. *Heart* **2017**, *103*, 1770–1779. [CrossRef]
75. Kamdar, F.; Garry, D.J. Dystrophin-Deficient Cardiomyopathy. *J. Am. Coll. Cardiol.* **2016**, *67*, 2533–2546. [CrossRef]
76. Faggioni, M.; Knollmann, B.C. Calsequestrin 2 and arrhythmias. *Am. J. Physiol.—Heart Circ. Physiol.* **2012**, *302*, H1250–H1260. [CrossRef]
77. Refaat, M.M.; Aouizerat, B.E.; Pullinger, C.R.; Malloy, M.; Kane, J.; Tseng, Z.H. Association of CASQ2 polymorphisms with sudden cardiac arrest and heart failure in patients with coronary artery disease. *Heart Rhythm.* **2014**, *11*, 646–652. [CrossRef]

Article

Characterization of a Read-through Fusion Transcript, BCL2L2-PABPN1, Involved in Porcine Adipogenesis

Jiyuan Zhu [1], Zewei Yang [1], Wanjun Hao [1], Jiaxin Li [1], Liang Wang [2], Jiqiao Xia [1], Dongjie Zhang [2], Di Liu [2,*] and Xiuqin Yang [1,*]

1   College of Animal Science and Technology, Northeast Agricultural University, Harbin 150030, China; zhujiyuan2008@163.com (J.Z.); yangzewei1997@163.com (Z.Y.); haowanjun1109@163.com (W.H.); ljxneau@163.com (J.L.); xiajiqiao365@163.com (J.X.)
2   Institute of Animal Husbandry, Heilongjiang Academy of Agricultural Sciences, Harbin 150086, China; wlwl448@163.com (L.W.); djzhang8109@163.com (D.Z.)
*   Correspondence: liudi1963@163.com (D.L.); xiuqinyang@neau.edu.cn (X.Y.); Tel.: +86-451-86677458 (D.L.); +86-451-55191738 (X.Y.)

Abstract: *cis*-Splicing of adjacent genes (*cis*-SAGe) has been involved in multiple physiological and pathological processes in humans. However, to the best of our knowledge, there is no report of *cis*-SAGe in adipogenic regulation. In this study, a *cis*-SAGe product, BCL2L2–PABPN1 (BP), was characterized in fat tissue of pigs with RT-PCR and RACE method. BP is an in-frame fusion product composed of 333 aa and all the functional domains of both parents. BP is highly conserved among species and rich in splicing variants. BP was found to promote proliferation and inhibit differentiation of primary porcine preadipocytes. A total of 3074/44 differentially expressed mRNAs (DEmRs)/known miRNAs (DEmiRs) were identified in porcine preadipocytes overexpressing BP through RNA-Seq analysis. Both DEmRs and target genes of DEmiRs were involved in various fat-related pathways with MAPK and PI3K-Akt being the top enriched. *PPP2CB, EGFR, Wnt5A* and *EHHADH* were hub genes among the fat-related pathways identified. Moreover, ssc-miR-339-3p was found to be critical for BP regulating adipogenesis through integrated analysis of mRNA and miRNA data. The results highlight the role of *cis*-SAGe in adipogenesis and contribute to further revealing the mechanisms underlying fat deposition, which will be conductive to human obesity control.

Keywords: adipogenesis; BCL2L2-PABPN1; chimeric RNA; *cis*-SAGe; genome-wide analysis; RNA-Seq

Citation: Zhu, J.; Yang, Z.; Hao, W.; Li, J.; Wang, L.; Xia, J.; Zhang, D.; Liu, D.; Yang, X. Characterization of a Read-through Fusion Transcript, BCL2L2-PABPN1, Involved in Porcine Adipogenesis. *Genes* **2022**, *13*, 445. https://doi.org/10.3390/genes13030445

Academic Editors: Katarzyna Piórkowska and Katarzyna Ropka-Molik

Received: 4 February 2022
Accepted: 26 February 2022
Published: 28 February 2022

Publisher's Note: MDPI stays neutral with regard to jurisdictional claims in published maps and institutional affiliations.

1. Introduction

Fat is a major factor affecting pig growth, development, and meat quality. Intramuscular fat (IMF) content is determinant of pork marbling and closely related to the juiciness, flavor and tenderness of pork. A suitable IMF content can bring a better taste and is important in improving pork quality [1]. However, back-fat thickness is negative related to lean meat yield [2]. The regulation of fat content and distribution in the body will bring major economic benefits to pig producers, which need to understand the mechanisms underlying fat deposition. Additionally, as an important endocrine organ, fat tissue plays key roles in maintaining body energy balance and glucose homeostasis [3], and is directly associated with some metabolic diseases, including diabetes and obesity. Pigs are similar to human beings in anatomy and physiology and have long been used as models in biomedical research [4–6]. Studies on adipogenesis in pigs will contribute to controlling metabolic diseases associated with fat.

It has been made clear that various transcription factors [7–9], signal transduction pathways [10–12], epigenetic factors [13,14], and functional RNAs [15,16] are involved in adipogenesis. However, adipogenesis is a complicated and precisely orchestrated process, and there are still many factors remaining to be identified before fully revealing the molecular mechanisms underlying adipogenesis. Chimeric RNA molecules are composed of exons from two independent genes. They can be produced by several mechanisms, including chromosome rearrangement, *cis*-splicing of adjacent genes (*cis*-SAGe), and *trans*-splicing [17]. *cis*-SAGe is cotranscription of adjacent genes coupled with intergenic splicing and forms read-through fusion transcripts [18,19]. Chimeric RNAs were first identified in tumor cells and once considered unique to tumors, which has focused researchers on their roles in carcinogenesis. They are involved in various tumors, and in some cases, can be used as diagnosis markers [20–22]. As research has progressed, chimeric RNAs have been found in normal tissues and can produce many fusion proteins, increasing greatly the complexity and diversity of the proteome. They can regulate gene expression, cell growth, vitality, and motility in normal physiological processes [23,24]. For example, PAX3–FOXO1 is needed for muscle lineage commitment [25,26], and DUS4L–BCAP29 is involved in neuronal differentiation [27].

The existing findings highlight the vital importance of chimeric RNAs and more researchers are paying attention to them. However, there are no studies on chimeric RNAs in adipogenesis in mammals. Here, we first identified a chimeric RNA produced by *cis*-SAGe, BCL2L2–PABPN1 (BP), in pigs and elucidated that it inhibited adipogenesis through MAPK and PI3K-Akt signaling pathways. The results highlight the role of read-through fusion transcripts in adipogenesis and contribute to further revealing the mechanisms underlying fat formation.

2. Materials and Methods

2.1. Animals, Nucleic Acid Isolation and cDNA Synthesis

All pigs were from the Institute of Animal Husbandry, Heilongjiang Academy of Agricultural Sciences (Harbin, China). The animal study was reviewed and approved by the Animal Care Committee of Northeast Agricultural University (Harbin, China). Fat tissues were obtained from 6-month-old Min and Yorkshire pigs raised in the same condition or from newborn Min pigs. RNA was extracted with TRIzol reagent (Invitrogen, Carlsbad, CA, USA) and reverse transcribed into cDNA with PrimeScriptTM 1st Strand cDNA Synthesis Kit (Takara, Dalian, China). In analysis of chimera formation, the reverse transcription (RT) primer was random 6 mers provided by the kit.

2.2. Hematoxylin and Eosin Staining and Triglyceride Assay

Adipose tissues were fixed in 4% paraformaldehyde solution, dehydrated in ethanol, and embedded in paraffin. 4 μm thickness section was sliced with HistoCore BIOCUT (Leica, Nussloch, Germany) and stained with hematoxylin and eosin (HE) for morphological analysis. Tissues from three pigs in each breed were used and more than five fields were chosen for morphological analysis. Adipocyte size was determined with Leica Application Suite V4 (Leica). Triglyceride (TG) contents were measured with an enzymatic TG assay kit (GPO-POD; Applygen, Beijing, China) according to the manufacturer's protocol.

2.3. Chimeric RNA Identification and cDNA Cloning

We previously obtained high-throughput paired-end RNA-seq data of fat tissues from Min and Yorkshire pigs [28], from which chimeric RNA was identified using ChimeraScan program [29] with the reference genome (*S. scrofa* 10.2) [30] using default parameters. Only read-through fusion candidates covering neighboring genes on the same strand of DNA were considered in this study. Other chimera candidates including inter-, intra-chromosomal, and adjacent ones were discarded.

Reverse transcription-polymerase chain reaction (RT-PCR) was used for validation of candidate BCL2L2-PABPN1 (BP), produced by B-cell lymphoma 2-like 2 protein (*BCL2L2*) and poly(A) binding protein nuclear 1 (*PABPN1*) genes, with specific primer pair, B1F/P1R, designed according to result of bioinformatic analysis and cDNA template from fat tissues. Another primer P2R designed according to porcine *PABPN1* mRNA (GenBank No. NM_001243548) was used to extend the chimera with B1F. The 5′ rapid amplification of cDNA ends (RACE) was used to clone the 5′ sequence using specific primers P1R and BPR, which is complementary to the junction of chimera with SMARTer RACE 5′/3′ kit (Takara). Primer P3F was used in the 3′ RACE reaction together with B1F.

Additionally, the 5′ RACE method was used to replenish the sequence of porcine *PABPN1* mRNA in which the outer and inner primers were P1R and P3R, respectively. P3R was complementary to exon 1 of porcine PABPN1. RT-PCR was performed with forward primer, P4F, complementary to the 5′ untranslated region (UTR) obtained and reverse primer, P4R, complementary to 3′ end of the chimera to verify the cDNA sequence of porcine *PABPN1*. Genomic structure was analyzed with BLAT program in UCSC genome (http://genome.ucsc.edu/, accessed on 9 March 2021). All primer sequences used in this study are listed in Table S1-1.

2.4. Primary Preadipocyte Isolation and Culture

Subcutaneous fat tissues were obtained after the newborn Min pigs were slaughtered, washed with sterile phosphate-buffered saline (PBS), and potentially contaminated muscle and connective tissue was carefully removed. After washing three times in PBS containing 1% penicillin–streptomycin (Invitrogen), fat tissues were cut into small pieces and digested with 0.1% type I collagenase (Invitrogen) for 40–50 min at 37 °C, then mixed with equal volumes of culture medium supplemented with penicillin–streptomycin and 10% fetal bovine serum (FBS) (Sigma, St. Louis, MO, USA), and filtered through 400-mesh filters. The filtrates were centrifuged at 1000 rpm for 5 min. The cell precipitation was resuspended with Dulbecco's modified Eagle's medium/Nutrient Mixture F-12 (DMEM/F12) containing 10% FBS and 1% penicillin–streptomycin. The medium was changed every 2 days until cells were grown to a desired density.

2.5. Preadipocyte Differentiation and Oil Red O Staining

To induce differentiation, DMEM/F12 was supplemented with 10% FBS, 0.5 mmol/L 3-isobytyl-1-methylxanthine, 1 μmol/L dexamethasone and 5 μg/mL insulin in which cells were incubated for 2 days. The cells were cultured with DMEM/F12 containing 10% FBS and 5 μg/mL insulin to maintain their differentiation until further analysis. The medium was changed every 2 days.

The differentiated adipocytes were stained with Oil Red O kit (Leagene, Beijing, China). The stained lipid droplets were viewed under a light microscope and photographed (Carl Zeiss AG, Jena, Germany). For quantification analysis, cellular Oil Red O was extracted with isopropanol and measured with optical absorbance at 510 nm.

To evaluate effects of BP on preadipocyte differentiation, overexpression vector of BP (pCMV-HA-BP) was constructed with pCMV-HA backbone at sites of *EcoR* I and *Kpn* I and transiently transfected with Lipofectamine 2000 (Invitrogen) according to manufacturer's protocol. At 24 h after transfection, cells were subjected to differentiation inducement.

2.6. Real-time Quantitative PCR

Real-time quantitative PCR (qPCR) was performed with TB Green® Premix Ex Taq™ reagent kit (Takara). The PCR volume and reaction program were set strictly according to the manufacturer's instructions. β-Actin was used as a reference and the relative expression level was analyzed with the $2^{-\Delta\Delta Ct}$ method [31].

2.7. Cell Counting Kit-8 Assay

Porcine preadipocytes were transiently transfected with pCMV-HA-BP or empty vector pCMV-HA for 24 h and further cultured until the CCK-8 assay was performed. In CCK-8 assays, cells were incubated with 10% CCK-8 (Beyotime, Shanghai, China) in complete medium for 2 h at 37 °C. The absorbance of cells was measured at 450 nm using a Tecan Microplate Reader Infinite F50 (Tecan GENios, Mannendorf, Switzerland).

2.8. Flow Cytometry

Porcine preadipocytes were inoculated in six-well plates at a density of 1×10^6 cells per well and cultured for 24 h. Cells were transfected with pCMV-HA-BP or empty vector and cultured for another 24 h. After digested with trypsin, cells were washed with PBS, and stained with cell cycle staining Kit (MultiSciences, Hangzhou, China). Then the cell cycle was analyzed with FACSCalibur Flow Cytometer (Becton Dickinson, Franklin Lakes, NJ, USA).

2.9. Illumina-Seq Library Construction and Sequencing

The recombinant adenoviruses were constructed using the AdEasy system (Hanbio, Shanghai, China) as described by He and colleagues (1998). BP CDS were inserted into the shuttle plasmid containing an enhanced green fluorescent protein (EGFP) and cytomegalovirus promoters using *Kpn* I and *Xho* I sites, and homologous recombination was performed in *Escherichia coli* BJ5183 with adenoviral backbone pAdEasy 1. The recombinant adenovirus plasmid was packaged in HEK-293A cells after linearized with *Pac* I. Preadipocytes were infected with adenovirus virions at multiplicity of infection (MOI) of 300. At 48 h post-infection, cells were collected for RNA-Seq with Illumina NovoSeq 6000 platform (Illumina, San Diego, CA, USA) by Geneseeq Technology (Nanjing, China) using pair-end sequencing strategy according to the manufacturer's protocols. Cells treated with empty adenovirus were used as a control. A total of six Ribo-Zero RNA-sequencing libraries including overexpressing and control groups were constructed, each with three replicates.

2.10. Genome-Wide mRNA Analysis

The raw reads were processed as described elsewhere [32]. High quality reads were mapped to reference genome of *S. scrofa* (11.1) using HISAT2 program [33]. Transcript abundances were quantified with StringTie software [34], and normalized with FPKM (Fragments per kilobase of transcript per million mapped reads) method across libraries. DESeq2 [35] was used to identify differentially expressed mRNAs (DEmRs) with an absolute $\log_2$-fold change ≥ 1 and $p < 0.05$. To functionally annotate DEmRs, Gene Ontology (GO) analysis was performed with Blast2GO with a cutoff E-value of 10^{-5}; Kyoto Encyclopedia of Genes and Genomes (KEGG) [36] pathway analysis was done using KEGG Orthology Based Annotation System (KOBAS 3.0) [37] with default parameters. Protein-protein interaction (PPI) network was constructed with STRING database (https://string-db.org, accessed on 3 August 2021) and visualized with Cytoscape software (version 3.8.2).

2.11. Genome-Wide miRNA Analysis

The raw reads were filtered as described elsewhere [32]. Briefly, clean reads were first obtained, and then non-coding RNA (rRNA, tRNA, scRNA, snRNA, snoRNA, etc.) and those reads aligned to exon, intron, and repeat sequences were removed. The remained clean reads were searched against miRbase database (Release 22.1) to characterize known miRNAs in *S. scrofa*. Furthermore, novel miRNAs were predicted with MiRDeep2 [38]. The miRNA expression level was normalized to transcripts per million (TPM), and the DESeq2 software [35] were used to characterize differentially expressed miRNA (DEmiRs) with absolute $\log_2$-fold change ≥ 1 and $p < 0.05$. Target genes were predicted with miRanda program [39,40].

2.12. Statistical Analysis

All experiments were performed at least three independent times, each with three repeats. Representative data of one experiment were given as mean $\pm$ standard error (SE). Data were analyzed with SPSS19.0 software (SPSS; Chicago, IL, USA) and Student's *t*-test was used to analyze differences between two groups. $p < 0.05$ (indicated with *) was considered statistically significant, and $p < 0.01$ (**) was considered very significant.

3. Results

3.1. Identification of Read-through Chimeric RNA Associated with Fat Content

Three subcutaneous fat tissues from neck, back, and hip were analyzed. Morphological analysis showed no difference between these tissues of Min and Yorkshire pigs (Figure 1A). There is significant difference in the average adipocyte area ($p < 0.05$) and TG content ($p < 0.05$) in subcutaneous tissues between Yorkshire and Min pigs except for TG content in hip fat, indicating that the subcutaneous fat deposition is different between Min and Yorkshire pigs (Figure 1B,C). Thus, backfat tissues from Min and Yorkshire pigs were used to screen chimeric RNAs related to fat deposition.

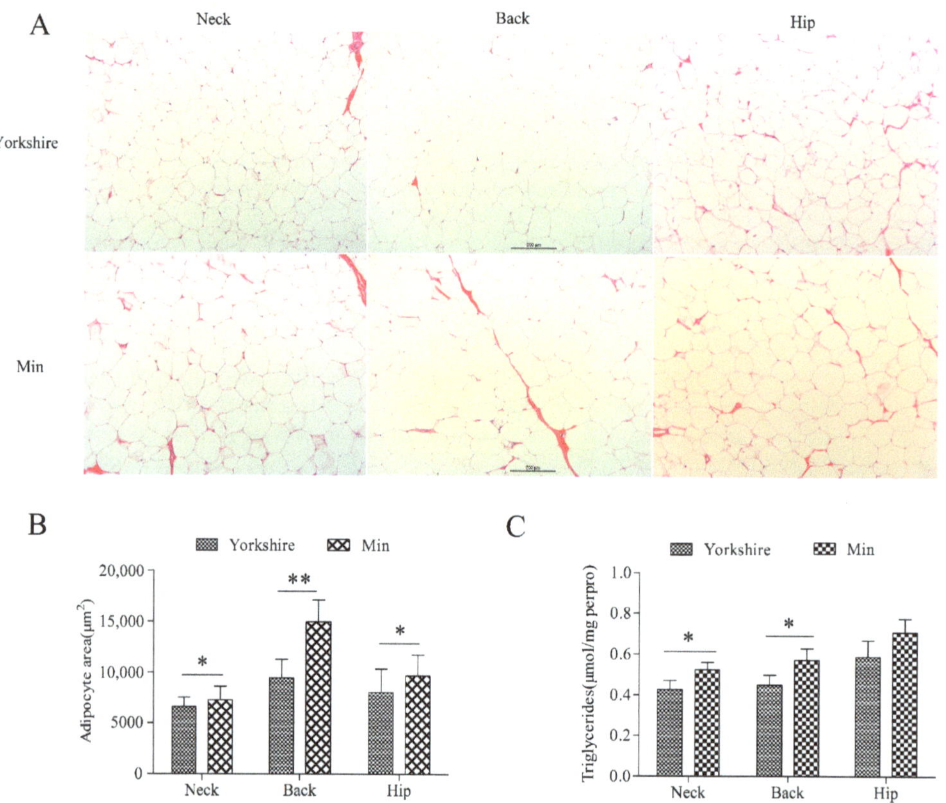

Figure 1. Comparison of subcutaneous adipose tissues between Yorkshire and Min pigs. (**A**) Morphological analysis with HE staining. (**B**) Adipocyte area of subcutaneous adipose tissues. The bar is 200 μm. (**C**) Triglyceride contents in subcutaneous adipose tissues. The data are shown as mean ± standard error. * and ** indicate $p < 0.05$, and $p < 0.01$, respectively.

The bioinformatic analysis of data obtained previously [28] called 77 read-throughs that presented in all of the four samples by ChimeraScan (Table S2). Most of them had low score values which mean total fragments supporting chimera, indicating humble expression levels of read-throughs (Figure 2A). Among the read-throughs identified, one tag named BP presented with differential score between fat tissues of Min and Yorkshire pigs. Furthermore, fragments spanning breakpoint junction accounted for a very large proportion in the total fragments of BP (Figure 2B). Through RT-PCR and sequencing analysis, the existence of BP was confirmed (Figure 2C).

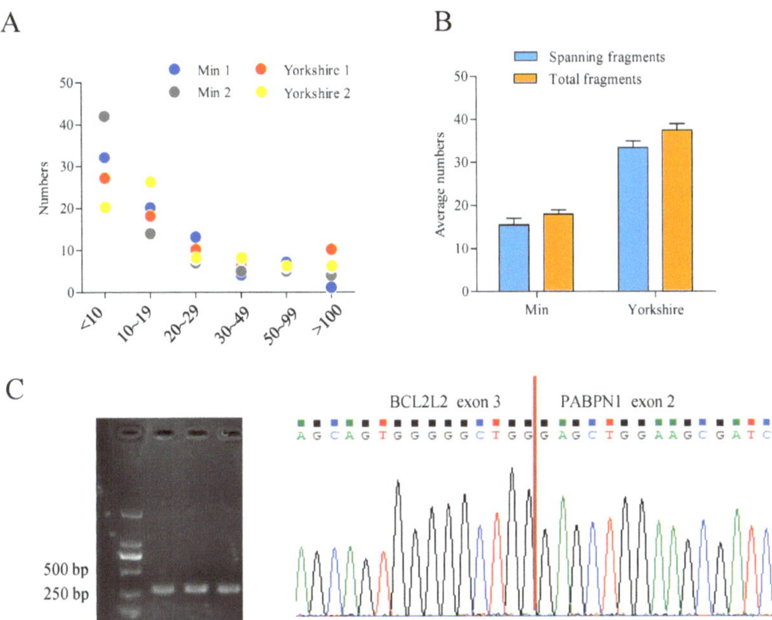

Figure 2. Characterization of porcine BCL2L2-PABPN1 (BP). (**A**) Score distribution of read-throughs identified. (**B**) Fragment distribution of BP in Min and Yorkshire pigs determined by ChimeraScan. (**C**) Confirmation of BP with RT-PCR and sequencing.

3.2. cDNA Cloning of BCL2L2–PABPN1

RACE analysis showed that porcine BP cDNA (V1) was 2097 bp in length and contained a complete CDS of 1002 bp, a 5′ UTR of 205 bp and a 3′ UTR of 890 bp. There was a typical poly(A) signal, AATAAA, at the 3′ end. It was located on chromosome 7 and composed of nine exons as revealed by BLAT program. The CDS spanned exons 3–9 with the first two exons and 8 bp of exon 3 comprising the 5′ UTR.

Seven alternative splicing (AS) variants of BP, named V2–V8, were obtained using 5′ RACE methods. V2 and V3 were formed by alternative 5′ splice sites (SSs) of exon 1, resulting in partial sequences of intron 1 being retained and having the same CDS as isoform V1. V4–V8 were absent of the start codon owing to the use of alternative 3′ SSs of exon 3 and thus could not be translated into a polypeptide. There were abundant alternative SSs in the first three exons of BP (Figure 3A). The sequences were deposited in GenBank under accession Nos. MH795109 for V1, and MW654158-64 for V2–V8.

Additionally, there was only an CDS sequence of porcine *PABPN1* deposited in Gen-Bank (No. NM_001243548) with a small 3′ UTR of 63 bp, 5′ UTR of 8 bp, and absence of poly(A) signal (Figure 3A). Through cloning the cDNA of BP, a fusion product of *BCL2L2* and *PABPN1* genes, we have obtained the complete 3′ sequence of porcine *PABPN1* mRNA. Using 5′ RACE, the 5′ UTR of *PABPN1* was obtained and subsequent RT-PCR with primers complementary to 5′ UTR of *PABPN1* and the end of the last exon of BP, respectively, confirmed the sequence of *PABPN1*.

The *PABPN1* cDNA was 1997 bp in length with 186 bp of 5′ UTR, 921 bp of CDS encoding a polypeptide of 306 aa, and 890 bp of 3′ UTR (GenBank accession No. MH795126). The porcine PABPN1 protein was completely identical to that in humans (NM004643), and thus contained three major domains as its counterpart in humans [41]: an acidic N-terminal domain containing a stretch of 10 alanines and a coiled-coiled domain (CCD, spanning 119–146 aa), a single ribonucleoprotein-type RNA recognition motif (RRM, 161–257 aa), and a basic arginine-rich C-terminal region (258–306 aa).

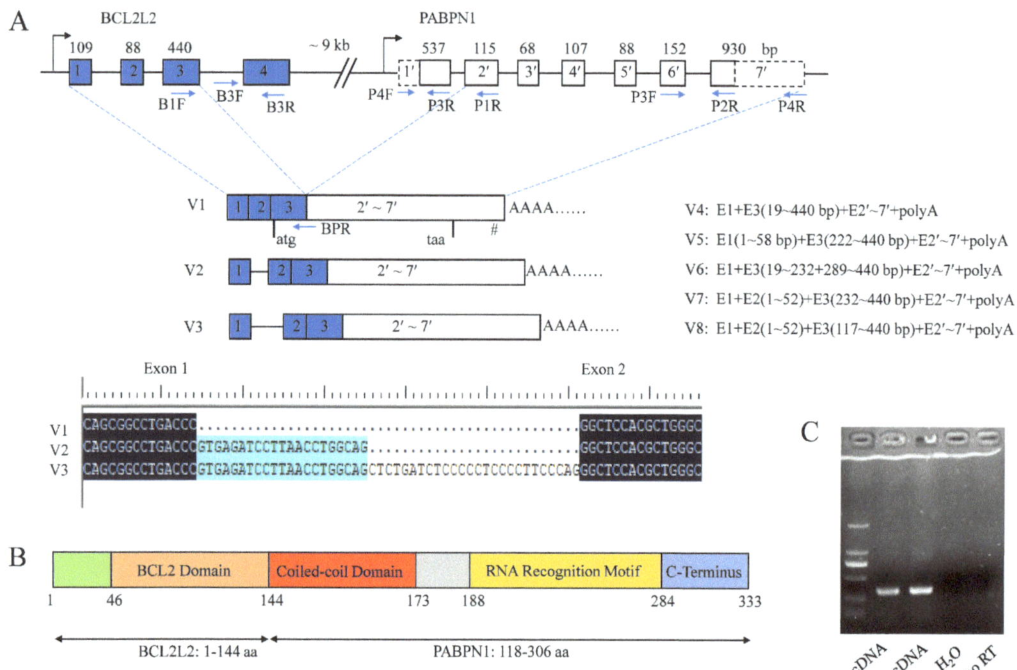

Figure 3. Characterization of porcine BCL2L2-PABPN1 (BP). (**A**) Genomic and mRNA structure of porcine BP. Boxes and lines indicate exons and introns, respectively. Dotted boxes in PABPN1 indicate that the sequences are first obtained here. Figures over the boxes indicate the length of the corresponding exons, while those in the boxes indicate exon No. Primer locations are shown with arrows. E, exon. I, intron. # indicates the position of poly(A) signal. Start and stop codons are shown under the boxes. (**B**) Motifs in the polypeptide of BP. (**C**) Identification of mechanisms underlying BP formation with RT-PCR. Templates were showed below. No RT, no reverse transcriptase control.

Porcine BP is an in-frame fusion product with a molecular weight of 37.2 kDa and a pI of 8.58. The predicted polypeptide was composed of 333 aa with the first 144 aa from BCL2L2 (1–144 aa of BCL2L2) and the last 189 aa from PABPN1 (118–306 aa of PABPN1). Porcine BCL2L2 was composed of 193 aa and contained a functional BCL2 domain in the 46–144 aa as revealed by Blastn program. Thus, BP has all functional domains of both parents except for a stretch of 10 alanines in the N-terminal region of PABPN1 (Figure 3B).

3.3. Mechanisms Underlying BCL2L2–PABPN1 Formation

BP comprised nine exons with exons 1–3 from the first three exons of *BCL2L2* and 4–9 from the last six exons of *PABPN1*. BLAT analysis showed that *BCL2L2* and *PABPN1* were composed of four and seven exons, respectively. The configuration of BP, involving the second-to-last exon in the former gene joining to the second exon in the latter gene, was the most common type of *cis*-SAGe [42–44]. Additionally, *BCL2L2* was adjacent to *PABPN1* on porcine chromosome 7 with a distance smaller than 10 kb (~9 kb) and had the same transcription direction, which is another characteristic of the formation of *cis*-SAGe [45]. These make BP a candidate for *cis*-SAGe.

To confirm its transcriptional read-through nature, RT-PCR was used to detect primary mRNA as described by Qin and coworkers [43] in which RT primer (P1R) was annealed to the second exon of *PABPN1*, and PCR primers (B3F/R) were complementary to the last intron and the last exon of *BCL2L2*, respectively (Figure 3A). To avoid DNA contamination, total RNA was digested with DNase I and no reverse transcriptase control was used. The fragment was successfully amplified (Figure 3C), indicating that the transcript ran from *BCL2L2* to *PABPN1* and BP was a product of *cis*-SAGe.

There is a read-through product (NM_001199864) of *BCL2L2* and *PABPN1* in humans. Both the human and pig sequences had very high identities in aa, CDS, even in the complete cDNA. Additionally, there were also homologs in various orders of Mammalia including Primates, Cetacea, Even-toed ungulates, Chiropter, Carnivora, and even in *Ornithorhynchus anatinus*, one of the oldest mammals. Except for in humans, no sequences were described as read-through origin, but the identities were more than 95% between the polypeptide of porcine BP and any of the homologs (Table S1-2). These indicated that BP was highly conserved in evolution and might have critical roles in life.

3.4. Effects of BCL2L2–PABPN1 on Preadipocyte Proliferation and Differentiation

BP mRNA level had a tendency to increase during preadipocyte proliferation but the changes were not significant ($p > 0.05$) (Figure 4A). During preadipocyte differentiation, the expression of BP was increased significantly ($p < 0.05$) since 4 days after induction compared with non-induced cells (Figure 4B). Real-time PCR and Western blot analysis showed that BP expression was increased effectively in preadipocytes transfected with the plasmids pCMV-HA-BP. CCK-8 assay showed that BP overexpression increased cell number compared to the cells transfected with empty vector, with the highest level ($p < 0.01$) at 4 days post-transfection (Figure 4C). Flow cytometry analysis showed that the number of G2-phase preadipocytes was increased significantly ($p < 0.01$) in groups overexpressing BP (Figure 4D). Overexpression of BP resulted in a decrease of lipid droplets compared to the control cells at 6-, 8- and 10-days post-induction as revealed with Oil Red O staining (Figure 4E). The results indicated that BP promoted proliferation and inhibited differentiation of preadipocytes.

3.5. Genome-Wide Identification of mRNAs Involved in BCL2L2-PABNP1 Regulation

RNA-Seq technology was used to explore mechanisms of BP on adipogenesis in cells transfected with adenoviruses expressing BP at a condition of MOI 300 and 48 h which was predetermined with fluorescence microscope and qPCR analysis (Figure S1). An average number of 104,791,770 and 128,873,418 raw reads were obtained in control and treatment groups, respectively. After removal of low-quality and adaptor containing reads, 103,766,967 and 127,695,699 clean reads were obtained. In these clean data, the Q30 content was more than 92.67%. A total of 3074 DEmRs were obtained by RNA-Seq analysis, among which 1476 were upregulated and 1598 were downregulated in cells overexpressing BP compared with control groups (Figure 5A, Table S3-1). Eleven DEmRs were selected randomly to validate RNA-Seq data with qPCR, and consistent results were obtained (Figure 5B).

To highlight the function of DEmRs, GO and KEGG analysis were performed. GO analysis revealed that the DEmRs were involved in multiple categories in molecular functions, cellular component, and biological processes (Figure 5C). The major biological processes enriched included fatty acid beta-oxidation, 2-oxoglutarate metabolic process, and 2-oxoglutarate metabolic process, etc. KEGG analysis performed on all DEmRs revealed that various fat-related pathways, such as MAPK, TGF-β, Wnt, PI3K-Akt, and Fatty acid metabolism, etc. were significantly enriched. When the up- and downregulated DEmRs were subjected to KEGG analysis separately, the pathways enriched suggested different roles between them. The upregulated DEmRs were mainly involved in fat metabolism-related pathways including fatty acid metabolism, fat digestion and absorption, arachidonic acid metabolism, butanoate metabolism, propanoate metabolism, and steroid biosynthesis; additionally, two signaling pathways associated with adipogenesis, PPAR and FoxO, were enriched by upregulated DEmRs. Downregulated DEmRs were enriched in some adipogenesis-related signaling pathways including MAPK, PI3K-Akt, Wnt, TGF-beta, insulin, Hippo, and cAMP signaling pathways, with MAPK and PI3K-Akt being the top two enriched pathways except for human papillomavirus infection whose enrichment might be associated with adenovirus infection (Figure 5D,E). These results indicated that DEmRs induced by BP overexpression were involved in fat deposition, which confirmed the role of BP in adipogenesis and highlighted the underlying mechanisms.

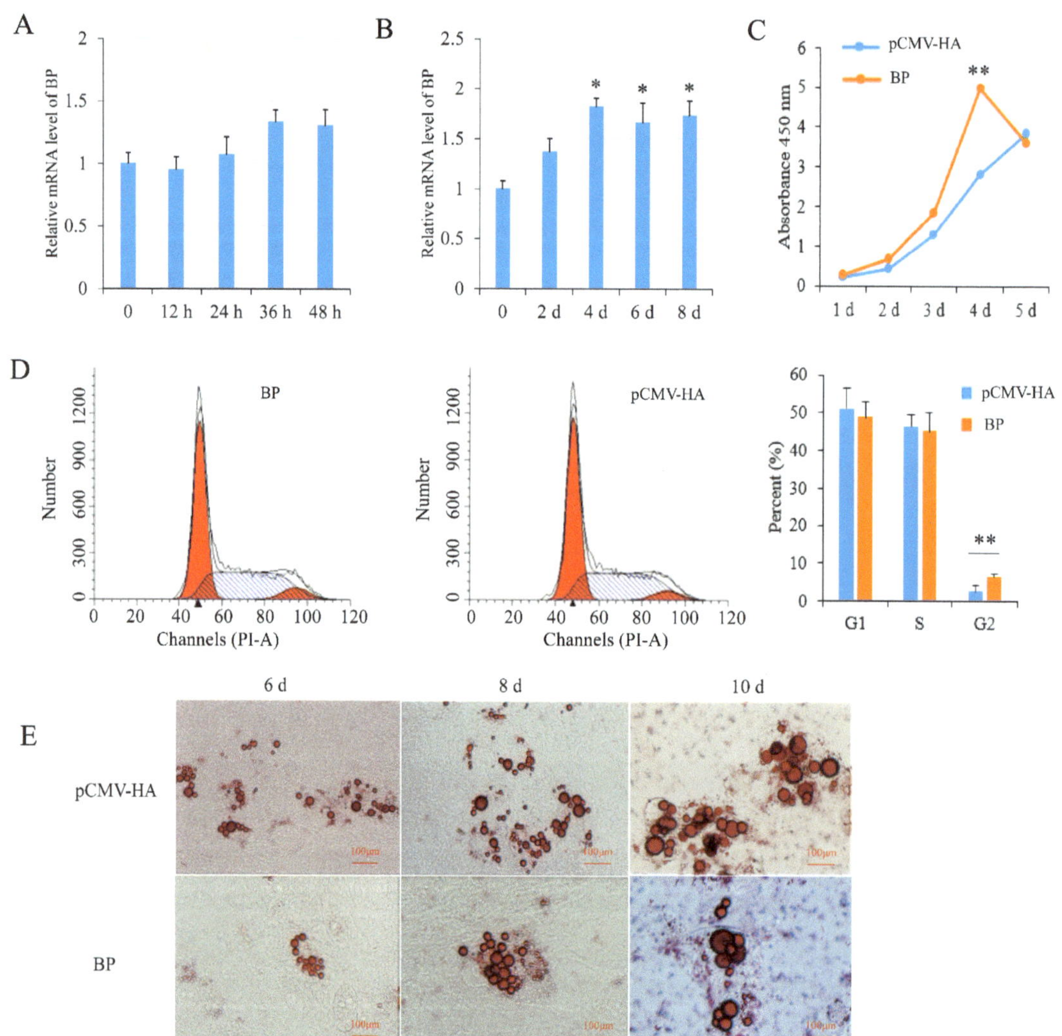

Figure 4. Effects of BCL2L2-PABPN1 (BP) on subcutaneous preadipocyte proliferation and differentiation. (**A**) Expression of BP during porcine preadipocyte proliferation determined with real-time PCR. (**B**) Expression of BP during porcine preadipocyte differentiation determined with real-time PCR. (**C**) Effect of BP on cell proliferation measured by CCK-8 assay. (**D**) Effects of BP on cell cycle measured with flow cytometry. (**E**) Oil red O staining of differentiated adipocyte. * and ** indicates $p < 0.05$ and $p < 0.01$, respectively, compared with control groups. The bar is 100 µm.

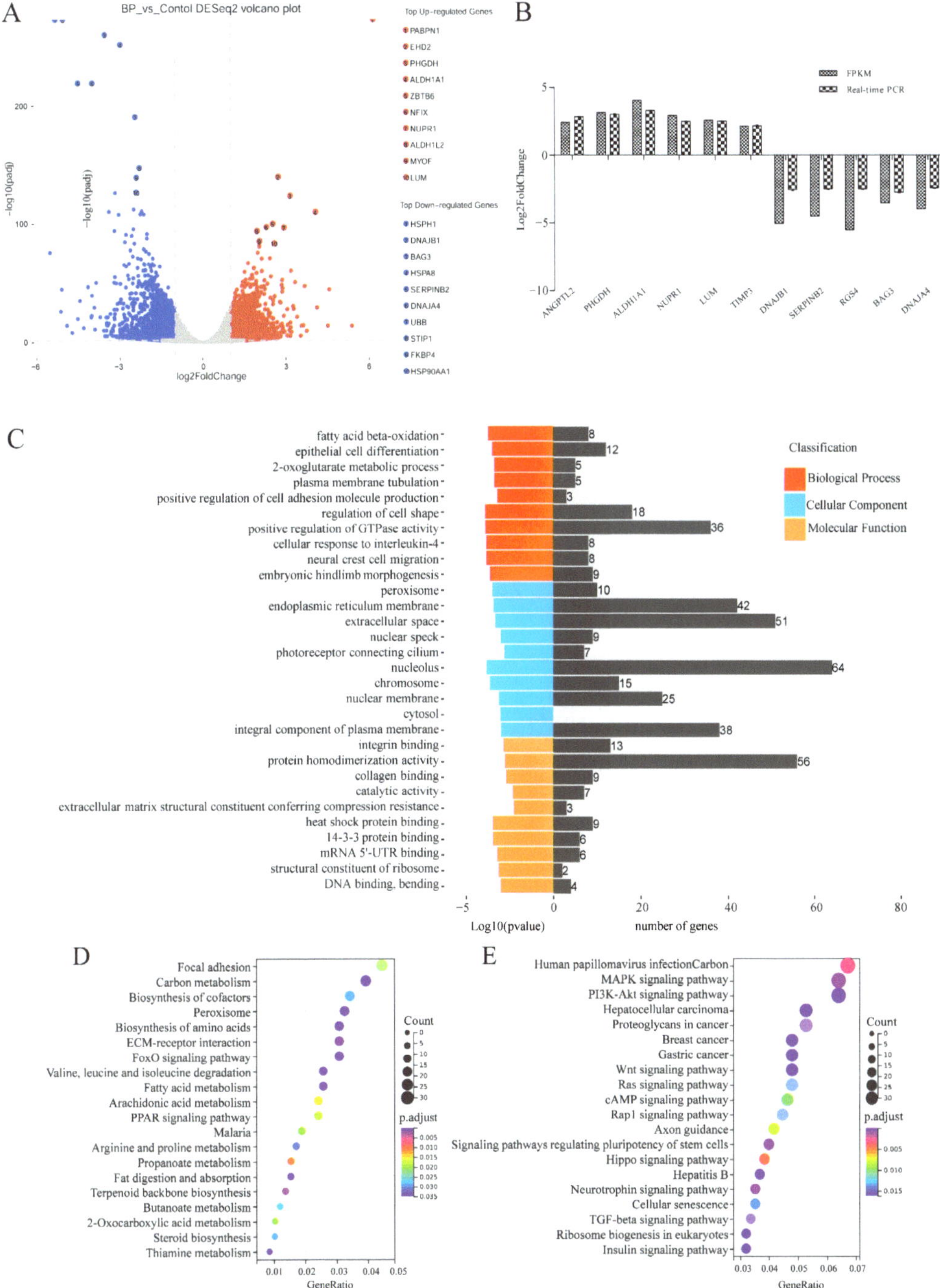

Figure 5. Characterization of differentially expressed mRNAs (DEmRs) induced by BCL2L2-PABPN1. (**A**) Volcano plot of DEmRs. (**B**) Validation of RNA-Seq data with real-time PCR method. (**C**) GO enrichment analysis of DEmRs. (**D**) KEGG pathway analysis of upregulated DEmRs. (**E**) KEGG pathway analysis of downregulated DEmRs.

The above 15 fat-related pathways involved 212 DEmRs of which 73 were upregulated and 139 were downregulated in response to BP treatment (Figure 6A, Table S3-2). PPI network analysis showed that these DEmRs were grouped into two separate modules. In module 1 the top three key nodes were PPP2CB, EGFR, and Wnt5A with a degree of 11, 9, and 9, respectively. Both PPP2CB and EGFR are involved in MAPK and PI3K-Akt signaling pathways. In module 2, EHHADH, ACAA 2, and ALDH6A1 were the top three key nodes with a degree of 10, 6, and 5, respectively (Figure 6B).

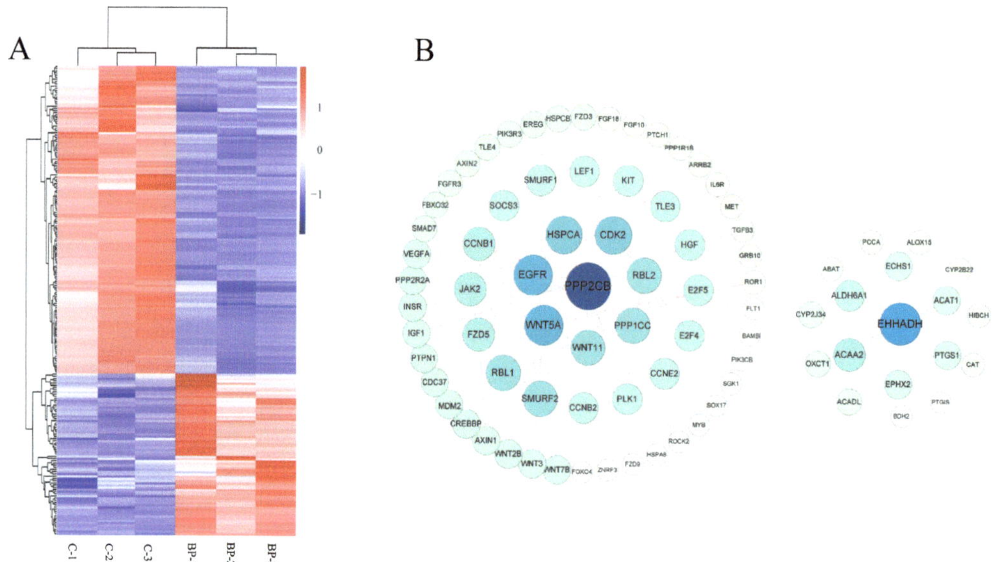

Figure 6. Characterization of differentially expressed mRNAs (DEmRs) involved in fat-related pathways. (**A**) Heatmap cluster of DEmRs involved in fat-related pathways. (**B**) The protein-protein interaction network of DEmRs involved in fat-related pathways. The size of the circle indicates the degree of interaction between the genes. The network was constructed with score > 0.9, FDR stringency = 1 percent, and disconnected nodes were hide.

3.6. Genome-Wide Identification of miRNAs Involved in BCL2L2-PABNP1 Regulation

To further characterize the mechanisms underlying the regulation of BP on adipogenesis, transcriptomic miRNA alteration induced by BP was analyzed using Illumina RNA sequencing. An average number of 25,583,601 and 18,115,912 clean reads (96.01% and 95.51% of raw reads) of which the average percentage of miRNAs was 63.72 and 65.80 were obtained from control and BP-treated groups, respectively. On the basis of *S. scrofa* genome (11.1), a total of 1987 unique miRNAs were identified (Table S4-1). The lengths of the miRNAs were mainly distributed in 19–24 nt in all six groups, with a maximum of 22 nt (Figure 7A). The expression level and count distribution of the total miRNAs in each sample was shown in Figure 7B. Compared with control groups, 44 known miRNAs were identified as differentially expressed miRNAs (DEmiRs) including 35 upregulated and nine downregulated during BP treatment (Figure 7C, Table S4-2). The expression of six DEmiRs were validated by real-time PCR, and consistent results were obtained between qPCR and RNA-Seq (Figure 7D).

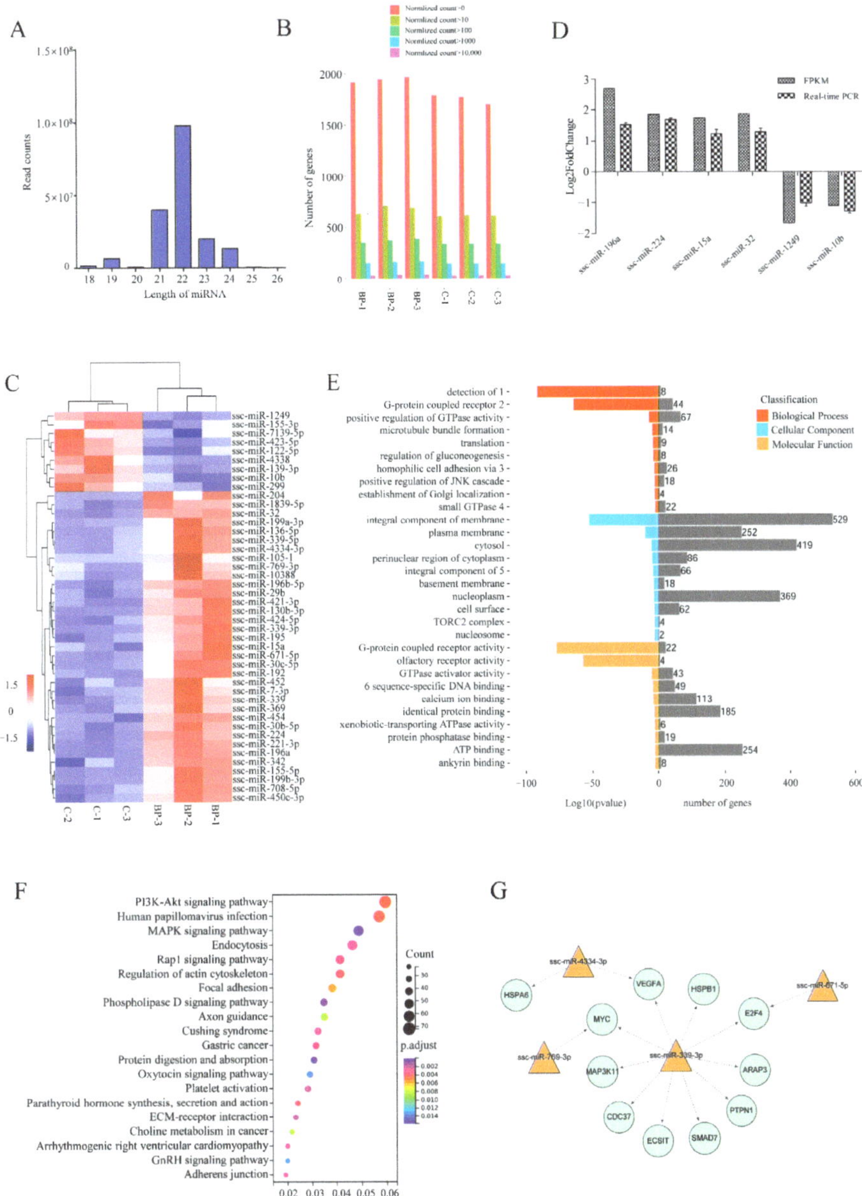

Figure 7. Characterization of differentially expressed miRNAs (DEmiRs) induced by BCL2L2-PABPN1. (**A**) Length distribution of miRNAs among six samples. (**B**) Barplot profiling the expression levels of the miRNAs in each sample. (**C**) Heatmap cluster of known DemiRs. (**D**) Validation of Illumina data with real-time PCR. (**E**) GO enrichment of target genes of DEmiRs. 1, detection of chemical stimulus involved in sensory perception of smell; 2, G-protein coupled receptor signaling pathway; 3, homophilic cell adhesion via plasma membrane adhesion molecules; 4, small GTPase mediated signal transduction; 5, integral component of plasma membrane; 6, RNA polymerase II core promoter proximal region sequence-specific DNA binding. (**F**) KEGG enrichment of target genes of DEmiRs. (**G**) Fat-related miRNA-mRNA interaction network.

A total of 4064 putative target mRNAs were predicted for these DEmiRs (Table S4-3), and top 10 target mRNAs of each DEmiRs were selected according to total score for GO and KEGG analysis. GO enrichment analysis showed that target genes were enriched in some categories of biological process such as positive regulation of GTPase activity, regulation of gluconeogenesis, and positive regulation of JNK cascade (Figure 7E). Two of the top three KEGG pathways enriched by target genes of DEmiRs were PI3K-Akt and MAPK signaling pathways both of which are important regulators of adipogenesis. Additionally, phospholipase D, Rap 1, and regulation of actin cytoskeleton signaling pathways were also significantly enriched (Figure 7F).

3.7. Integrated Analysis of mRNA and miRNA Data

To highlight DEmiRs involved in the regulation of BP on adipogenesis, integrated analysis was performed between known DEmiRs and 212 DEmRs involved in 15 fat-related pathways, in which those DEmRs identified as target genes of known DEmiRs and had a negatively correlated expression levels with the paired DEmiRs were selected for further analysis. A total of 11 differentially expressed target mRNAs (DETmRs) and four paired DEmiRs were obtained (Table S4-4). These fat-related DETmRs and the paired DEmiRs constituted a network in which ssc-miR-339-3p was critical (Figure 7G). Five of the 10 genes regulated by ssc-miR-339-3p, *MYC*, *VEGFA*, *MAP3K11*, *HSPB1* and *ECSIT*, are involved in MAPK signaling pathway.

4. Discussion

Chimeric RNAs were traditionally believed to be produced by chromosome rearrangement and unique to carcinogenesis until recent discoveries of RNA *trans*-splicing and *cis*-SAGe. It has been found that chimeric RNAs are expressed in noncancerous cells and tissues and involved in normal physiological process such as muscle lineage commitment [25,26] and neuronal differentiation [27]. However, there is no report on chimeric RNAs in fat formation in mammals.

Here, through analyzing our previous paired-end high-throughput sequencing data from backfat tissues of Min and Yorkshire pigs [28], chimeric RNA BP was characterized and the full-length cDNA was cloned in pigs using RT-PCR and RACE. Additionally, BP formation was identified as *cis*-SAGe. To the best of our knowledge, this is the first report on *cis*-SAGe in pigs. Min pig is a local breed in Northeast China with abundant fat content, while Yorkshire pigs have high lean meat percentage owing to long extensive breeding. BP was differentially characterized between fats from the two breeds. The role of BP in fat formation was thus expected and confirmed. The results extend the function of chimeric RNAs to adipogenesis, a normal physiological process that chimeric RNAs have not been involved in.

During cloning, we obtained seven AS variants of BP; some of which occurred in the 5′ UTR. The AS patterns identified here include exon skipping, alternative 5′ and 3′ SSs. It has been shown that exons closer to the intergenic region of two parental genes have lower conservation than those farther from the region and tend to be alternatively spliced in the formation of read-through chimeric RNA [44,46]. In the present study, AS variants were by-products in 5′ RACE cloning of BP and we did not focus on AS characterization. Additionally, the reverse primer of 5′ RACE was complementary to the junction of the two parents, resulting in inability to identify variants alternatively spliced in this region. Thereafter, there should be more AS variants remaining to be identified and more AS patterns might be present in BP. In a previous report analyzing formation and structures of *cis*-SAGe chimeric RNA [46], 20 and 23 transcript variants were obtained from ZNF343–SNPRB and COX17–POPDC2, respectively. These indicate that *cis*-SAGe is also rich in AS like regular pre-mRNAs.

As a nuclear poly(A) RNA binding protein, PABPN1 plays a key role in polyadenylation. It can directly interact with poly(A) polymerase (PAP) through the CCD domain leading to stimulation of the processivity of PAP [47], while binding to poly(A) RNA via the single RRM domain with a contribution from the C-terminal region [48]. BP contained all the functional domains of PABPN1 including CCD, RRM and C-terminal region implicated in polyadenylation, which provides a structural basis for polyadenylation. This suggests a role of BP in post-transcriptional regulation through mediating poly(A) tailing.

Because BP was identified in fat tissues, we focused on its role in adipogenesis and found that it can promote preadipocyte proliferation and inhibit differentiation. Mechanisms underlying the regulation of BP on adipogenesis were then analyzed with RNA-Seq, and genome-wide DEmRs and DEmiRs were characterized in preadipocytes overexpressing BP. Both DEmRs and target genes of DEmiRs were significantly involved in MAPK and PI3K-Akt, two of the important signaling pathways regulating adipogenesis. Various recent studies showed that PI3K-Akt pathway positively regulated the adipogenesis [49–51]. However, the role of MAPK signaling pathway in adipogenesis was bi-directional. Some reports demonstrated that activation of the MAPK pathway phosphorylates PPARγ, an adipogenic marker, and thus opposed adipogenesis [52–54]. While studies on effects of genes such as pigment epithelium-derived factor, a newly identified adipokine, miR-145 and lncRNA 332443 on adipogenesis revealed a positive role of MAPK signaling pathway during differentiation process [55,56]. These indicates that the role of MAPK in adipogenesis is complicated and versatile. Nevertheless, we showed that MAPK and PI3K-Akt were important for the regulation of BP on adipogenesis.

5. Conclusions

In this study, a read-through fusion transcript BP was first characterized in pigs. The deduced polypeptide contains the main functional domains of both parents, BCL2L2 and PABPN1, and highly conserved among species including *Ornithorhynchus anatinus*. BP was found to inhibit differentiation of primary porcine preadipocytes, and MAPK and PI3K-Akt were identified as the key signaling pathways affected by BP in which *PPP2CB* and *EGFR* were the hub genes. Additionally, ssc-miR-339-3p was critical for BP regulating adipogenesis. The results highlight the role of chimeric RNA in adipogenesis in mammals.

Supplementary Materials: The following are available online at https://www.mdpi.com/article/10.3390/genes13030445/s1, Figure S1: Optimum condition for transfection of adenovirus, Table S1: Primers and identities of BP among species, Table S2: Read-throughs expressed in all of the four samples, Table S3: differentially expressed mRNAs in response to BP treatment, Table S4: miRNAs identified in this study.

Author Contributions: Conceptualization, X.Y.; resources, D.L. and J.X.; investigation, J.Z., Z.Y., J.L. and L.W.; validation, J.Z. and W.H.; data curation, J.Z. and Z.Y.; visualization, J.X. and D.Z.; writing—original draft preparation, J.Z. and X.Y.; writing—reviewing and editing, X.Y.; funding acquisition, X.Y. and D.L. All authors have read and agreed to the published version of the manuscript.

Funding: This research was funded by the National Natural Science Foundation of China (31741114) and Heilongjiang science fund for distinguished youth scholars (JQ2020C005).

Institutional Review Board Statement: The animal study was reviewed and approved by the Animal Care Committee of Northeast Agricultural University (protocol code NEAUEC20200201).

Informed Consent Statement: Not applicable.

Data Availability Statement: All the relevant data are provided along with the manuscript as Supplementary Files. The RNA-Seq data associated with this study have been submitted to the NCBI SRA database (accession number GSE192412).

Conflicts of Interest: The authors declare no conflict of interest.

References

1. Fortin, A.; Robertson, W.M.; Tong, A.K.W. The eating quality of Canadian pork and its relationship with intramuscular fat. *Meat Sci.* **2005**, *69*, 297–305. [CrossRef]
2. Hoa, V.; Seo, H.; Seong, P.; Cho, S.; Kang, S.; Kim, Y.; Moon, S.; Choi, Y.; Kim, J.; Seol, K. Back-fat thickness as a primary index reflecting the yield and overall acceptance of pork meat. *Anim. Sci. J.* **2021**, *92*, e13515. [CrossRef] [PubMed]

3. Rosen, E.D.; Spiegelman, B.M. Adipocytes as regulators of energy balance and glucose homeostasis. *Nature* **2006**, *444*, 847–853. [CrossRef] [PubMed]

4. Lunney, J.K. Advances in swine biomedical model genomics. *Int. J. Biol. Sci.* **2007**, *3*, 179–184. [CrossRef] [PubMed]

5. Ross, J.W.; Fernandez de Castro, J.P.; Zhao, J.; Samuel, M.; Walters, E.; Rios, C.; Bray-Ward, P.; Jones, B.W.; Marc, R.E.; Wang, W.; et al. Generation of an inbred miniature pig model of retinitis pigmentosa. *Investig. Ophthalmol. Vis. Sci.* **2012**, *53*, 501–507. [CrossRef]

6. Stachowiak, M.; Szczerbal, I.; Switonski, M. Genetics of adiposity in large animal models for human obesity-studies on pigs and dogs. *Prog. Mol. Biol. Transl. Sci.* **2016**, *140*, 233–270. [CrossRef]

7. Lefterova, M.I.; Lazar, M.A. New developments in adipogenesis. *Trends Endocrinol. Metab.* **2009**, *20*, 107–114. [CrossRef]

8. Siersbæk, R.; Nielsen, R.; Mandrup, S. Transcriptional networks and chromatin remodeling controlling adipogenesis. *Trends Endocrinol. Metab.* **2012**, *23*, 56–64. [CrossRef]

9. Lefterova, M.I.; Haakonsson, A.K.; Lazar, M.A.; Mandrup, S. PPARγ and the global map of adipogenesis and beyond. *Trends Endocrinol. Metab.* **2014**, *25*, 293–302. [CrossRef]

10. Garcés, C.; Ruiz-Hidalgo, M.J.; Font de Mora, J.; Park, C.; Miele, L.; Goldstein, J.; Bonvini, E.; Porrás, A.; Laborda, J. Notch-1 controls the expression of fatty acid-activated transcription factors and is required for adipogenesis. *J. Biol. Chem.* **1997**, *272*, 29729–29734. [CrossRef]

11. Bennett, C.N.; Ross, S.E.; Longo, K.A.; Bajnok, L.; Hemati, N.; Johnson, K.W.; Harrison, S.D.; MacDougald, O.A. Regulation of Wnt signaling during adipogenesis. *J. Biol. Chem.* **2002**, *277*, 30998–31004. [CrossRef] [PubMed]

12. Nishimoto, S.; Nishida, E. MAPK signalling: ERK5 versus ERK1/2. *EMBO Rep.* **2006**, *7*, 782–786. [CrossRef]

13. Qimuge, N.; He, Z.; Qin, J.; Sun, Y.; Wang, X.; Yu, T.; Dong, W.; Yang, G.; Pang, W. Overexpression of DNMT3A promotes proliferation and inhibits differentiation of porcine intramuscular preadipocytes by methylating p21 and PPARg promoters. *Gene* **2019**, *696*, 54–62. [CrossRef]

14. Abdalla, B.A.; Li, Z.; Ouyang, H.; Jebessa, E.; Sun, T.; Yu, J.; Cai, B.; Chen, B.; Nie, Q.; Zhang, X. A novel Dnmt3a1 transcript inhibits adipogenesis. *Front. Physiol.* **2018**, *9*, 1270. [CrossRef]

15. Greither, T.; Wenzel, C.; Jansen, J.; Kraus, M.; Wabitsch, M.; Behre, H.M. MiR-130a in the adipogenesis of human SGBS preadipocytes and its susceptibility to androgen regulation. *Adipocyte* **2020**, *9*, 197–205. [CrossRef] [PubMed]

16. Zhang, T.; Liu, H.; Mao, R.; Yang, H.; Zhang, Y.; Zhang, Y.; Guo, P.; Zhan, D.; Xiang, B.; Liu, Y. The lncRNA RP11-142A22.4 promotes adipogenesis by sponging miR-587 to modulate Wnt5β expression. *Cell Death Dis.* **2020**, *11*, 475. [CrossRef] [PubMed]

17. Zhuo, J.; Jing, X.; Du, X.; Yang, X.Q. Generation of chimeric RNAs by cis-splicing of adjacent genes (cis-SAGe) in mammals. *Yi Chuan* **2018**, *40*, 145–154. [CrossRef]

18. Communi, D.; Suarez-Huerta, N.; Dussossoy, D.; Savi, P.; Boeynaems, J.M. Cotranscription and intergenic splicing of human P2Y11 and SSF1 genes. *J. Biol. Chem.* **2001**, *276*, 16561–16566. [CrossRef]

19. Varley, K.E.; Gertz, J.; Roberts, B.S.; Davis, N.S.; Bowling, K.M.; Kirby, M.K.; Nesmith, A.S.; Oliver, P.G.; Grizzle, W.E.; Forero, A.; et al. Recurrent read-through fusion transcripts in breast cancer. *Breast Cancer Res. Treat.* **2014**, *146*, 287–297. [CrossRef]

20. Missiaglia, E.; Williamson, D.; Chisholm, J.; Wirapati, P.; Pierron, G.; Petel, F.; Concordet, J.; Thway, K.; Oberlin, O.; Pritchard-Jones, K.; et al. PAX3/FOXO1 fusion gene status is the key prognostic molecular marker in rhabdomyosarcoma and significantly improves current risk stratification. *J. Clin. Oncol.* **2012**, *30*, 1670–1677. [CrossRef]

21. Jothi, M.; Mal, M.; Keller, C.; Mal, A.K. Small molecule inhibition of PAX3-FOXO1 through AKT activation suppresses malignant phenotypes of alveolar rhabdomyosarcoma. *Mol. Cancer Ther.* **2013**, *12*, 2663–2674. [CrossRef] [PubMed]

22. Loupe, J.M.; Miller, P.J.; Ruffin, D.R.; Stark, M.W.; Hollenbach, A.D. Inhibiting phosphorylation of the oncogenic PAX3-FOXO1 reduces alveolar rhabdomyosarcoma phenotypes identifying novel therapy options. *Oncogenesis* **2015**, *4*, e145. [CrossRef] [PubMed]

23. Tang, Y.; Guan, F.; Li, H. Case Study: The Recurrent Fusion RNA DUS4L-BCAP29 in Noncancer Human Tissues and Cells. *Methods Mol. Biol.* **2020**, *2079*, 243–258. [CrossRef] [PubMed]

24. Tang, Y.; Qin, F.; Liu, A.; Li, H. Recurrent fusion RNA DUS4L-BCAP29 in non-cancer human tissues and cells. *Oncotarget* **2017**, *8*, 31415–31423. [CrossRef] [PubMed]

25. Yuan, H.; Qin, F.; Movassagh, M.; Park, H.; Golden, W.; Xie, Z.; Zhang, P.; Sklar, J.; Li, H. A chimeric RNA characteristic of rhabdomyosarcoma in normal myogenesis process. *Cancer Discov.* **2013**, *3*, 1394–1403. [CrossRef]

26. Xie, Z.; Tang, Y.; Su, X.; Cao, J.; Zhang, Y.; Li, H. PAX3-FOXO1 escapes miR-495 regulation during muscle differentiation. *RNA Biol.* **2019**, *16*, 144–153. [CrossRef]

27. Tang, Y.; Ma, S.; Wang, X.; Xing, Q.; Huang, T.; Liu, H.; Li, Q.; Zhang, Y.; Zhang, K.; Yao, M.; et al. Identification of chimeric RNAs in human infant brains and their implications in neural differentiation. *Int. J. Biochem. Cell Biol.* **2019**, *111*, 19–26. [CrossRef]

28. Wang, W.T.; Zhang, D.J.; Liu, Z.G.; Peng, F.G.; Wang, L.; Fu, B.; Wu, S.H.; Li, Z.Q.; Guo, Z.H.; Liu, D. Identification of differentially expressed genes in adipose tissue of min pig and large white pig using RNA-seq. *Acta Agr. Scand. Sect. A—Anim. Sci.* **2018**, *68*, 73–80. [CrossRef]

29. Iyer, M.K.; Chinnaiyan, A.M.; Maher, C.A. ChimeraScan: A tool for identifying chimeric transcription in sequencing data. *Bioinformatics* **2011**, *27*, 2903–2904. [CrossRef]

30. Warr, A.; Affara, N.; Aken, B.; Beiki, H.; Bickhart, D.M.; Billis, K.; Chow, W.; Eory, L.; Finlayson, H.A.; Flicek, P.; et al. An improved pig reference genome sequence to enable pig genetics and genomics research. *Gigascience* **2020**, *9*, giaa051. [CrossRef]

31. Livak, K.J.; Schmittgen, T.D. Analysis of relative gene expression data using real-time quantitative PCR and the $2^{-\Delta\Delta CT}$ Method. *Methods* **2001**, *25*, 402–408. [CrossRef] [PubMed]

32. Wu, Q.; Ning, X.; Sun, L. Megalocytivirus Induces Complicated Fish Immune Response at Multiple RNA Levels Involving mRNA, miRNA, and circRNA. *Int. J. Mol. Sci.* **2021**, *22*, 3156. [CrossRef] [PubMed]

33. Kim, D.; Langmead, B.; Salzberg, S.L. HISAT: A fast spliced aligner with low memory requirements. *Nat. Methods* **2015**, *12*, 357–360. [CrossRef] [PubMed]

34. Trapnell, C.; Williams, B.A.; Pertea, G.; Mortazavi, A.; Kwan, G.; van Baren, M.J.; Salzberg, S.L.; Wold, B.J.; Pachter, L. Transcript assembly and quantification by RNA-Seq reveals unannotated transcripts and isoform switching during cell differentiation. *Nat. Biotechnol.* **2010**, *28*, 511–515. [CrossRef] [PubMed]

35. Love, M.I.; Huber, W.; Anders, S. Moderated estimation of fold change and dispersion for RNA-seq data with DESeq2. *Genome Biol.* **2014**, *15*, 550. [CrossRef] [PubMed]

36. Kanehisa, M.; Goto, S. KEGG: Kyoto encyclopedia of genes and genomes. *Nucleic Acids Res.* **2000**, *28*, 27–30. [CrossRef]

37. Wu, J.; Mao, X.; Cai, T.; Luo, J.; Wei, L. KOBAS server: A web-based platform for automated annotation and pathway identification. *Nucleic Acids Res.* **2006**, *34*, W720–W724. [CrossRef]

38. Friedländer, M.R.; Mackowiak, S.D.; Li, N.; Chen, W.; Rajewsky, N. miRDeep2 accurately identifies known and hundreds of novel microRNA genes in seven animal clades. *Nucleic Acids Res.* **2012**, *40*, 37–52. [CrossRef]

39. Enright, A.J.; John, B.; Gaul, U.; Tuschl, T.; Sander, C.; Marks, D.S. MicroRNA targets in Drosophila. *Genome Biol.* **2003**, *5*, R1. [CrossRef]

40. Riffo-Campos, Á.L.; Riquelme, I.; Brebi-Mieville, P. Tools for Sequence-Based miRNA Target Prediction: What to Choose? *Int. J. Mol. Sci.* **2016**, *17*, 1987. [CrossRef]

41. Banerjee, A.; Apponi, L.H.; Pavlath, G.K.; Corbett, A.H. PABPN1: Molecular function and muscle disease. *FEBS J.* **2013**, *280*, 4230–4250. [CrossRef] [PubMed]

42. Prakash, T.; Sharma, V.K.; Adati, N.; Ozawa, R.; Kumar, N.; Nishida, Y.; Fujikake, T.; Takeda, T.; Taylor, T.D. Expression of conjoined genes: Another mechanism for gene regulation in eukaryotes. *PLoS ONE* **2010**, *5*, e13284. [CrossRef] [PubMed]

43. Qin, F.; Song, Z.; Babiceanu, M.; Song, Y.; Facemire, L.; Singh, R.; Adli, M.; Li, H. Discovery of CTCF-sensitive Cis-spliced fusion RNAs between adjacent genes in human prostate cells. *PLoS Genet.* **2015**, *11*, e1005001. [CrossRef]

44. Funnell, T.; Tasaki, S.; Oloumi, A.; Araki, S.; Kong, E.; Yap, D.; Nakayama, Y.; Hughes, C.S.; Cheng, S.G.; Tozaki, H.; et al. CLK-dependent exon recognition and conjoined gene formation revealed with a novel small molecule inhibitor. *Nat. Commun.* **2017**, *8*, 7. [CrossRef] [PubMed]

45. Akiva, P.; Toporik, A.; Edelheit, S.; Peretz, Y.; Diber, A.; Shemesh, R.; Novik, A.; Sorek, R. Transcription-mediated gene fusion in the human genome. *Genome Res.* **2006**, *16*, 30–36. [CrossRef]

46. Kim, R.N.; Kim, A.; Choi, S.H.; Kim, D.S.; Nam, S.H.; Kim, D.W.; Kim, D.W.; Kang, A.; Kim, M.Y.; Park, K.H.; et al. Novel mechanism of conjoined gene formation in the human genome. *Funct. Integr. Genom.* **2012**, *12*, 45–61. [CrossRef]

47. Kerwitz, Y.; Kühn, U.; Lilie, H.; Knoth, A.; Scheuermann, T.; Friedrich, H.; Schwarz, E.; Wahle, E. Stimulation of poly(A) polymerase through a direct interaction with the nuclear poly(A) binding protein allosterically regulated by RNA. *EMBO J.* **2003**, *22*, 3705–3714. [CrossRef] [PubMed]

48. Kühn, U.; Nemeth, A.; Meyer, S.; Wahle, E. The RNA binding domains of the nuclear poly(A)-binding protein. *J. Biol. Chem.* **2003**, *278*, 16916–16925. [CrossRef]

49. Wang, T.; Zhong, D.; Qin, Z.; He, S.; Gong, Y.; Li, W.; Li, X. miR-100-3p inhibits the adipogenic differentiation of hMSCs by targeting PIK3R1 via the PI3K/AKT signaling pathway. *Aging* **2020**, *12*, 25090–25100. [CrossRef]

50. Liu, X.; Li, Z.; Liu, H.; Zhu, Y.; Xia, D.; Wang, S.; Gu, R.; Zhang, P.; Liu, Y.; Zhou, Y. Flufenamic Acid Inhibits Adipogenic Differentiation of Mesenchymal Stem Cells by Antagonizing the PI3K/AKT Signaling Pathway. *Stem Cells Int.* **2020**, *2020*, 1540905. [CrossRef]

51. Savova, M.S.; Vasileva, L.V.; Mladenova, S.G.; Amirova, K.M.; Ferrante, C.; Orlando, G.; Wabitsch, M.; Georgiev, M.I. Ziziphus jujuba Mill. leaf extract restrains adipogenesis by targeting PI3K/AKT signaling pathway. *Biomed. Pharmacother.* **2021**, *141*, 111934. [CrossRef]

52. Adams, M.; Reginato, M.J.; Shao, D.; Lazar, M.A.; Chatterjee, V.K. Transcriptional activation by peroxisome proliferator-activated receptor gamma is inhibited by phosphorylation at a consensus mitogen-activated protein kinase site. *J. Biol. Chem.* **1997**, *272*, 5128–5132. [CrossRef] [PubMed]

53. Camp, H.S.; Tafuri, S.R. Regulation of peroxisome proliferator-activated receptor gamma activity by mitogen-activated protein kinase. *J. Biol. Chem.* **1997**, *272*, 10811–10816. [CrossRef] [PubMed]

54. Hu, E.; Kim, J.B.; Sarraf, P.; Spiegelman, B.M. Inhibition of adipogenesis through MAP kinase-mediated phosphorylation of PPARgamma. *Science* **1996**, *274*, 2100–2103. [CrossRef] [PubMed]

55. Wang, M.; Wang, J.J.; Li, J.; Park, K.; Qian, X.; Ma, J.X.; Zhang, S.X. Pigment epithelium-derived factor suppresses adipogenesis via inhibition of the MAPK/ERK pathway in 3T3-L1 preadipocytes. *Am. J. Physiol. Endocrinol. Metab.* **2009**, *297*, E1378–E1387. [CrossRef] [PubMed]

56. Xiao, F.; Tang, C.Y.; Tang, H.N.; Wu, H.X.; Hu, N.; Li, L.; Zhou, H.D. Long Non-coding RNA 332443 Inhibits Preadipocyte Differentiation by Targeting Runx1 and p38-MAPK and ERK1/2-MAPK Signaling Pathways. *Front. Cell Dev. Biol.* **2021**, *9*, 663959. [CrossRef]

Article

Additive and Dominance Genomic Analysis for Litter Size in Purebred and Crossbred Iberian Pigs

Houssemeddine Srihi [1], José Luis Noguera [2], Victoria Topayan [3], Melani Martín de Hijas [4], Noelia Ibañez-Escriche [3], Joaquim Casellas [4], Marta Vázquez-Gómez [4], María Martínez-Castillero [1], Juan Pablo Rosas [5] and Luis Varona [1,*]

[1] Departamento de Anatomía, Embriología y Genética Animal, Facultad de Veterinaria, Instituto Agrolimentario de Aragón (IA2), 50013 Zaragoza, Spain; houssemsrihi@unizar.es (H.S.); mmartinezcastillero@gmail.com (M.M.-C.)

[2] Area de Producció Animal, Centre UdL-IRTA, 25198 Lleida, Spain; JoseLuis.Noguera@irta.cat

[3] Departamento de Ciència Animal, Universitat Politècnica de València, 46071 Valencia, Spain; victoriatopayan@gmail.com (V.T.); noeibes@dca.upv.es (N.I.-E.)

[4] Departament de Ciència Animal i dels Aliments, Universitat Autònoma de Barcelona, 08193 Barcelona, Spain; melani.mhv@gmail.com (M.M.d.H.); joaquim.casellas@uab.cat (J.C.); Marta.Vazquez@uab.cat (M.V.-G.)

[5] Programa de Mejora Genética "Castúa", INGA FOOD S.A. (Nutreco), Avda. A Rúa, 2—Bajo Edificio San Marcos, 06200 Almendralejo, Spain; juan.rosas@nutreco.com

* Correspondence: lvarona@unizar.es; Tel.: +34-876-554209

check for updates

Citation: Srihi, H.; Noguera, J.L.; Topayan, V.; Martín de Hijas, M.; Ibañez-Escriche, N.; Casellas, J.; Vázquez-Gómez, M.; Martínez-Castillero, M.; Rosas, J.P.; Varona, L. Additive and Dominance Genomic Analysis for Litter Size in Purebred and Crossbred Iberian Pigs. *Genes* **2022**, *13*, 12. https://doi.org/10.3390/genes13010012

Academic Editor: Katarzyna Piórkowska

Received: 30 November 2021
Accepted: 17 December 2021
Published: 22 December 2021

Publisher's Note: MDPI stays neutral with regard to jurisdictional claims in published maps and institutional affiliations.

Abstract: INGA FOOD S. A., as a Spanish company that produces and commercializes fattened pigs, has produced a hybrid Iberian sow called CASTÚA by crossing the Retinto and Entrepelado varieties. The selection of the parental populations is based on selection criteria calculated from purebred information, under the assumption that the genetic correlation between purebred and crossbred performance is high; however, these correlations can be less than one because of a GxE interaction or the presence of non-additive genetic effects. This study estimated the additive and dominance variances of the purebred and crossbred populations for litter size, and calculated the additive genetic correlations between the purebred and crossbred performances. The dataset consisted of 2030 litters from the Entrepelado population, 1977 litters from the Retinto population, and 1958 litters from the crossbred population. The individuals were genotyped with a GeneSeek® GGP Porcine70K HDchip. The model of analysis was a 'biological' multivariate mixed model that included additive and dominance SNP effects. The estimates of the additive genotypic variance for the total number born (TNB) were 0.248, 0.282 and 0.546 for the Entrepelado, Retinto and Crossbred populations, respectively. The estimates of the dominance genotypic variances were 0.177, 0.172 and 0.262 for the Entrepelado, Retinto and Crossbred populations. The results for the number born alive (NBA) were similar. The genetic correlations between the purebred and crossbred performance for TNB and NBA—between the brackets—were 0.663 in the Entrepelado and 0.881 in Retinto poplulations. After backsolving to obtain estimates of the SNP effects, the additive genetic variance associated with genomic regions containing 30 SNPs was estimated, and we identified four genomic regions that each explained >2% of the additive genetic variance in chromosomes (SSC) 6, 8 and 12: one region in SSC6, two regions in SSC8, and one region in SSC12.

Keywords: pig; Iberian; additive; dominance; genetic correlation; crossbreeding; genomic selection

1. Introduction

The Iberian pig breed is one of the porcine populations that has the highest meat quality [1]. Historically, Iberian pig production was developed extensively with purebred varieties, which took advantage of the *Dehesa* environment in southwestern Spain. In recent decades, however, many traditional production systems have been substituted with intensive production systems that use crossbreeding with Duroc populations to improve growth and efficiency [2]. The norms that regulate Iberian pig production [3] obligate farmers crossing Iberian and Duroc varieties to cross boars from the Duroc variety and sows from the Iberian variety. Prolificacy, which is lower than that of white pig populations, is the major limitation in the intensive production of crossbred pigs from Iberian dams [4]. The INGA FOOD, S.A. company has developed a crossbreeding scheme between two Iberian varieties (Retinto–R- and Entrepelado–E-) that has created a hybrid sow called CASTUA–ER, which has an important heterosis effect in prolificacy [5]. In addition, the company has been developing a breeding scheme for increasing litter size through selection in the parental Retinto and Entrepelado populations.

Theoretically, the optimal strategy for the selection of purebreds for crossbred performance is Recurrent Reciprocal Selection [6]; however, it has not been routinely used in pig breeding because it involves a delay in the generation interval. In fact, purebred parental populations are selected based on selection criteria calculated from purebred phenotypic information, and under the assumption that the genetic correlation between purebred and crossbred performance is high [7]. Those genetic correlations can be imperfect (<1) because of genotype-by-environment (GxE) interactions and the presence of non-additive genetic effects [7].

Genomic information facilitates the analysis of crossbreeding data, even if genotyped and phenotyped individuals are not directly related [8], by the definition of an additive-dominance genotypic model that provides estimates of genotype x environmental interactions through genotypic correlations. In addition, the estimates of genotypic and dominance variances can be used to estimate the additive genetic correlation between purebred and crossbred performances. Backsolving, as proposed by Wang et al. [9], provides an estimate of the SNP effects and allows us to calculate the amount of additive genetic variance associated with each genomic region in purebred and crossbred performances.

This study estimated the additive and dominance genotypic variances and covariances, which were used to calculate the additive and dominance genetic variances and the genetic correlations between purebred and crossbred performances in the Retinto and Entrepelado populations. In addition, the distribution of the additive genetic variance within the autosomal genome for purebred and crossbred performance was quantified.

2. Materials and Methods

The phenotypic data included the number of piglets born alive (NBA) and the total number born (TNB) for 306 Entrepelado and 313 Retinto purebred sows, and for 333 crossbred (Entrepelado x Retinto) sows when crossed with Entrepelado, Retinto or Duroc boars (Table 1).

Table 1. Number of records (and number of sows between brackets) and the mean ± standard deviation of the number born alive and the total number born in Entrepelado, Crossbred and Retinto populations.

	Entrepelado	Crossbred	Retinto
N [1] (NS) [2]	2030 (306)	1958 (333)	1977 (313)
NBA [3]	7.75 ± 1.85	8.57 ± 2.27	8.07 ± 2.07
TNB [4]	8.02 ± 1.89	8.80 ± 2.29	8.33 ± 2.11

[1] N: number of records. [2] NS: number of sows. [3] NBA: number born alive. [4] TNB: total number born.

All of the sows were genotyped with the GeneSeek® GPP Porcine 70K HDchip (Illumina Inc., San Diego, CA, USA). Filtering excluded genotypes that had a minor allele frequency < 0.05 and an SNP call rate < 0.90 in the overall population. From that, 34,316 SNP markers were used to build the genomic relationship matrices with our own developed software in the R environment [10]. The missing genotypes were replaced with their expectation.

The model of analysis assumed that the phenotypic values of individuals (y) (TNB and NBA) are explained by the (biological) additive (u) and dominance (v) effects of the SNPs, and a covariate (c) with the average homozygosity (f), the systematic effects (b)

order of parity (1, 2, 3, and >3), the sire of service breed (Entreplado, Retinto, or Duroc) and herd-year-season (122 levels). Phenotypic data were generated in three herds, and herd-year-season effects were defined every 3 months. The sow permanent environmental effects (s) with 2030, 1958 and 1977 levels for the Entrepelado, Retinto and Crossbred populations, and the residuals (e), were as follows:

$$
\begin{bmatrix} y_E \\ y_R \\ y_{ER} \end{bmatrix} = \begin{bmatrix} f_E c_E \\ f_R c_R \\ f_{RE} c_{RE} \end{bmatrix} + \begin{bmatrix} X_E & 0 & 0 \\ 0 & X_R & 0 \\ 0 & 0 & X_{RE} \end{bmatrix} \begin{bmatrix} b_E \\ b_R \\ b_{ER} \end{bmatrix} + \begin{bmatrix} T_E & 0 & 0 \\ 0 & T_R & 0 \\ 0 & 0 & T_{RE} \end{bmatrix} \begin{bmatrix} s_E \\ s_R \\ s_{ER} \end{bmatrix} + \begin{bmatrix} T_E & 0 & 0 \\ 0 & T_R & 0 \\ 0 & 0 & T_{RE} \end{bmatrix} \begin{bmatrix} u_E \\ u_R \\ u_{ER} \end{bmatrix} + \begin{bmatrix} T_E & 0 & 0 \\ 0 & T_R & 0 \\ 0 & 0 & T_{RE} \end{bmatrix} \begin{bmatrix} v_E \\ v_R \\ v_{ER} \end{bmatrix} + \begin{bmatrix} e_E \\ e_R \\ e_{ER} \end{bmatrix}
$$

where X and T are the corresponding incidence matrices. Following Vitezica et al. [8], u and v can be described in terms of the vectors of additive (a) and dominance (d) SNP genotypic effects as follows:

$$
\begin{bmatrix} u_E \\ u_R \\ u_{RE} \end{bmatrix} = \begin{bmatrix} Za_E \\ Za_R \\ Za_{ER} \end{bmatrix} \text{ and } \begin{bmatrix} v_E \\ v_R \\ v_{RE} \end{bmatrix} = \begin{bmatrix} Wd_E \\ Wd_R \\ Wd_{ER} \end{bmatrix}
$$

The matrices $Z = (z_1 \ldots \ldots z_m)$ and $W = (w_1 \ldots \ldots w_m)$ are equal to 1, 0, -1 and 0, 1, 0 for SNP genotypes A_1A_1, A_1A_2 and A_2A_2, respectively.

The covariance across individual genotypic additive (u) and dominance (v) effects is

$$
cov \begin{bmatrix} u_E \\ u_R \\ u_{RE} \end{bmatrix} = G_o \bigotimes G \text{ and } cov \begin{bmatrix} v_E \\ v_R \\ v_{RE} \end{bmatrix} = D_o \bigotimes D
$$

with

$$
G_o = \begin{bmatrix} \sigma^2_{U_E} & \sigma_{U_E U_R} & \sigma_{U_E U_{ER}} \\ \sigma_{U_E U_R} & \sigma^2_{U_R} & \sigma_{U_R U_{ER}} \\ \sigma_{U_E U_{ER}} & \sigma_{U_R U_{ER}} & \sigma^2_{U_{ER}} \end{bmatrix} \text{ and } D_o = \begin{bmatrix} \sigma^2_{V_E} & \sigma_{V_E V_R} & \sigma_{V_E V_{ER}} \\ \sigma_{V_E V_R} & \sigma^2_{V_R} & \sigma_{V_R V_{ER}} \\ \sigma_{V_E V_{ER}} & \sigma_{V_R V_{ER}} & \sigma^2_{V_{ER}} \end{bmatrix}
$$

and

$$
G = \frac{ZZ'}{\{tr[ZZ']/n\}} \text{ and } D = \frac{WW'}{\{tr[WW']/n\}}
$$

The variance components were estimated by REML [11] through the EM-REML algorithm using *remlf90* software [12] and, in order to obtain the average information matrix, we used one extra iteration with *airemlf90*. Additive and dominance variance components were calculated in each of the populations (E, R, and ER) as follows:

$$
\begin{bmatrix} \hat{\sigma}^2_{aE} \\ \hat{\sigma}^2_{aR} \\ \hat{\sigma}^2_{aER} \end{bmatrix} = \begin{bmatrix} \dfrac{\hat{\sigma}^2_{U_E}}{\{tr[ZZ']/n\}} \\ \dfrac{\hat{\sigma}^2_{U_R}}{\{tr[ZZ']/n\}} \\ \dfrac{\hat{\sigma}^2_{U_{ER}}}{\{tr[ZZ']/n\}} \end{bmatrix} \text{ and } \begin{bmatrix} \hat{\sigma}^2_{dE} \\ \hat{\sigma}^2_{dR} \\ \hat{\sigma}^2_{dER} \end{bmatrix} = \begin{bmatrix} \dfrac{\hat{\sigma}^2_{D_E}}{\{tr[WW']/n\}} \\ \dfrac{\hat{\sigma}^2_{D_R}}{\{tr[WW']/n\}} \\ \dfrac{\hat{\sigma}^2_{D_{ER}}}{\{tr[WW']/n\}} \end{bmatrix} \tag{1}
$$

The additive (σ^2_A) and dominance (σ^2_D) genetic variances of the purebred populations were calculated as follows:

$$
\hat{\sigma}^2_{A_E} = \sum_{i=1}^{n} 2\hat{p}_{Ei}\hat{q}_{Ei}\hat{\sigma}^2_{a_E} + 2\hat{p}_{Ei}\hat{q}_{Ei}(\hat{q}_{Ei} - \hat{p}_{Ei})^2\hat{\sigma}^2_{d_E} \tag{2}
$$

$$
\hat{\sigma}^2_{D_E} = \sum_{i=1}^{n} (2\hat{p}_{Ei}\hat{q}_{Ei})^2\hat{\sigma}^2_{d_E} \tag{3}
$$

$$
\hat{\sigma}^2_{A_R} = \sum_{i=1}^{n} 2\hat{p}_{Ri}\hat{q}_{Ri}\hat{\sigma}^2_{a_R} + 2\hat{p}_{Ri}\hat{q}_{Ri}(\hat{q}_{Ri} - \hat{p}_{Ri})^2\hat{\sigma}^2_{d_R} \tag{4}
$$

$$
\hat{\sigma}^2_{D_R} = \sum_{i=1}^{n} (2\hat{p}_{Ri}\hat{q}_{Ri})^2\hat{\sigma}^2_{d_R} \tag{5}
$$

where $\hat{p}_{Xi}$ and $\hat{q}_{Xi}$ are the raw estimates of the allelic frequencies for A_1 and A_2 at the *ith* SNP marker and the $X = \{E, R \text{ or } ER\}$ population, respectively. The estimates of the contributions to the additive variance in the crossbred population from the Entrepelado ($\sigma^2_{A_{ER(E)}}$) and Retinto ($\sigma^2_{A_{ER(R)}}$) were obtained by [8] as follows:

$$\hat{\sigma}^2_{A_{ER(E)}} = \sum_{i=1}^{n} 2\hat{p}_{Ri}\hat{q}_{Ri}\hat{\sigma}^2_{a_{ER}} + 2\hat{p}_{Ri}\hat{q}_{Ri}(\hat{q}_{Ei} - \hat{p}_{Ei})^2\hat{\sigma}^2_{d_{ER}} \tag{6}$$

$$\hat{\sigma}^2_{A_{ER(R)}} = \sum_{i=1}^{n} 2\hat{p}_{Ei}\hat{q}_{Ei}\hat{\sigma}^2_{a_{ER}} + 2\hat{p}_{Ei}\hat{q}_{Ei}(\hat{q}_{Ri} - \hat{p}_{Ri})^2\hat{\sigma}^2_{d_{ER}} \tag{7}$$

Following Vitezica et al. [8], the additive variance in the crossbred population was the average of the two values resulting from Equations (6) and (7), as follows:

$$\hat{\sigma}^2_{A_{ER}} = \frac{1}{2}\hat{\sigma}^2_{A_{ER(E)}} + \frac{1}{2}\hat{\sigma}^2_{A_{ER(R)}} \tag{8}$$

The estimate of the dominance variance of the crossbred population [8] was calculated as follows:

$$\hat{\sigma}^2_{D_{ER}} = \sum_{i=1}^{n} 4\hat{p}_{Ei}\hat{q}_{Ei}\hat{p}_{Ri}\hat{q}_{Ri}\hat{\sigma}^2_{d_{ER}} \tag{9}$$

With these estimates, the heritabilities (h_X^2) and dominance ratios (d_X^2) in the purebred ($X = E,R$) and crossbred ($X = ER$) populations were obtained by:

$$\hat{h}_X^2 = \hat{\sigma}^2_{A_X} / \left(\hat{\sigma}^2_{A_X} + \hat{\sigma}^2_{D_X} + \hat{\sigma}^2_{S_X} + \hat{\sigma}^2_{E_X}\right) \tag{10}$$

$$\hat{d}_X^2 = \hat{\sigma}^2_{D_X} / \left(\hat{\sigma}^2_{A_X} + \hat{\sigma}^2_{D_X} + \hat{\sigma}^2_{S_X} + \hat{\sigma}^2_{E_X}\right) \tag{11}$$

where $\hat{\sigma}^2_{S_X}$ and $\hat{\sigma}^2_{E_X}$ are the estimates of the sow permanent environmental and residual variance in the $X = \{E,R,ER\}$ population.

The covariance between purebred and crossbred additive genetic effects in the Entrepelado ($\sigma_{A_E A_{ER(E)}}$) and Retinto ($\sigma_{A_R A_{ER(R)}}$) populations were as follows:

$$\hat{\sigma}_{A_E A_{ER(E)}} = \sum_{i=1}^{n} 2\hat{p}_{Ei}\hat{q}_{Ei}\hat{\sigma}_{a_E a_{ER}} + 2\hat{p}_{Ei}\hat{q}_{Ei}(\hat{q}_{Ei} - \hat{p}_{Ei})(\hat{q}_{Ri} - \hat{p}_{Ri})\hat{\sigma}_{d_E d_{ER}} \tag{12}$$

$$\hat{\sigma}_{A_E A_{ER(R)}} = \sum_{i=1}^{n} 2\hat{p}_{Ri}\hat{q}_{Ri}\hat{\sigma}_{a_R a_{ER}} + 2\hat{p}_{Ri}\hat{q}_{Ri}(\hat{q}_{Ri} - \hat{p}_{Ri})(\hat{q}_{Ei} - \hat{p}_{Ei})\hat{\sigma}_{d_R d_{ER}} \tag{13}$$

with

$$\begin{bmatrix} \hat{\sigma}_{a_E a_{ER}} \\ \hat{\sigma}_{a_R a_{ER}} \end{bmatrix} = \begin{bmatrix} \frac{\hat{\sigma}_{u_E u_{ER}}}{\{tr[\mathbf{ZZ'}]/n\}} \\ \frac{\hat{\sigma}_{u_R u_{ER}}}{\{tr[\mathbf{ZZ'}]/n\}} \end{bmatrix} \text{ and } \begin{bmatrix} \hat{\sigma}_{d_E d_{ER}} \\ \hat{\sigma}_{d_R d_{ER}} \end{bmatrix} = \begin{bmatrix} \frac{\hat{\sigma}_{v_E v_{ER}}}{\{tr[\mathbf{WW'}]/n\}} \\ \frac{\hat{\sigma}_{v_R v_{ER}}}{\{tr[\mathbf{WW'}]/n\}} \end{bmatrix} \tag{14}$$

Therefore, the genetic correlations between the purebred and crossbreed breeding values in the Entrepelado and Retinto populations were computed as follows:

$$\hat{r}_{A_E A_{ER(E)}} = \frac{\hat{\sigma}_{A_E A_{ER(E)}}}{\sqrt{\hat{\sigma}^2_{A_E}\hat{\sigma}^2_{A_{ER(E)}}}} \text{ and } \hat{r}_{A_R A_{ER(R)}} = \frac{\hat{\sigma}_{A_R A_{ER(R)}}}{\sqrt{\hat{\sigma}^2_{A_R}\hat{\sigma}^2_{A_{ER(R)}}}} \tag{15}$$

The vector of the SNP additive effects ($\hat{a}_E$, $\hat{a}_R$ and $\hat{a}_{ER}$) was obtained by backsolving [9], as

$$\hat{a}_E = \frac{\hat{\sigma}^2_{a_E}}{\hat{\sigma}^2_{u_E}}\mathbf{Z}\mathbf{G}^{-1}\hat{u}_E, \ \hat{a}_R = \frac{\hat{\sigma}^2_{a_R}}{\hat{\sigma}^2_{u_R}}\mathbf{Z}\mathbf{G}^{-1}\hat{u}_R \text{ and } \hat{a}_{ER} = \frac{\hat{\sigma}^2_{a_{ER}}}{\hat{\sigma}^2_{u_{ER}}}\mathbf{Z}\mathbf{G}^{-1}\hat{u}_{ER} \tag{16}$$

and the vector of the SNP dominance effects ($\hat{d}_E$, $\hat{d}_R$ and $\hat{d}_{ER}$) was as follows:

$$\hat{d}_E = \frac{\hat{\sigma}^2_{d_E}}{\hat{\sigma}^2_{v_E}}\mathbf{W}\mathbf{D}^{-1}\hat{v}_E, \ \hat{d}_R = \frac{\hat{\sigma}^2_{d_R}}{\hat{\sigma}^2_{v_R}}\mathbf{W}\mathbf{D}^{-1}\hat{v}_R \text{ and } \hat{d}_{ER} = \frac{\hat{\sigma}^2_{d_{ER}}}{\hat{\sigma}^2_{v_{ER}}}\mathbf{W}\mathbf{D}^{-1}\hat{v}_{ER} \tag{17}$$

With those, the genetic additive variances ($\sigma^2_{A_E(k)}$, $\sigma^2_{A_R(k)}$, $\sigma^2_{A_{ER(E)}(k)}$ and $\sigma^2_{A_{ER(R)}(k)}$) explained by the kth segment of the genome were calculated as follows:

$$\hat{\sigma}^2_{A_E(k)} = \sum_{i=1}^{n(k)} 2\hat{p}_{Ei}\hat{q}_{Ei}\hat{a}^2_{E_i} + 2\hat{p}_{Ei}\hat{q}_{Ei}(\hat{q}_{Ei} - \hat{p}_{Ei})^2\hat{d}^2_{E_i} \tag{18}$$

$$\hat{\sigma}^2_{A_R(k)} = \sum_{i=1}^{n(k)} 2\hat{p}_{Ri}\hat{q}_{Ri}\hat{a}^2_{R_i} + 2\hat{p}_{Ri}\hat{q}_{Ri}(\hat{q}_{Ri} - \hat{p}_{Ri})^2\hat{d}^2_{R_i} \tag{19}$$

$$\hat{\sigma}^2_{A_{ER(E)}(k)} = \sum_{i=1}^{n(k)} 2\hat{p}_{Ri}\hat{q}_{Ri}\hat{a}^2_{ER_i} + 2\hat{p}_{Ri}\hat{q}_{Ri}(\hat{q}_{Ei} - \hat{p}_{Ei})^2\hat{d}^2_{ER_i} \tag{20}$$

$$\hat{\sigma}^2_{A_{ER(R)}(k)} = \sum_{i=1}^{n(k)} 2\hat{p}_{Ei}\hat{q}_{Ei}\hat{a}^2_{ER_i} + 2\hat{p}_{Ei}\hat{q}_{Ei}(\hat{q}_{Ri} - \hat{p}_{Ri})^2\hat{d}^2_{ER_i} \tag{21}$$

where $n(k)$ is the number of SNP markers within the kth segment, which was set to 30 after testing several number of the SNP markers (20, 30 and 40). In order to identify the genes within the genomic regions that explained >2.0% of the total genetic variance, we used the biomart tool (www.ensembl.org (accessed on 10 October 2021)).

3. Results and Discussion

The results based on TNB and NBA were similar, which was expected because these two traits have a high genetic correlation [13], and the raw correlation between them in the analyzed dataset was 0.94; therefore, we focused on the results with the TNB, and the results for NBA are presented as Supplementary Information (Tables S1–S3 and Figure S1). The REML estimates of the additive genotypic (co) variances are shown in Tables 2 and 3 in TNB and NBA.

Table 2. REML estimates ± standard error (SE) of the additive genotypic (co)variances for the total number born (TNB).

	Entrepelado	**Crossbred**	**Retinto**
Entrepelado	0.248 ± 0.161	0.259 ± 0.178	0.200 ± 0.135
Crossbred	-	0.546 ± 0.268	0.388 ± 0.170
Retinto	-	-	0.282 ± 0.146

Table 3. REML estimates ± standard error (SE) of the dominance genotypic (co)variances for the total number born (TNB).

	Entrepelado	**Crossbred**	**Retinto**
Entrepelado	0.177 ± 0.165	0.212 ± 0.171	0.166 ± 0.152
Crossbred	-	0.262 ± 0.210	0.202 ± 0.179
Retinto	-	-	0.172 ± 0.199

The additive genotypic variance was higher in the crossbred populations than it was in the purebred populations. This may be due to scale effects, as the phenotypic variation in also greater. In addition, the estimates of the genotypic covariances between purebreds (Entrepelado and Retinto) and the crossbred population were all high and positive, and they corresponded to additive genotypic correlations of 0.704 ($0.259/\sqrt{0.248 \times 0.546}$) between Entrepelado and Crossbred pigs, 0.988 ($0.388/\sqrt{0.546 \times 0.282}$) between Retinto and Crossbred pigs, and 0.756 ($0.200/\sqrt{0.248 \times 0.282}$) between the two purebreds. These results indicated that the genotype x environmental interaction was small, and the additive genotypic correlations were similar to those obtained by Vitezica et al. [8] in white pig populations. The REML estimates of the dominance genotypic (co)variances ranged from 0.170 (Retinto) to 0.265 (Crossbred) (Table 3).

The estimates of the dominance genotypic covariances were all positive, and reflected genotypic dominance correlations >0.95. The analysis provided the REML estimates of the sow permanent and residual effects (Table 4). The residual variance (σ_E^2) is greater in the crossbred population than in purebreds, consistently with the greater phenotypic variation. In contrast, the estimate of the sow environmental variance (σ_R^2) was very low in the crossbred population.

Table 4. REML estimates ± standard error (SE) of the permanent environmental and residual variances for the total number born (TNB) in the Entrepelado, Crossbred and Retinto populations.

Variances [1]	Entrepelado	Crossbred	Retinto
σ_S^2	0.191 ± 0.105	0.009 ± 0.029	0.268 ± 0.120
σ_E^2	2.810 ± 0.099	4.467 ± 0.155	3.534 ± 0.128

[1] σ_S^2: Sow permanent environmental variance; σ_E^2: residual variance.

The additive and dominance genotypic (co) variances were used to calculate the additive and dominance genetic variances in the purebred populations based on expressions (1) to (5) (Table 3). The estimates of the additive genetic variances were 0.170 (Entrepelado) and 0.150 (Retinto), and the estimates of the dominance genetic variances were 0.074 (Entrepelado) and 0.056 (Retinto). The heritability estimates were calculated using Equation (10); they were 0.052 (Entrepelado) and 0.037 (Retinto), which were within the range or slightly lower than those of white pigs [13–15] and in the same [5] or other Iberian [16,17] populations. The dominance ratios were obtained from Equation (11), and were 0.023 for Entrepelado and 0.014 for Retinto. They were smaller than the heritabilities, but their ratios with them were approximately 40%, which was higher than those reported for white pig populations [8,18] for litter size and similar to the results of Tusell et al. [19] in other swine traits.

We used Equations (6) and (7) to calculate the additive variances for crossbred performance in the purebred populations, which were 0.413 (Entrepelado) and 0.293 (Retinto). Therefore, the additive genetic variance in the crossbred population was the average of the two (0.353), which was higher than the additive genetic variances in the purebred populations, which were similar to the results of Vitezica et al. [8] with regard to litter size, and to the results of Tusell et al. [19] for other pig traits. Nevertheless, Xiang et al. [20] found the opposite in a cross between Landrace and Yorkshire breeds (0.86 and 0.54 in purebreds and 0.28 in crossbreds). In the present study, the dominance genetic variance in the crossbred population (0.079) was calculated based on the Equation (8), which was similar to the dominance genetic variance in the purebreds; however, its ratio with the additive genetic variances was lower (22%). Given those variance components, the heritability and dominance ratio estimates in the crossbred population were 0.072 and 0.016, respectively.

In addition, the additive genetic correlations between purebred and crossbred performances in the Entrepelado and Retinto populations were calculated based on expressions (12) to (15), which were 0.663 in Entrepelado and 0.881 in Retinto populations. Those correlations were within the range of the estimates summarized by Wientjes and Calus [7], and suggest that the efficiency of the selection for increased crossbred performance by selecting for purebred performance will be more effective in Retinto than in Entrepelado pigs.

We used Equations (16) and (17) to calculate the additive and dominance genotypic effects associated with each of the 34,316 SNP markers, which were used in Equations (18)–(21) to calculate the proportion of the additive genetic variance that was explained by segments of 30 consecutive SNPs (Figure 1). The distribution of the additive variance explained by segments of 20 and 40 SNP markers were similar, and are presented as supplementary information (Figures S2 and S3).

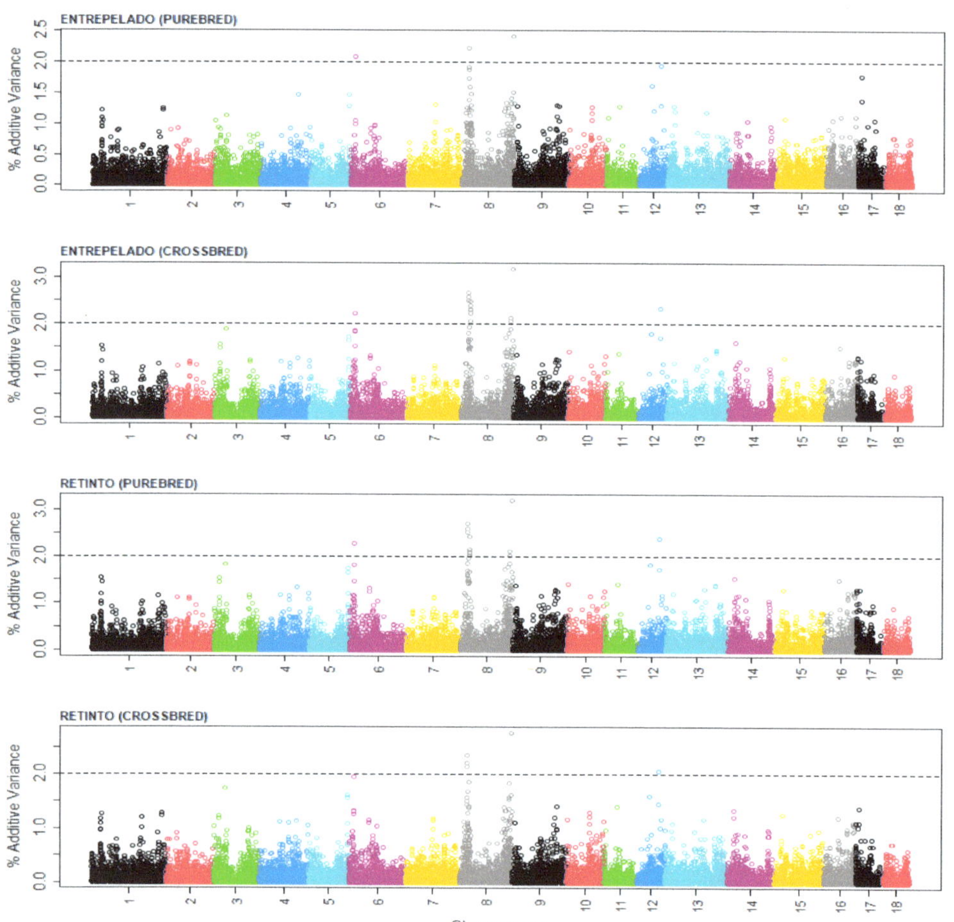

Figure 1. Distribution of the percentage of the additive genetic variance explained by genomic segments of 30 SNPs within the autosomal genome of purebred and crossbred performance for the total number born (TNB) in the Entrepelado and Retinto varieties. Black: chromosomes 1, 9 and 17; red: chromosomes 2, 10 and 18; green: chromosomes 3 and 11; deep blue: chromosomes 4 and 12; blue: chromosomes 5 and 13; purple: chromosomes 6 and 14; yellow: chromosomes 7 and 15; grey: chromosomes 8 and 16.

The figure presents the distribution of the additive variance along the autosomal chromosomes in the Entrepelado and Retinto populations, and for the purebred and crossbred performance. Four genomic regions can be highlighted; each explained >2% of the additive genetic variance in at least one of the populations. The SNPs at the center of each of the genomic regions that explained the highest amount of additive genetic variance, and the genes in the Sus_Scrofa 11.1. genomic map that were within 1 Mb downstream or upstream, are presented in Table 5.

Table 5. SNPs at the center of each of the four genomic regions that explained > 2% of the additive genetic variance in at least one of the populations, and the genes located within 1 Mb downstream or upstream.

SNP [1]	SSC [2]	bp [3]	Genes
rs326244568	6	7,597,405	*BCO1, PKD1L2, GCSH, ATMIN, CENPN, CDYL2, DYNLRB2*
rs81401202	8	11,585,865	*CD38, FGFBP1, PROM1, TAPT1, LDB2*
rs81406142	8	137,540,516	*CFAP299, FGF5, PRDM8, ANTXR2*
rs345468811	12	46,079,417	*TAOK1, ABHD15, TP53I13, GIT1, ANKRD1, CORO6, EFCAB5, NSRP1, SLC6A4, BLMH, TMIGD1, CPD, GOSR1*

[1] SNP: single nucleotide polymorphism, [2] SSC: Sus Scrofa chromosome, [3] bp: base pair.

Among those genes, several can be proposed as candidate genes to explain the additive genetic variation. The genomic region surrounding bp 7,597,405 in SSC6 included BCO1 (*β-Carotene Oxygenase 1*), which encodes an enzyme that catalyzes the breakdown of provitamin A and provides retinoids for embryogenesis [21,22]. Furthermore, the GCSH (*Glycine Cleavage System H*) protein plays an important role in embryonic viability [23].

Two genomic regions were identified in SSC8 around bp 11,585,865 and bp 137,540,516. Among the genes within those regions, PRDM8 (*PR/SET Domain 8*) is involved in the neurogenesis [24] of the FGF5 (*Fibroblast Growth Factor 5*), a member of the fibroblast growth factor family that is involved in several biological processes, including embryonic development, cell growth, and morphogenesis [25,26].

The genomic region around bp 46,079,417 in SSC12 contains, among others, the GIT1 (*G protein-coupled receptor kinase interactor 1*) gene, which plays a role in spine morphogenesis [27], the NSRP1 (*Nuclear Speckle Splicing Regulatory Protein 1*) development process, and in utero embryonic development [28], and ANKRD1 (*Ankyrin Repeat Domain 1*), which is involved in neuron projection development [29].

The Gene Ontology (GO) terms for the biological processes for the proposed candidate genes are presented as Supplementary Table S4.

4. Conclusions

(1) The additive genetic variance and the heritabilities were higher in the crossbred than those in the purebred populations, (2) the genetic correlation between purebred and crossbred performances were higher in Retinto than they were in Entrepelado pigs, and (3) the additive genetic variances were heterogeneously distributed throughout the autosomal genome, and four genomic regions in SSC6, SSC8, and SSC12 with several candidate genes were identified.

Supplementary Materials: The following supporting information can be downloaded at: https://www.mdpi.com/article/10.3390/genes13010012/s1. Table S1: REML estimates ± the standard error (SE) of the additive genotypic (co)variances for the number born alive (NBA). Table S2: REML estimates ± the standard error (SE) of the dominance genotypic (co)variances for the number born alive (NBA). Table S3: REML estimates ± the standard error (SE) of the permanent environmental and residual variances for the number born alive (NBA). Table S4: GO (Gene Ontology) terms for the biological process of the proposed candidate genes. Figure S1: Distribution of the percentage of the additive genetic variance explained by genomic segments of 30 SNPs within the autosomal genome of the purebred and crossbred performance for the number born alive (NBA) in the Entrepelado and Retinto varieties. Figure S2: Distribution of the percentage of the additive genetic variance explained by genomic segments of 20 SNPs within the autosomal genome of purebred and crossbred performance for the total number born (TNB) in the Entrepelado and Retinto varieties. Figure S3: Distribution of the percentage of the additive genetic variance explained by genomic segments of 40 SNPs within the autosomal genome of the purebred and crossbred performance for the total number born (TNB) in the Entrepelado and Retinto varieties.

Author Contributions: Conceptualization, L.V.; methodology, H.S. and L.V.; software, L.V.; formal analysis, H.S., V.T. and L.V.; data curation, J.L.N., J.P.R., M.M.d.H. and N.I.-E.; writing—original draft preparation, H.S. and L.V.; writing—review and editing M.M.d.H., M.V.-G., N.I.-E., M.M.-C. and J.C.;

supervision, L.V., N.I.-E., J.C. and J.L.N.; project administration, J.L.N., J.C. and N.I.-E.; funding acquisition, J.L.N., J.C. and N.I.-E. All authors have read and agreed to the published version of the manuscript.

Funding: The research was partially funded by grants CGL2016-80155-R and PID2020-114705RB-I00 (Ministerio de Ciencia e Innovación), and IDI-20170304 (CDTI). Srihi received funding from the European Union's H2020 research and innovation program under a Marie Sklodowska-Curie grant agreement, No. 801586.

Institutional Review Board Statement: The protocols were approved by the Ethical and Animal Care Committee at IRTA (Institute of Agrifood Research and Technology, Caldes de Montbui, Spain; IRTA-22032012). It was performed according to the Spanish Policy for Animal Protection (RD 53/2013), which meets the European Union Directive 2010/63/UE on the protection of research animals.

Informed Consent Statement: Not applicable.

Data Availability Statement: The dataset used in this study will be available upon reasonable request to the corresponding author (lvarona@unizar.es).

Acknowledgments: We thank the company Inga Food SA (Almendralejo, Spain) and its technicians (E. Magallón, L. Muñoz, P. Díaz, and D. Iniesta) for their cooperation and technical support.

Conflicts of Interest: The authors declare no conflict of interest.

References

1. Serra, X.; Gil, F.; Pérez-Enciso, M.; Oliver, M.A.; Vázquez, J.M.; Gispert, M.; Díaz, I.; Moreno, F.; Latorre, R.; Noguera, J.L. A comparison of carcass, meat quality and histochemical characteristics of Iberian (Guadyerbas line) and Landrace pigs. *Livest. Prod. Sci.* **1998**, *56*, 215–223. [CrossRef]
2. Serrano, M.P.; Valencia, D.G.; Nieto, M.; Lázaro, R.; Mateos, G.G. Influence of sex and terminal sire line on performance and carcass and meat quality of Iberian pigs reared under intensive production systems. *Meat Sci.* **2008**, *78*, 420–428. [CrossRef]
3. Ministry of Agriculture, Food and Environment. *Boletín Oficial del Estado Real Decreto 4/2014 de 10 enero, por el que se aprueba la norma de calidad para la carne, el jamón, la paleta y la caña de lómo ibérico*; Ministry of Agriculture, Food and Environment: Madrid, Spain, 2014; pp. 1569–1585.
4. Silio, L.; Rodriguez, M.C.; Rodrigañez, J.; Toro, M.A. La selección de cerdos ibéricos. In *Porcino Ibérico: Aspectos Claves*; Buxade, C., Daza, A., Eds.; Mundi-Press: Madrid, Spain, 2001; pp. 125–149.
5. Noguera, J.L.; Ibáñez-Escriche, N.; Casellas, J.; Rosas, J.P.; Varona, L. Genetic parameters and direct, maternal and heterosis effects on litter size in a diallel cross among three commercial varieties of Iberian pig. *Animal* **2019**, *13*, 2765–2772. [CrossRef] [PubMed]
6. Comstock, R.E.; Robinson, H.F.; Harvey, P.H. A Breeding Procedure Designed to Make Maximum Use of Both General and Specific Combining Ability. *Agron. J.* **1949**, *41*, 360–367. [CrossRef]
7. Wientjes, Y.C.J.; Calus, M.P.L. BOARD INVITED REVIEW: The purebred-crossbred correlation in pigs: A review of theory, estimates, and implications. *J. Anim. Sci.* **2017**, *95*, 3467–3478. [CrossRef]
8. Vitezica, Z.G.; Varona, L.; Elsen, J.-M.; Misztal, I.; Herring, W.; Legarra, A. Genomic BLUP including additive and dominant variation in purebreds and F1 crossbreds, with an application in pigs. *Genet. Sel. Evol.* **2016**, *48*, 6. [CrossRef] [PubMed]
9. Wang, H.; Misztal, I.; Aguilar, I.; Legarra, A.; Muir, W.M. Genome-wide association mapping including phenotypes from relatives without genotypes. *Genet. Res.* **2012**, *94*, 73–83. [CrossRef] [PubMed]
10. R Core Team. *R: A Language and Environment for Statistical Computing*; R Foundation for Statistical Computing: Vienna, Austria, 2021; Available online: http://www.R-project.org/ (accessed on 1 November 2021).
11. Patterson, H.D.; Thompson, R. Recovery of Inter-Block Information when Block Sizes are Unequal. *Biometrika* **1971**, *58*, 545–554. [CrossRef]
12. Lourenco, D.; Legarra, A.; Tsuruta, S.; Masuda, Y.; Aguilar, I.; Misztal, I. Single-step genomic evaluations from theory to practice: Using snp chips and sequence data in blupf90. *Genes* **2020**, *11*, 790. [CrossRef]
13. Bidanel, J.P. Biology and genetics of reproduction. In *The Genetics of the Pig*; CABI: Wallingford, UK, 2011; pp. 218–241.
14. Putz, A.M.; Tiezzi, F.; Maltecca, C.; Gray, K.A.; Knauer, M.T. Variance component estimates for alternative litter size traits in swine. *J. Anim. Sci.* **2015**, *93*, 5153–5163. [CrossRef]
15. Ogawa, S.; Konta, A.; Kimata, M.; Ishii, K.; Uemoto, Y.; Satoh, M. Estimation of genetic parameters for farrowing traits in purebred Landrace and Large White pigs. *Anim. Sci. J.* **2019**, *90*, 23–28. [CrossRef]
16. Fernández, A.; Rodrigáñez, J.; Zuzúarregui, J.; Rodríguez, M.C.; Silió, L. Genetic parameters for litter size and weight at different parities in Iberian pigs. *Span. J. Agric. Res.* **2008**, *6*, 98–106. [CrossRef]
17. García-Casco, J.M.; Fernández, A.; Rodríguez, M.C.; Silió, L. Heterosis for litter size and growth in crosses of four strains of Iberian pig. *Livest. Sci.* **2012**, *147*, 1–8. [CrossRef]
18. Culbertson, M.S.; Mabry, J.W.; Misztal, I.; Gengler, N.; Bertrand, J.K.; Varona, L. Estimation of Dominance Variance in Purebred Yorkshire Swine. *J. Anim. Sci.* **1998**, *76*, 448–451. [CrossRef]

19. Tusell, L.; Gilbert, H.; Vitezica, Z.G.; Mercat, M.J.; Legarra, A.; Larzul, C. Dissecting total genetic variance into additive and dominance components of purebred and crossbred pig traits. *Animal* **2019**, *13*, 2429–2439. [CrossRef] [PubMed]
20. Xiang, T.; Christensen, O.F.; Vitezica, Z.G.; Legarra, A. Genomic evaluation by including dominance effects and inbreeding depression for purebred and crossbred performance with an application in pigs. *Genet. Sel. Evol.* **2016**, *48*, 92. [CrossRef] [PubMed]
21. Quadro, L.; Giordano, E.; Costabile, B.K.; Nargis, T.; Iqbal, J.; Kim, Y.; Wassef, L.; Hussain, M.M. Interplay between β-carotene and lipoprotein metabolism at the maternal-fetal barrier. *Biochim. Biophys. Acta—Mol. Cell Biol. Lipids* **2020**, *1865*, 158591. [CrossRef]
22. Wassef, L.; Spiegler, E.; Quadro, L. Embryonic phenotype, β-carotene and retinoid metabolism upon maternal supplementation of β-carotene in a mouse model of severe vitamin A deficiency. *Arch. Biochem. Biophys.* **2013**, *539*, 223–229. [CrossRef]
23. Leung, K.Y.; De Castro, S.C.P.; Galea, G.L.; Copp, A.J.; Greene, N.D.E. Glycine Cleavage System H Protein Is Essential for Embryonic Viability, Implying Additional Function Beyond the Glycine Cleavage System. *Front. Genet.* **2021**, *12*, 625120. [CrossRef] [PubMed]
24. Kinameri, E.; Inoue, T.; Aruga, J.; Imayoshi, I.; Kageyama, R.; Shimogori, T.; Moore, A.W. Prdm proto-oncogene transcription factor family expression and interaction with the Notch-Hes pathway in mouse neurogenesis. *PLoS ONE* **2008**, *3*, e3859. [CrossRef]
25. Haub, O.; Goldfarb, M. Expression of FGF5 gene in mouse embryo. *Development* **1991**, *406*, 397–406. [CrossRef]
26. Beer, H.D.; Bittner, M.; Niklaus, G.; Munding, C.; Max, N.; Goppelt, A.; Werner, S. The fibroblast growth factor binding protein is a novel interaction partner of FGF-7, FGF-10 and FGF-22 and regulates FGF activity: Implications for epithelial repair. *Oncogene* **2005**, *24*, 5269–5277. [CrossRef] [PubMed]
27. Segura, I.; Essmann, C.L.; Weinges, S.; Acker-Palmer, A. Grb4 and GIT1 transduce ephrinB reverse signals modulating spine morphogenesis and synapse formation. *Nat. Neurosci.* **2007**, *10*, 301–310. [CrossRef] [PubMed]
28. Kim, Y.D.; Lee, J.Y.; Oh, K.M.; Araki, M.; Araki, K.; Yamamura, K.I.; Jun, C.D. NSrp70 is a novel nuclear speckle-related protein that modulates alternative pre-mRNA splicing in vivo. *Nucleic Acids Res.* **2011**, *39*, 4300–4314. [CrossRef] [PubMed]
29. Stam, F.J.; MacGillavry, H.D.; Armstrong, N.J.; De Gunst, M.C.M.; Zhang, Y.; Van Kesteren, R.E.; Smit, A.B.; Verhaagen, J. Identification of candidate transcriptional modulators involved in successful regeneration after nerve injury. *Eur. J. Neurosci.* **2007**, *25*, 3629–3637. [CrossRef] [PubMed]

Article

SESN3 Inhibited SMAD3 to Relieve Its Suppression for *MiR-124*, Thus Regulating Pre-Adipocyte Adipogenesis

Weimin Lin [†], Jindi Zhao [†], Mengting Yan, Xuexin Li, Kai Yang, Wei Wei, Lifan Zhang and Jie Chen *

College of Animal Science and Technology, Nanjing Agricultural University, Nanjing 210095, China;
2018205002@njau.edu.cn (W.L.); 2019105005@stu.njau.edu.cn (J.Z.); 2018105035@njau.edu.cn (M.Y.);
2020105004@stu.njau.edu.cn (X.L.); 2019105037@stu.njau.edu.cn (K.Y.); wei-wei-4213@njau.edu.cn (W.W.);
lifanzhang@njau.edu.cn (L.Z.)
* Correspondence: jiechen@njau.edu.cn; Tel.: +86-18759141669
† The authors have contributed equally to this work.

Abstract: Sestrin-3, together with the other two members Sestrin-1 and Sestrin-2, belongs to the Sestrin family. The Sestrin protein family has been demonstrated to be involved in antioxidative, metabolic homeostasis, and even the development of nonalcoholic steatohepatitis (NASH). However, the adipogenic regulatory role of SESN3 in adipogenesis still needs to be further explored. In this study, we demonstrated SESN3 inhibited porcine pre-adipocyte proliferation, thus suppressing its adipogenesis. Meanwhile, SESN3 has been demonstrated to inhibit Smad3 thus protecting against NASH. Further, for our previous study, we found *mmu-miR-124* involved in 3T3-L1 cell adipogenesis regulation. In this study, we also identified that *ssc-miR-124* inhibited porcine pre-adipocyte proliferation, thus suppressing its adipogenesis, and the SMAD3 was an inhibitor of *ssc-miR-124* by binding to its promoter. Furthermore, the *ssc-miR-124* targeted porcine *C/EBPα* and *GR* and thus inhibited pre-adipocyte adipogenesis. In conclusion, SESN3 inhibited SMAD3, thus improving *ssc-miR124*, and then suppressed *C/EBPα* and *GR* to regulate pre-adipocytes adipogenesis.

Keywords: SESN3; SMAD3; *ssc-miR-124-3p*; pig; adipogenesis

Citation: Lin, W.; Zhao, J.; Yan, M.; Li, X.; Yang, K.; Wei, W.; Zhang, L.; Chen, J. SESN3 Inhibited SMAD3 to Relieve Its Suppression for *MiR-124*, Thus Regulating Pre-Adipocyte Adipogenesis. *Genes* **2021**, *12*, 1852. https://doi.org/10.3390/genes12121852

Academic Editors: Katarzyna Piórkowska and Katarzyna Ropka-Molik

Received: 16 October 2021
Accepted: 22 November 2021
Published: 23 November 2021

Publisher's Note: MDPI stays neutral with regard to jurisdictional claims in published maps and institutional affiliations.

1. Introduction

Overweight and obesity lead to a batch of diseases including but not limited to metabolic disease, hypertension, cardiovascular disease, and type II diabetes (T2D), which becomes one of the major threats to human health [1,2]. As a basic unit of adipose tissue, adipocyte differentiation and proliferation lead to adipose tissue expansion, and excessive adipose tissue cause obesity-related metabolic syndromes. Hence, adipocytes are emerging as a significant target in the treatment of obesity-related metabolic syndromes in the clinical setting [3]. Meanwhile, the molecular mechanisms of adipocyte differentiation and proliferation are of significant and escalating biomedical interest.

According to the classical extensive consensus on adipogenesis, hundreds of factors and genes involve enormous and complicated interaction networks to control this progress. Among them is the peroxisome proliferator activated receptor (PPAR) family, containing PPARα and PPARβ/δ, which are numbers of the nuclear receptor superfamily of ligand-activated transcription factors, and important adipogenic regulators [4]. In particular, PPARγ has been widely confirmed to play a critical role in adipocyte development, such as insulin signaling pathway, which is involved in regulating insulin sensitivity [5–7]. Apart from that, the CCAAT/enhancer binding protein (C/EBP) family is another key adipogenic regulator, including C/EBPα/β/γ/δ/ε and ζ, which can bind to the promoter of target genes through the CCAAT element, and form both homodimers and heterodimers to regulate their targets' expression level [8,9]. Moreover, C/EBPα and PPARγ activate with each other in mammalian cells to co-regulate adipogenesis. Meanwhile, it has been proven that there is a broad overlap in both of their transcriptional target genes [10]. The C/EBPα and

PPARγ mainly played regulatory role within the early and middle stage of adipogenesis. However, for the middle to late stage of adipogenesis, the fusing of small lipid droplets to mature lipid droplets causes the expansion of adipocytes. Numerous regulators are involved in this progress, including glucose transporter 4 (GLUT4), lipoprotein lipase (LPL), stearyl-CoA-desaturase (SCD) and fatty acid synthase (FAS), and so on [11,12].

MicroRNA (miRNA) is an important part of ncRNAs that has been demonstrated involving a series of biological progress, including adipogenesis. Generally, miRNAs are a kind of small molecular with ~21 nucleotides ncRNA, by targeting in either 3′untranslated region (3′UTR) of coding genes, or ncRNA, for instance, long non-coding RNAs (lncRNAs), circleRNAs (circRNA) and pseudogenes, to play the post-transcriptional role [13–15]. A series of miRNAs have been proven involving adipogenesis regulation. For instance, *miR-27* inhibited lipoprotein lipase (LPL) [16], and *miR-130* suppressed adipocytes adipogenesis by targeting *PPARγ* [17]. *MiR-103* was one of miRNAs that improved 3T3-L cell differentiation by targeting *MEF2D* and activating the Akt/mTOR signal pathway [18]. The other adipogenic miRNAs contained but was not limited to *miR-34* [19], *miR-17* [20], *miR-124* [21], *miR-144* [22], and so on.

Sestrin (SESN) proteins have been demonstrated as multifunctional ones by their biochemical characterization. Among them, Sestrin 1 and 2 share the conserved domain structure, whereas the structure of Sestrin 3 differs from other eukaryotic proteins, including Sestrin 1 and 2 [23]. Sestrin is able to activate adenosine monophosphate-activated protein kinase (AMPK) and the mammalian target of rapamycin kinase complex 2 (mTORC2,) but inhibits mammalian target of rapamycin kinase complex 1 (mTORC1) [24]. SESN1 and SESN2 are regarded as the sensor to leucine for the mTORC1 regulation [25,26]. Though SESN1/2/3 shared part common functions in AMPK and mTOR regulation, whereas there is a hierarchy of their leucine binding ability: Generally, SESN1 is the highest, while SESN3 is the lowest [24]. Moreover, SESN3 has been demonstrated to protect against diet-induced non-alcoholic steatohepatitis (NASH) in mice by directly inhibiting SMAD3 through protein–protein interaction, thus suppressing the TGFβ-SMAD3 signal pathway [24]. However, the role of SESN3 in adipogenesis still needs to be further explored. Hence, in this paper, we tried to study the adipogenic role of SESN3 in pre-adipocytes adipogenesis.

2. Materials and Methods

2.1. Experiment Animals

The animals in this study were 7-days-old Erhualian piglets, and they all came from the Changzhou Erhualian Pig Production Cooperation (Changzhou, Jiangsu, China).

All animal experiments including the pre-adipocyte collection were approved and reviewed by the Animal Ethics Committee of Nanjing Agricultural University (STYK (Su) 2011-0036).

2.2. Cell Culture, Transfection and Differentiation

Adipose tissue, isolated from the porcine back subcutaneous, was cut with scissors into 1-mm³ pieces, and we used 1 mg/mL of collagenase type I (Invitrogen, Carlsbad, CA, United States). The sample was kept in a 37 °C, 50 rpm/min shaking bath for over 2 h to digest. We added growth medium (89% Dulbecco's modified Eagle's medium/Ham's F-12 (DMEM-F12), 10% fetal bovine serum and 1% penicillin-streptomycin) 1.5 times the volume of the total digestion product to stop the digestion progress. The pre-adipocytes were collected with a 200-mm nylon mesh, filtrated, and were obtained via differential centrifugation. The pre-adipocytes were cultured in the growth medium at 37 °C with 5% CO_2. The medium was replaced with a fresh one every 2 days.

When a density of 85% confluence, the cell was cultured in 6- or 12-well plates, following the protocol transfection purpose plasmids or oligonucleotides into the cell using Lipofectamine 3000 (Invitrogen, Shanghai, China). The plasmids and oligonucleotides are shown in Supplementary Table S1.

The DIM inducer comprised of 2.5 mM of dexamethasone, 8.6 mM of insulin, 0.1 mM of 3-isobutyl-1 methylxanthine (IBMX), 1% penicillin-streptomycin, and 10% FBS in Dulbecco's modified Eagle's medium/Ham's-high glucose (DMEM-HG) (Sigma–Aldrich, Shanghai, China) and was used to induce for adipogenic differentiation. After 4 days' induction of adipogenic differentiation, the medium was replaced with maintenance medium comprised of 8.9 mM of insulin and 10% FBS-DMEM-HG until day 8. Every 2 days, we replaced it with a fresh medium.

2.3. RNA Isolation, Library Preparation, and RT-PCR

We used the Trizol reagent (TaKaRa, Dalian, China) to isolate the total RNA of the cell or tissue. The cDNA libraries of mRNA and miRNA were either reverse-transcribed by the PrimeScript^{TM} RT Master Mix (TaKaRa, Dalian, China), or miRNA 1st Strand cDNA Synthesis Kit (by stem-loop) (Vazyme, Nanjing, China), respectively. The AceQ Universal SYBR qPCR Master Mix (Vazyme, Nanjing, China) and the miRNA Universal SYBR qPCR Master Mix (Vazyme, Nanjing, China) were used to detect Quantitative real-time PCR (q-PCR) of mRNA and miRNA, respectively. The relative level of RNA expression was normalized to *GAPDH* and miR-17 [27–29] expression levels using the $2^{-\Delta\Delta Ct}$ method. Every sample was detected in triplicate. Primers were shown in Supplementary Table S1.

2.4. Oil Red O Staining and Triglyceride Assay

We used 4% paraformaldehyde (Biosharp, Hefei, China) to fix differentiated porcine pre-adipocytes for 30 min after gently washing three times with fresh $1\times$ PBS. We washed the fixed cells three times with $1\times$ PBS, using the 60% saturated oil red O to stain for 30 min (Sigma–Aldrich, Shanghai, China).

Subsequently, we captured images of the cells using a Zeiss Axiovert 40 CFL inverted microscope (Thornwood, NY, United States).

Quantitation of total triglyceride was carried out via eluting oil red O in pure isopropanol and absorbance was measured at the 510-nm wavelength.

2.5. Cell Counting Kit-8 (CCK-8) Assay

Pre-adipocytes were seeded in a 96-well plate for 24 h, adding 10 µL of CCK-8 reagent (UE, Shanghai, China) to each well and incubating the for 4 h in the cell culture incubator. The absorbance was measured at 450 nm with a microplate reader. The cell viability was calculated with the following equation: (OD treatment—OD blank). Each group had nine independent replicates ($n = 9$).

2.6. Cell Division Assay by EdU Incubation

The pre-adipocytes were incubated for 2 h with YF$^{®}$ 488 Click-iT EdU (50 µM) (UE, Shanghai, China). Then, we washed the pre-adipocytes with $1\times$ PBS, fixed them in 4% paraformaldehyde for 30 min, incubated them with 2 mg/mL glycine solution to neutralize the residual paraformaldehyde for 5 min, washed them with 3% BSA (configurating with $1\times$ PBS) thrice, and then incubated them with 0.5% Triton X-100 for 20 min. Finally, we washed them with 3% BSA (configurating with $1\times$ PBS) thrice and incubated them with Click-iT EdU work solution following protocol for 30 min. Subsequently, we washed them with $1\times$ PBS thrice, finally incubating with $1\times$ Hoechst33342 for 30 min. The representative images were captured via confocal microscopy (LSM700META, Zeiss, Oberkochen, Germany).

2.7. Luciferase Reporter Assay

The porcine pre-adipocytes were cultured in 12- or 24-well plates until the density was of 85% confluence. The pGL3 basic vector either contained the promoter of *Sesn3* or *ssc-miR-124* transfected into pre-adipocytes, by using Lipofectamine 3000 (Invitrogen Shanghai, China). Furthermore, the pmirGLO vector contained the 3'UTR of *GR* gene, which co-transfected with *ssc-miR-124* mimics/NC or *ssc-miR-124* inhibitors/NC, respectively.

The Dual-Luciferase Reporter Assay System (Promega, Madison, WI, United States) was used to quantify Firefly and Renilla luciferase activity.

2.8. Chromatin Immunoprecipitation Assay PCR (ChIP-PCR)

We used the ChIP Assay Kit (Boytime, Nanjing, China) following the manufacturer's instruction to detect FoxO1 protein binding with the promoter of *Sesn* and SMAD3 binding to the promoter of *ssc-miR-124*, respectively. Briefly, 1% formaldehyde was used to cross-link pre-adipocytes at 37 °C for 10 min, quenching the cross-linking by $1\times$ glycine for 5 min at room temperature. We used the ultrasonic cell smash machine VCX750 (Sonics, United States) to smash the pre-adipocytes and to obtain DNA fragments between 200 and 1000 bp as verified by ethidium bromide electrophoresis. For immunoprecipitation, 5 μg of antibody against FoxO1 (cat 383312 ZEN BIO) and SMAD3 (bs-3484R, Bioss) was incubated with 100 mL of cellular lysis at 4 °C overnight. Protein A Agarose/SalmonSperm DNA at was incubated at 4 °C for 2 h to isolate the immunoprecipitated complexes, and then washed as following: low salt wash buffer, high salt wash buffer, LiCl wash buffer once, and TE buffer twice. The final PCR analysis subjected to ChIP DNA and specific primers are shown in Supplementary Table S1.

2.9. Western Blotting

Briefly, total protein was extracted from porcine pre-adipocytes by using radioimmunoprecipitation assay (RIPA) lysis buffer (Beyotime, Jiangsu, China), following the protocol, and then quantified the protein by BCA Protein Assay kit (Beyotime, Jiangsu, China). We loaded the protein sample (1.5 μg/well) into a 12% SDS-PAGE gel (Zoman, Beijing, China). After its electrophoresis, SDS-PAGE was transferred to a PVDF membrane (Millipore, Billerica, MA, United States). Subsequently, we used the $1\times$ TBST containing 5% bovine serum albumin (BSA) to block the transferred membranes for 2 h at room temperature, followed by overnight primary antibody incubation (ZenBioScience, Chengdu, China) at 4 °C. On the second day, the immunoblot membranes were washed thrice with $1\times$ TBST for 30 min, and horseradish peroxidase conjugated secondary antibody was used to incubate for 2 h at room temperature. Using the ECL Chemiluminescence Detection Kit (Vazyme, Nanjing, China) to develop the blots, and using the VersaDoc 4000 MP system (Bio-Rad) to photograph.

2.10. Bioinformatics Analysis

Four kinds of online software were used to predict the *ssc-miR-124* targets, respectively, including miRanda (http://cbio.mskcc.org/microrna_data/miRanda-aug2010.tar.gz, accessed on 27 March 2010), miRmap (https://mirmap.ezlab.org/, accessed on 9 January 2013), TargetScan (http://www.targetscan.org/, accessed on 15 March 2018), and PITA (http://genie.weizmann.ac.il/pubs/mir07/mir07_data.html, accessed on 23 September 2007). RNAhybrid (http://bibiserv.techfak.uni-bielefeld.de/rnahybrid/submission.html, accessed on 18 September 2017) was used to predict the binding site and affinity between C/EBPα, GR, and ssc-miR-124. The precursor and mature sequences of ssc-miR-124 were obtained from miRBase (http://www.mirbase.org/, accessed on 7 June 2018).

PromoterScan 2 (http://bimas.dcrt.nih.gov:80/molbio/proscan, accessed on 4 February 1999) and JASPAR (http://jaspar.genereg.net/, accessed on 1 January 2013) were used for promoter and transcription factor prediction. Image J was used to quantify the results of C/EBPα and GR Western blotting, and the relative fluorescence intensity (EdU/DAPI).

2.11. Statistical Analysis

SPSS software (21.0 version, IBM, United States) was used to carry out the statistical analysis. All data were presented as means $\pm$ standard error of mean (s.e.m). To compare the average difference between the groups, a two-tailed Student's *t*-test was used. $P < 0.05$ was regarded as a statistically significant difference.

3. Results

3.1. Identification of the SESN3 Proliferative Role in the Pre-Adipocyte

We constructed the pcDNA3.1-Sesn3-CDS vector to explore its proliferative role in pre-adipocytes, then transfected them into pre-adipocytes for 48 h and detected their proliferative efficiency by EdU incubation and CCK-8 detection. The confocal fluorescence detection result showed the positive EdU level of SESN3 transfection groups was down-regulated compared with control group (Figure 1A,B). Moreover, CCK-8 detection resulted in further identified SESN3-inhibited pre-adipocyte proliferation (Figure 1C).

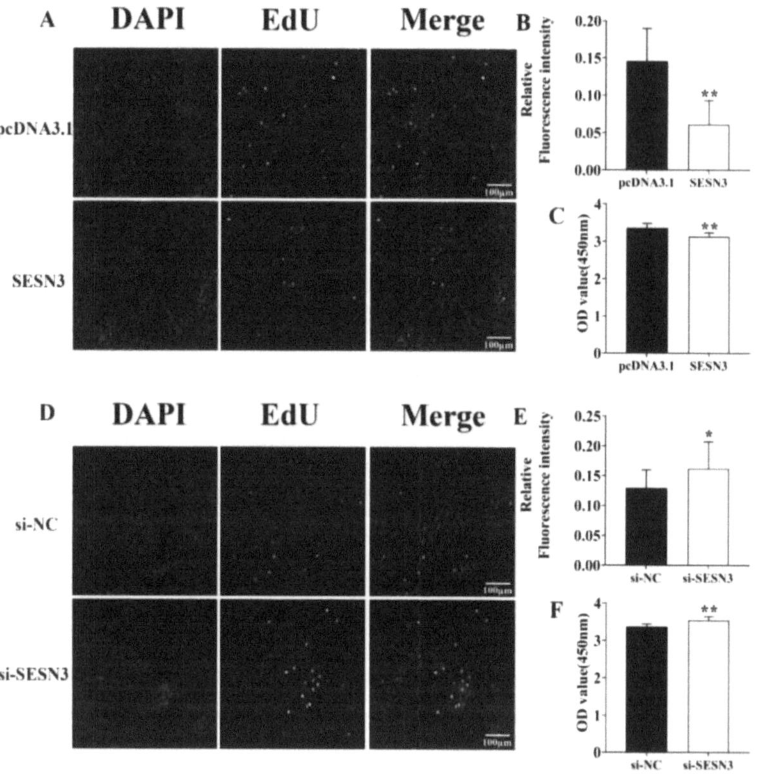

Figure 1. SESN3 inhibited pre-adipocyte proliferation. (**A,D**) Detection of pre-adipocyte positive EdU level after transfecting pcDNA3.1-Sesn3-CDS vector (**A**) or si-Sesn3 (**D**) for 48 h. The proliferating nuclei were stained green with EdU (50 μM) for 4 h, while the nuclei of all cells were stained blue with Hoechst33342 for 30 min. The scale bar was 100 μm. (**B,E**) Quantification of the relative fluorescence intensity of EdU-positive cells (EdU/DAPI) using ImageJ software. (**C,F**) The Cell Counting Kit-8 assay (CCK-8) detection after transfecting pcDNA3.1-Sesn3-CDS vector (**A**) or si-Sesn3 (**D**) for 48 h (n = 9 per group). The wavelength of CCK-8 detection was 450 nm. Data are presented as mean ± s.e.m. * means $P < 0.05$, ** means $P < 0.01$.

Besides, we designed the specific small interference RNA of *Sesn3* (si-Sesn3), then transfected them into pre-adipocyte for 48 h. After EdU incubation, the confocal fluorescence detection result showed the positive EdU level of si-Sesn3 transfection group was upregulated compared with NC group (Figure 1D,E). The CCK-8 result showed that si-Sesn3 increased pre-adipocyte proliferation (Figure 1F).

Above results demonstrated that SESN3 inhibited pre-adipocyte proliferation.

3.2. Sesn3 Inhibits Pre-Adipocyte Adipogenesis

Due to its suppression proliferative role in pre-adipocytes, we transfected the pcDNA3.1-Sesn3-CDS into pre-adipocytes to confirm its adipogenic role (Figure 2A). We found that except for the *PPARγ*, other adipogenic marker genes including *C/EBPα*, *C/EBPβ*, and *Fabp4* in Sesn3-overexpressed group were downregulated compared with the control group (Figure 2B). Pre-adipocytes were stimulated with DIM for 8 days after transfection, and we found that pre-adipocyte differentiation was suppressed in the pcDNA3.1-Sesn3-CDS-transfected group (Figure 2C,D). Moreover, si-Sesn3 was further transfected into the pre-adipocyte (Figure 2E), we found that *C/EBPα*, *C/EBPβ*, and *PPARγ* expression were upregulated in the si-Sesn3-transfected group (Figure 2F). After stimulating with DIM for 8 days, pre-adipocyte differentiation was improved in the si-Sesn3 transfection group (Figure 2G,H). Hence, these results identified that SESN3 inhibited pre-adipocyte adipogenesis.

3.3. SESN3 Inhibits SMAD3 to Reduce Its Suppressing Effect on ssc-miR-124 Transcription

According to the report, SESN3 has been demonstrated to inhibit SMAD3, thus suppressing the TGFβ-SMAD3 signal pathway [24]. Thus, we respectively transfected pcDNA3.1-Sesn3-CDS and si-Sesn3 into pre-adipocyte for 48 h, then detected the expression patterns of *Smad3*. We found that Sens3 inhibited *Smad3* expression, which was consistent with the report (Figure 3A–D). Our previous study identified that *mmu-miR-124* played an important role in 3T3-L1 adipogenesis [21]; hence, we predicted the *ssc-miR-124* promoter and further explored whether the SMAD3 regulated *ssc-miR-124* transcription by binding to promoter. We predicted 4 binding sites, and the dual-luciferase reporter assay system result showed that the predicted binding site 3 was the most suppressive efficiency one (Figure 3E,F). We further confirmed this result via the ChIP-PCR (Figure 3G). Besides, the pcDNA3.1-Smad3-CDS vector was transfected to pre-adipocyte, and the result showed that Smad3 suppressed the *ssc-miR-124* expression level (Figure 3H).

3.4. Identification of ssc-miR-124 Proliferative Role in the Pre-Adipocyte

To identify the role of *ssc-miR-124* in pre-adipocytes proliferation, we transfected them into pre-adipocytes for 48 h. The confocal fluorescence detection result showed that compared with NC group, the positive EdU level of *ssc-miR-124* mimicking the transfection group was downregulated (Figure 4A,B), and the CCK-8 detection result further proved that *ssc-miR-124* mimics inhibited pre-adipocyte proliferation (Figure 4C).

Besides, we also transfected the *ssc-miR-124* inhibitors into pre-adipocyte for 48 h. The confocal fluorescence detection result identified that the positive EdU level of *ssc-miR-124* inhibitors was upregulated compared with NC groups (Figure 4D,E). Moreover, the CCK-8 result further confirmed that *ssc-miR-124* inhibitors improved pre-adipocyte proliferation (Figure 4F).

The above results further demonstrated that *ssc-miR-124* impaired pre-adipocytes proliferation.

3.5. Ssc-miR-124 Inhibits Pre-Adipocyte Adipogenesis

To identify the adipogenic role of *ssc-miR-124* in porcine pre-adipocytes, we transfected *ssc-miR-124* mimics and *ssc-miR-124* inhibitors into porcine pre-adipocyte, respectively (Figure 5A,E), and detected the expression level of adipogenic marker genes, including *C/EBPα/β*, *PPARγ*, and *Fabp4*. The result showed that *ssc-miR-124* inhibited these genes' expression (Figure 5B,F). To confirm the inhibitory role of *ssc-miR-124* in pre-adipocyte adipogenesis, we further stimulated pre-adipocyte with DIM for 8 days after transfecting *ssc-miR-124* mimics or inhibitors into pre-adipocyte for 48 h. The results identified that *ssc-miR-124* mimics inhibited pre-adipocyte differentiation (Figure 5C,D), and *ssc-miR-124* inhibitors promoted pre-adipocytes differentiation (Figure 5G,H).

To sum up, we found *ssc-miR-124* suppressed porcine pre-adipocytes adipogenesis.

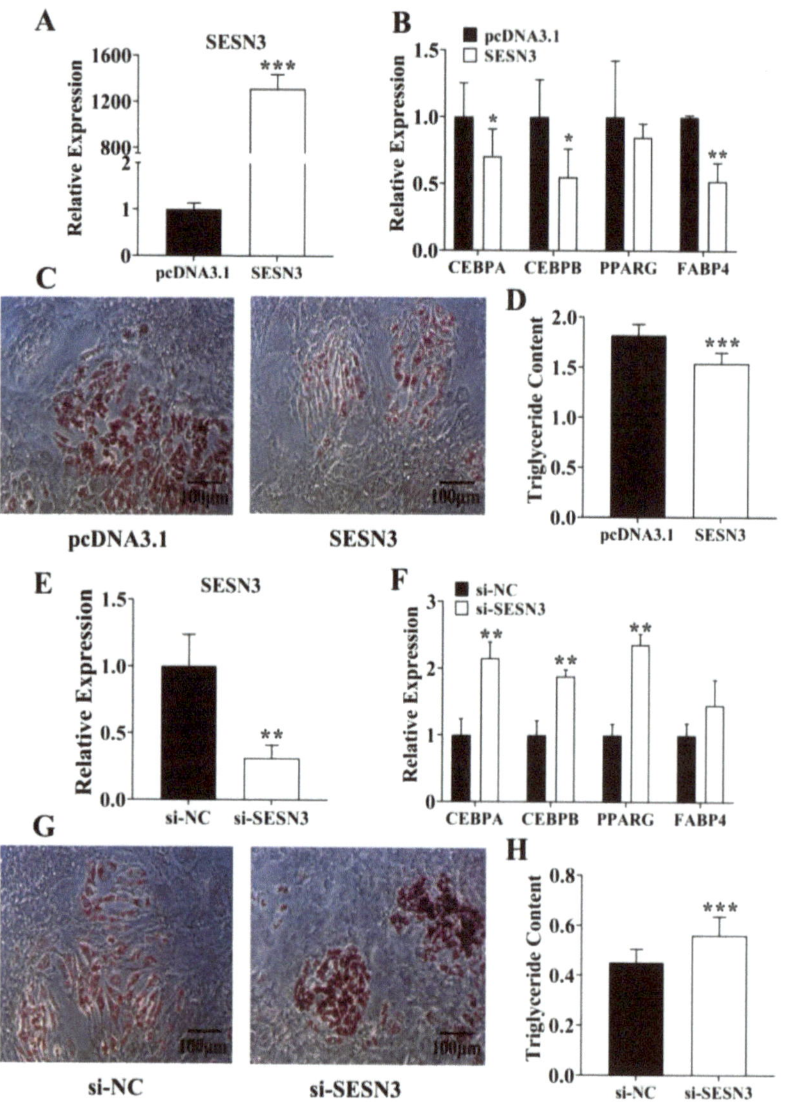

Figure 2. SESN3 inhibited pre-adipocyte adipogenesis. (**A**) *Sesn3* was overexpressed in pre-adipocytes after transfecting pcDNA3.1-Sesn3-CDS vector. (**B**) The expression of adipogenic marker genes after transfecting pcDNA3.1-Sesn3-CDS vector into pre-adipocyte for 48 h. (**C,D**) The pre-adipocyte oil red O staining (**C**) and quantitation of total triglyceride result (**D**) after transfecting pcDNA3.1-Sesn3-CDS vector. (**E**) *Sesn3* expression was inhibited after transfecting si-Sesn3. (**F**) The expression of adipogenic marker genes after transfecting si-Sesn3 into pre-adipocyte for 48 h. (**G,H**) The pre-adipocyte oil red O staining (**G**) and quantitation of total triglyceride result (**H**) after transfecting si-Sesn3. Data are presented as mean ± s.e.m. * means $P < 0.05$, ** means $P < 0.01$, and *** means $P < 0.001$.

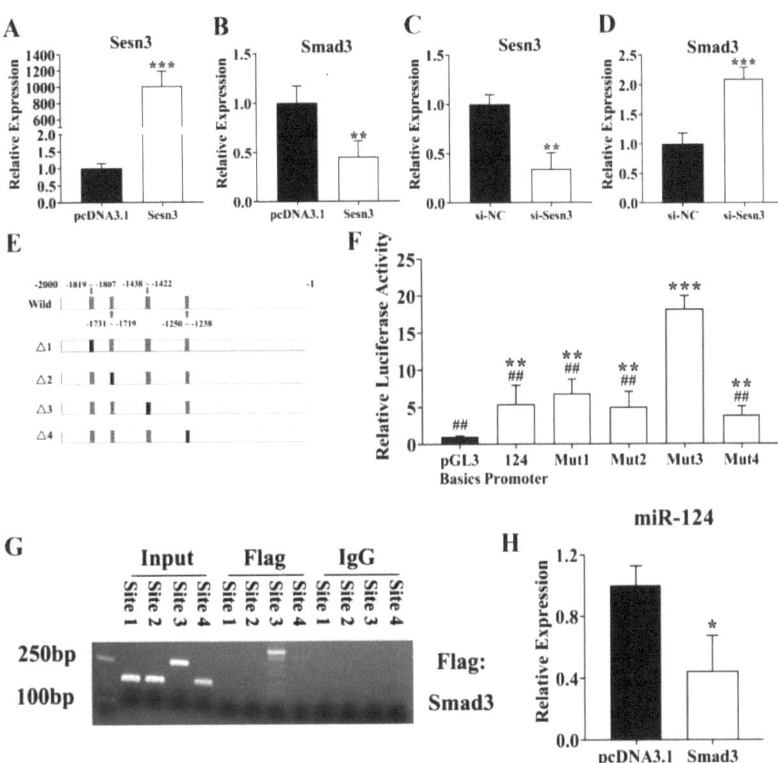

Figure 3. SMAD3 suppressed *ssc-miR-124* transcription by binding to its promoter. (**A,B**) Relative expression of *Sesn3* and *Smad3* after transfecting pcDNA3.1-Sesn3-CDS into pre-adipocyte. (**C,D**) Relative repression of *Sesn3* and *Smad3* after transfecting si-Sesn3 into pre-adipocyte. (**E**) The schematic diagram of SMAD3 binding sites in *ssc-miR-124* promoter. The red box meant wild binding sites, and black boxes were mutated ones. (**F**) Relative luciferase activity analysis in pre-adipocytes after pGL3 basic vector that contained the sequence with wild binding site or mutated one transfection. (**G**) ChIP-PCR detection of the binding of SMAD3 in *ssc-miR-124* promoter region. (**H**) The miRNA real-time PCR analysis of *ssc-miR-124* expression level in pre-adipocyte after transfecting pcDNA3.1-Smad3-CDS for 48 h. Data are presented as mean ± s.e.m. * means $P < 0.05$, ** means $P < 0.01$, and *** means $P < 0.001$, which shows the data compared with the control group, ## represented significant difference between promoter Mut3 and other five groups, and means $P < 0.01$

3.6. Ssc-miR-124 Inhibits C/EBPα and GR by Targeting Their 3′UTR

Ssc-miR-124 targets were predicted by 4 main target prediction tools. After the comprehensive analysis, we found 18 genes were shared among 4 prediction tools (Figure 6A), among them *C/EBPα* and *GR*. Hence, we further analyzed the *C/EBPα* and *GR* and respectively predicted 3 *ssc-miR-124* binding sites in *C/EBPα* gene 3′UTR and 2 *ssc-miR-124* binding sites in *GR* gene 3′UTR (Figure 6B). We constructed the pmirGLO vector with *C/EBPα* and *GR* gene 3′UTR which contained the predicted *ssc-miR-124* binding sites and their mutated binding sites (Figure 6B). The dual-luciferase reporter assay results identified that *ssc-miR-124* targeted *C/EBPα* and *GR* gene (Figure 6C,D). We further transfected *ssc-miR-124* mimics into pre-adipocyte for 48 h, and western bolting results showed that *ssc-miR-124* inhibited C/EBPα and the GR protein level (Figure 6E,F). Besides, we also transfected *ssc-miR-124* inhibitors into pre-adipocytes, and western bolting results identified that *ssc-miR-124* inhibitors increased C/EBPα and GR protein levels (Figure 6G,H).

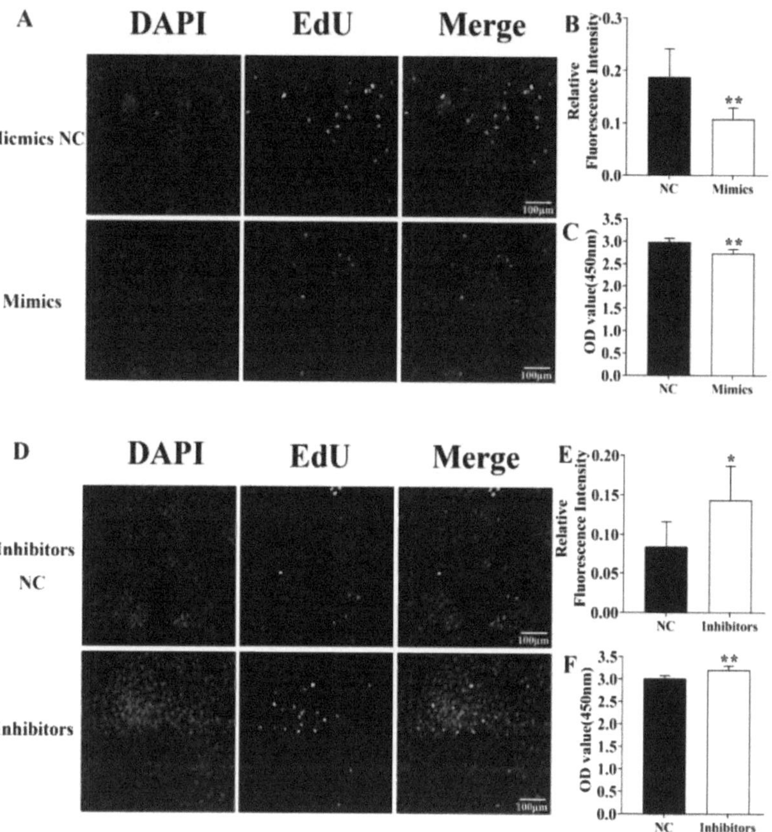

Figure 4. *Ssc-miR-124* inhibited pre-adipocyte proliferation. (**A,D**) Detection of pre-adipocyte positive EdU level after transfecting *ssc-miR-124* mimics (**A**) or *ssc-miR-124* inhibitors (**D**) for 48 h. The proliferating nuclei were stained green with EdU (50 μM) for 4 h, while the nuclei of all cells were stained blue with Hoechst33342 for 30 min. The scale bar was 100 μm. (**B,E**) Quantification of the relative fluorescence intensity of EdU-positive cells (EdU/DAPI) by ImageJ software. (**C,F**) The Cell Counting Kit-8 assay (CCK-8) detection after transfecting *ssc-miR-124* mimics (**A**) or *ssc-miR-124* inhibitors (**D**) for 48 h (n = 9 per group). The wavelength of CCK-8 detection was 450 nm. Data are presented as mean ± s.e.m. * means $P < 0.05$, ** means $P < 0.01$.

The above results further proved that *ssc-miR-124* suppressed porcine *C/EBPα* and *GR* by targeting their 3′UTR, and thus inhibiting their protein expression.

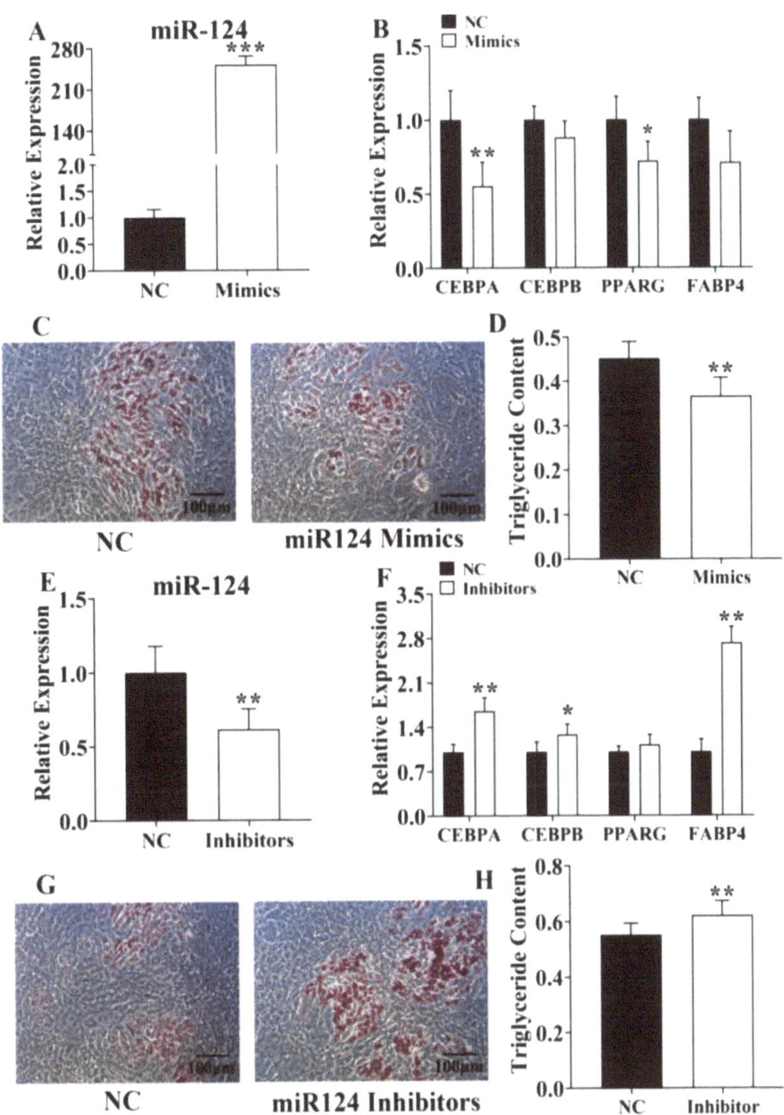

Figure 5. *Ssc-miR-124* inhibited pre-adipocyte adipogenesis. (**A**) *Ssc-miR-124* was overexpressed in pre-adipocyte by transfecting *ssc-miR-124* mimics. (**B**) The real-time analysis of adipogenic marker genes in pre-adipocyte after transfecting *ssc-miR-124* mimics for 48 h. (**C,D**) The pre-adipocytes oil red O staining (**C**) and quantitation of total triglyceride result (**D**) after transfecting *ssc-miR-124* mimics. (**E**) *Ssc-miR-124* was inhibited in pre-adipocyte after transfecting *ssc-miR-124* inhibitors. (**F**) The real-time analysis of adipogenic marker genes after transfecting *ssc-miR-124* inhibitors for 48 h. (**G,H**) The pre-adipocyte oil red O staining (**G**) and quantitation of total triglyceride result (**H**) after transfecting *ssc-miR-124* inhibitors. Data are presented as mean $\pm$ s.e.m. * means $P < 0.05$, ** means $P < 0.01$, and *** means $P < 0.001$.

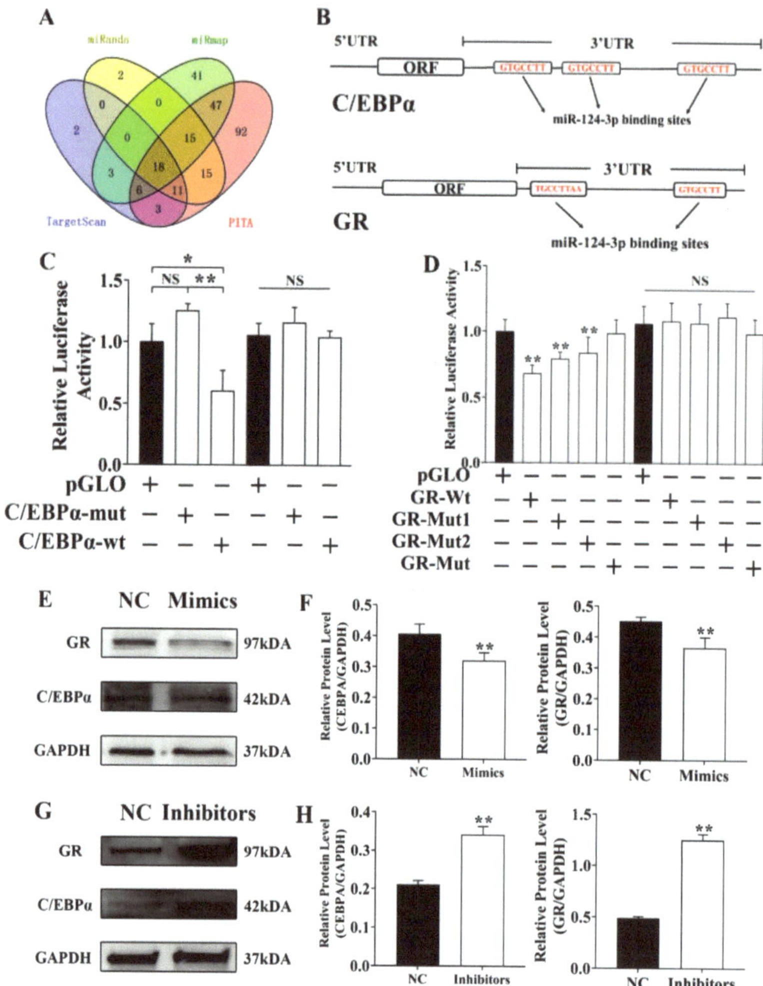

Figure 6. *Ssc-miR-124* suppressed *GR* and *C/EBPα* by targeting their 3′UTR. (**A**) The Venn diagram of *ssc-miR-124* targets by 4 prediction software. (**B**) The schematic diagram of *ssc-miR-124* miRNA response elements (MREs) of *C/EBPα* and *GR* genes. (**C,D**) Relative luciferase activity of constructed pmirGLO vector that contained the *C/EBPα* 3′UTR and *GR* 3′UTR, which respectively contained *ssc-miR-124* wild or mutated binding site sequences. (**E,F**) Western blotting result of C/EBPα and GR in pre-adipocyte after transfecting *ssc-miR-124* mimics for 48 h (**E**), and the relative quantification of their WB results used ImageJ software (**F**). (**G,H**) Western blotting results of C/EBPα and GR in pre-adipocyte after transfecting ssc-miR-124 inhibitors for 48 h (**G**), and the relative quantification of their WB results using ImageJ software (**H**). Data are presented as mean ± s.e.m. * means $P < 0.05$, ** means $P < 0.01$.

4. Discussion

In general, Sestrin was regarded as a versatile anti-aging molecule, which was considered an important component of antioxidant defense, and was transcriptionally induced upon oxidative damage through diverse transcription factors, for example p53, Nrf2, AP-1, and FoxOs [23]. The excessive reactive oxygen species (ROS) was considered correlated with the aging so far, although an appropriate level of ROS was necessary for physiological homeostasis. Besides, other promoters of aging include nutritional overabundance and obesity. It is known that the mechanistic target of rapamycin complex 1 (mTORC1) is one of major regulators that mediates the nutritional effect on aging; meanwhile, including mammals, either genetic or pharmacological inhibition of mTORC1 extends longevity and health span in most organisms [30,31]. The Sestrin family contains three paralogues in vertebrates, they were Sestrin 1–3, respectively [32], specially, Sestrin 1 and Sestrin 2 have been demonstrated tp reduce ROS and suppress mTORC1 activity [33]. Therefore, the suppression of ROS or mTORC1 was treated as a therapy among fat accumulation, insulin resistance, muscle degeneration, cardiac dysfunction, mitochondrial pathologies and tumorigenesis that was led by genetic depletion of Sestrin in many animal models, for example, worms [34], flies [35], and mice [36]. However, Sestrin 1 and Sestrin 2 were highly conserved between their protein structure, while the structure of Sestrin 3 differed from other eukaryotic proteins, including Sestrin 1 and 2 [23]. By report, chronic activation of p70 S6 kinase (S6K) and mTORC1 in response to hypernutrition contributed to obesity-associated metabolic pathologies including hepatosteatosis and insulin resistance. Furthermore, the ablation of the Sestrin 2 and 3, especially Sestrin 2, provoked hepatic mTORC1-S6K activation and insulin resistance [36]. On the other hands, some reports have been identified that the Sesn2 induced by a high-fat diet in the liver and skeletal muscle of mice, whereas Sesn1 was decreased in liver, and Sesn3 was decreased in liver and adipose tissue. In the skeletal muscle of mice and the leg muscle biopsies of human diabetics, Sesn3 was increased [37,38], which would offer parts of explanation for the adipogenic phenotype we observed. However, the adipogenic role of Sestrin 3 among porcine pre-adipocyte differentiation and proliferation still needs to be further explored.

We first identified the macroscopically adipogenic role of SESN3, and we found SESN3 inhibited pre-adipocyte proliferation thus suppressed its adipogenesis (Figures 1 and 2). Although cell proliferation and differentiation are usually incompatible within the same cell cycle, generally, it is necessary to stop proliferation then initiate differentiation. However, our results showed that Sesn3 inhibited pre-adipocyte proliferation and differentiation, which seemingly was not consistent with that. However, as the basic unit of adipose tissue, the lipid droplet enrichment of single adipocyte or increasing the number of adipocytes both lead to adipogenesis development. Meanwhile, we detected pre-adipocyte differentiation by transfecting vector or siRNA for 48 h, as well as after 8 days of induction, when we detected their proliferative effect was at the end of transfection. Hence, we consider that SESN3 regulated pre-adipocyte adipogenesis through inhibiting its proliferation effect, and then further suppressing differentiation. There are some reports that also concur this idea: For instance, the inhibition effect of HMGB2 and KDM5 for pre-adipocytes thus suppressed adipogenesis [39–41].

Besides, SESN3 has been demonstrated to be in involved TGFβ-Smads signaling by inhibiting the SMADs family at both protein and mRNA expression levels, thus protecting against diet-induced non-alcoholic steatohepatitis (NASH) in mice [24]. Thus, we further detected its inhibition for SMAD3 by respectively transfecting pcDNA3.1-Sesn3-CDS and si-Sesn3. Our results further identified that Sesn3 inhibited Smad3 within the porcine pre-adipocytes (Figure 3).

We demonstrated that *mmu-miR-124* played an important role in 3T3-L cell adipogenesis; however, the adipogenic molecular mechanism of *ssc-miR-124* for pigs still needs to be further identify [21]. We found 4 predicted binding sites of SMAD3 in the *ssc-miR-124* promoter region, and the results of the dual-luciferase reporter system and ChIP-PCR further identified that the predicted binding site3 was a positive site (Figure 3C–E). To identify the regulatory role of SMAD3 for *ssc-miR-124*, results showed that SMAD3 inhibited *ssc-miR-124* expression (Figure 3F).

To understand the adipogenic role of *ssc-miR-124* in pigs, we respectively detected its proliferative and differentiated effect for porcine pre-adipocyte (Figures 4 and 5). As was our prediction, *ssc-miR-124* promoted pre-adipocyte proliferation, thus improving its differentiation, which was consistent with the SESN3 adipogenic role (Figures 1 and 2), as well as being consistent with our previous report of *mmu-miR-124* regulation for 3T3-L1 cell [21].

Based on the post-transcriptional regulatory role of miRNA, we further explored ssc-miR-124 underlying molecular mechanisms. We identified two key adipogenic factors, GR and C/EBPα, which were targets of *ssc-miR-124*. As one of the most important adipogenic regulators, C/EBPα has been demonstrated interacting with PPARγ to regulate insulin signal pathway that playing the crucial role among adipogenesis [5,6,42,43]. For *GR*, also known as the *NR3C1* gene, it was a nuclear receptor that activated by ligand binding. Glucocorticoid (GC) was one of the key signals that were capable of inducing adipocyte differentiation, while glucocorticoid sensitivity of a cell largely depended on its GR levels [21,44,45].

Our results demonstrated that *ssc-miR-124* targeted *GR* 3′UTR and *C/EBPα* 3′UTR and inhibited their mRNA and protein levels, thus suppressing those two regulators from regulating adipogenesis among pre-adipocytes (Figure 6).

5. Conclusions

We identified that SESN3 inhibited SMAD3, thus improving *ssc-miR-124* expression, and finally suppressing two adipogenic vital factors, *GR* and *C/EBPα*, to regulate pre-adipocyte adipogenesis.

Supplementary Materials: The following are available online at DOI: https://doi.org/10.6084/m9.figshare.16436838, Supplementary Table S1: The sequence used in the whole paper.

Author Contributions: W.L. and J.Z. conceived the idea, performed the analyses, and wrote the manuscript. W.L. and J.Z. implemented the package. X.L., M.Y. and K.Y. contributed to the data analysis and checked the manuscript. L.Z. and W.W. checked the finished manuscript. J.C. conceived the idea, supervised the project analysis, and contributed to the manuscript preparation, provided the funding support. All authors have read and agreed to the published version of the manuscript.

Funding: This work was supported by the National Natural Science Foundation of China (Grant Nos. 31872334), and the Joint Funds of the National Natural Science Foundation of China (Grant No. U20A2052), and the National Natural Science Foundation of China (Grant Nos. 31902132).

Institutional Review Board Statement: All animal experiments including the pre-adipocyte collection were approved and reviewed by the Animal Ethics Committee of Nanjing Agricultural University (STYK (Su) 2011-0036).

Data Availability Statement: The data that support the findings of this study are available from the corresponding author upon reasonable request.

Conflicts of Interest: The authors declare that the research was conducted in the absence of any commercial or financial relationships that could be construed as a potential conflict of interest.

References

1. Goran, M.I.; Ball, G.D.C.; Cruz, M.L. Obesity and Risk of Type 2 Diabetes and Cardiovascular Disease in Children and Adolescents. *J. Clin. Endocrinol. Metab.* **2003**, *88*, 1417–1427. [CrossRef] [PubMed]
2. Kopelman, P.G. Obesity as a medical problem. *Nat. Cell Biol.* **2000**, *404*, 635–643. [CrossRef] [PubMed]
3. Shen, L.; Li, Q.; Wang, J.; Zhao, Y.; Niu, L.; Bai, L.; Shuai, S.; Li, X.; Zhang, S.; Zhu, L. miR-144-3p Promotes Adipogenesis Through Releasing C/EBPα From Klf3 and CtBP. *Front. Genet.* **2018**, *9*, 677. [CrossRef] [PubMed]
4. Souza-Mello, V. Peroxisome proliferator-activated receptors as targets to treat non-alcoholic fatty liver disease. *World J. Hepatol.* **2015**, *7*, 1012–1019. [CrossRef] [PubMed]
5. Barroso, I.; Gurnell, M.; Crowley, V.; Agostini, M.; Schwabe, J.; Soos, M.; Maslen, G.; Williams, T.; Lewis, H.; Schafer, A.; et al. Dominant negative mutations in human PPARgamma associated with severe insulin resistance, diabetes mellitus and hypertension. *Nat. Commun.* **1999**, *402*, 880–883. [CrossRef] [PubMed]
6. Doney, A.; Fischer, B.; Leese, G.; Morris, A.; Palmer, C. Cardiovascular risk in type 2 diabetes is associated with variation at the PPARG locus: A Go-DARTS study. *Arterioscler. Thromb. Vasc. Biol.* **2004**, *24*, 2403–2407. [CrossRef]
7. Ortega-Molina, A.; Efeyan, A.; Lopez-Guadamillas, E.; Muñoz-Martin, M.; López, G.G.; Cañamero, M.; Mulero, F.; Pastor, J.; Martinez, S.; Romanos, E.; et al. Pten Positively Regulates Brown Adipose Function, Energy Expenditure, and Longevity. *Cell Metab.* **2012**, *15*, 382–394. [CrossRef]

8. Hamm, J.; Park, B.; Farmer, S. A role for C/EBPbeta in regulating peroxisome proliferator-activated receptor Gamma activity during adipogenesis in 3T3-L1 preadipocytes. *J. Biol. Chem.* **2001**, *276*, 18464–18471. [CrossRef]

9. Yu, H.; He, K.; Wang, L.; Hu, J.; Gu, J.; Zhou, C.; Lu, R.; Jin, Y. Stk40 represses adipogenesis through translational control of CCAAT/enhancer-binding proteins. *J. Cell Sci.* **2015**, *128*, 2881–2890. [CrossRef]

10. Lefterova, M.; Lazar, M. New developments in adipogenesis. *Trends Endocrinol. Metab. TEM* **2009**, *20*, 107–114. [CrossRef]

11. Student, A.; Hsu, R.; Lane, M. Induction of fatty acid synthetase synthesis in differentiating 3T3-L1 preadipocytes. *J. Biol. Chem.* **1980**, *255*, 4745–4750. [CrossRef]

12. Vu, D.; Ong, J.M.; Clemens, T.L.; Kern, P.A. 1,25-Dihydroxyvitamin D induces lipoprotein lipase expression in 3T3-L1 cells in association with adipocyte differentiation. *Endocrinology* **1996**, *137*, 1540–1544. [CrossRef]

13. Hamilton, A.J.; Baulcombe, D.C. A Species of Small Antisense RNA in Posttranscriptional Gene Silencing in Plants. *Science* **1999**, *286*, 950–952. [CrossRef]

14. Ketting, R.F.; Fischer, S.E.; Bernstein, E.; Sijen, T.; Hannon, G.J.; Plasterk, R.H. Dicer functions in RNA interference and in synthesis of small RNA involved in developmental timing in *C. elegans*. *Genes Dev.* **2001**, *15*, 2654–2659. [CrossRef]

15. Grishok, A.; Pasquinelli, A.E.; Conte, D.; Li, N.; Parrish, S.; Ha, I.; Baillie, D.L.; Fire, A.; Ruvkun, G.; Mello, C.C. Genes and mechanisms related to RNA interference regulate expression of the small temporal RNAs that control *C. elegans* developmental timing. *Cell* **2001**, *106*, 23–34. [CrossRef]

16. Zhang, M.; Wu, J.; Chen, W.; Tang, S.; Mo, Z.; Tang, Y.; Li, Y.; Wang, J.; Liu, X.; Peng, J.; et al. MicroRNA-27a/b regulates cellular cholesterol efflux, in-flux and esterification/hydrolysis in THP-1 macrophages. *Atherosclerosis* **2014**, *234*, 54–64. [CrossRef]

17. Pan, S.; Yang, X.; Jia, Y.; Li, R.; Zhao, R. Microvesicle-shuttled miR-130b reduces fat deposition in recipient primary cul-tured porcine adipocytes by inhibiting PPAR-γ expression. *J. Cell. Physiol.* **2014**, *229*, 631–639. [CrossRef]

18. Li, M.; Liu, Z.; Zhang, Z.; Liu, G.; Sun, S.; Sun, C. miR-103 promotes 3T3-L1 cell adipogenesis through AKT/mTOR signal pathway with its target being MEF2D. *Biol. Chem.* **2015**, *396*, 235–244. [CrossRef]

19. Fu, T.; Seok, S.; Choi, S.; Huang, Z.; Suino-Powell, K.; Xu, H.E.; Kemper, B.; Kemper, J.K. MicroRNA 34a Inhibits Beige and Brown Fat Formation in Obesity in Part by Suppressing Adipocyte Fibroblast Growth Factor 21 Signaling and SIRT1 Function. *Mol. Cell. Biol.* **2014**, *34*, 4130–4142. [CrossRef]

20. Han, H.; Gu, S.; Chu, W.; Sun, W.; Wei, W.; Dang, X.; Tian, Y.; Liu, K.; Chen, J. miR-17-5p Regulates Differential Expression of NCOA3 in Pig Intramuscular and Subcutaneous Adipose Tissue. *Lipids* **2017**, *52*, 939–949. [CrossRef]

21. Liu, K.; Zhang, X.; Wei, W.; Liu, X.; Tian, Y.; Han, H.; Zhang, L.; Wu, W.; Chen, J. Myostatin/SMAD4 signaling-mediated reg-ulation of miR-124-3p represses glucocorticoid receptor expression and inhibits adipocyte differentiation. *Am. J. Physiol.-Endocrinol. Metab.* **2019**, *316*, E635–E645.

22. Lin, W.; Tang, Y.; Zhao, Y.; Zhao, J.; Zhang, L.; Wei, W.; Chen, J. MiR-144-3p targets FoxO1 to reduce its regulation of adi-ponectin and promote adipogenesis. *Front. Genet.* **2020**, *11*, 1–12. [CrossRef] [PubMed]

23. Ho, A.; Cho, C.-S.; Namkoong, S.; Cho, U.-S.; Lee, J.H. Biochemical Basis of Sestrin Physiological Activities. *Trends Biochem. Sci.* **2016**, *41*, 621–632. [CrossRef] [PubMed]

24. Huang, M.; Kim, H.; Zhong, X.; Dong, C.; Zhang, B.; Fang, Z.; Zhang, Y.; Lu, X.; Saxena, R.; Liu, Y.; et al. Sestrin 3 Protects Against Diet-Induced Nonalcoholic Steatohepatitis in Mice Through Suppression of Transform-ing Growth Factor β Signal Transduction. *Hepatology* **2020**, *71*, 76–92. [CrossRef] [PubMed]

25. Xu, D.; Shimkus, K.; Lacko, H.; Kutzler, L.; Jefferson, L.; Kimball, S. Evidence for a role for Sestrin1 in mediating leu-cine-induced activation of mTORC1 in skeletal muscle. *Am. J. Physiol.-Endocrinol. Metab.* **2019**, *316*, E817–E828.

26. Wolfson, R.L.; Chantranupong, L.; Saxton, R.A.; Shen, K.; Scaria, S.M.; Cantor, J.R.; Sabatini, D.M. Sestrin2 is a leucine sensor for the mTORC1 pathway. *Science* **2015**, *351*, 43–48. [CrossRef] [PubMed]

27. Gu, Y.; Li, M.; Zhang, K.; Chen, L.; Jiang, A.-A.; Wang, J.; Lv, X.; Li, X. Identification of suitable endogenous control microRNA genes in normal pig tissues. *Anim. Sci. J.* **2011**, *82*, 722–728. [CrossRef]

28. Timoneda, O.; Balcells, I.; Córdoba, S.; Castelló, A.; Sánchez, A. Determination of reference microRNAs for relative quantification in porcine tissues. *PLoS ONE* **2012**, *7*, e44413.

29. Bae, I.-S.; Seo, K.-S.; Kim, S.H. Identification of endogenous microRNA references in porcine serum for quantitative real-time PCR normalization. *Mol. Biol. Rep.* **2018**, *45*, 943–949. [CrossRef]

30. Harrison, D.E.; Strong, R.; Sharp, Z.D.; Nelson, J.F.; Astle, C.M.; Flurkey, K.; Nadon, N.L.; Wilkinson, J.E.; Frenkel, K.; Carter, C.S.; et al. Rapamycin fed late in life extends lifespan in genetically heterogeneous mice. *Nature* **2009**, *460*, 392–395. [CrossRef]

31. Johnson, S.; Rabinovitch, P.S.; Kaeberlein, M. mTOR is a key modulator of ageing and age-related disease. *Nat. Cell Biol.* **2013**, *493*, 338–345. [CrossRef]

32. Lee, J.H.; Budanov, A.V.; Karin, M. Sestrins Orchestrate Cellular Metabolism to Attenuate Aging. *Cell Metab.* **2013**, *18*, 792–801. [CrossRef]

33. Budanov, A.V.; Karin, M. p53 Target Genes Sestrin1 and Sestrin2 Connect Genotoxic Stress and mTOR Signaling. *Cell* **2008**, *134*, 451–460. [CrossRef]

34. Yang, Y.-L.; Loh, K.-S.; Liou, B.-Y.; Chu, I.-H.; Kuo, C.-J.; Chen, H.-D.; Chen, C.-S. SESN-1 is a positive regulator of lifespan in Caenorhabditis elegans. *Exp. Gerontol.* **2013**, *48*, 371–379. [CrossRef]

35. Lee, J.H.; Budanov, A.V.; Park, E.J.; Birse, R.; Kim, T.E.; Perkins, G.A.; Ocorr, K.; Ellisman, M.H.; Bodmer, R.; Bier, E.; et al. Sestrin as a Feedback Inhibitor of TOR That Prevents Age-Related Pathologies. *Science* **2010**, *327*, 1223–1228. [CrossRef]

36. Lee, J.H.; Budanov, A.V.; Talukdar, S.; Park, E.J.; Park, H.L.; Park, H.-W.; Bandyopadhyay, G.; Li, N.; Aghajan, M.; Jang, I.; et al. Maintenance of Metabolic Homeostasis by Sestrin2 and Sestrin3. *Cell Metab.* **2012**, *16*, 311–321. [CrossRef]
37. Dong, X.C. The potential of sestrins as therapeutic targets for diabetes. *Expert Opin. Ther. Targets* **2015**, *19*, 1011–1015. [CrossRef]
38. Kim, M.; Kowalsky, A.H.; Lee, J.H. Sestrins in Physiological Stress Responses. *Annu. Rev. Physiol.* **2021**, *83*, 381–403. [CrossRef]
39. Brier, A.-S.B.; Loft, A.; Madsen, J.G.S.; Rosengren, T.; Nielsen, R.; Schmidt, S.F.; Liu, Z.; Yan, Q.; Gronemeyer, H.; Mandrup, S. The KDM5 family is required for activation of pro-proliferative cell cycle genes during adipocyte differentiation. *Nucleic Acids Res.* **2017**, *45*, 1743–1759. [CrossRef]
40. Sun, J.-M.; Ho, C.-K.; Gao, Y.; Chong, C.-H.; Zheng, D.-N.; Zhang, Y.-F.; Yu, L. Salvianolic acid-B improves fat graft survival by promoting proliferation and adipogenesis. *Stem Cell Res. Ther.* **2021**, *12*, 1–16. [CrossRef]
41. Chen, K.; Zhang, J.; Liang, F.; Zhu, Q.; Cai, S.; Tong, X.; He, Z.; Liu, X.; Chen, Y.; Mo, D. HMGB2 orchestrates mitotic clonal expansion by binding to the promoter of C/EBPβ to facilitate adipogenesis. *Cell Death Dis.* **2021**, *12*, 666. [CrossRef]
42. Qadir, A.S.; Woo, K.M.; Ryoo, H.-M.; Baek, J.-H. Insulin suppresses distal-less homeobox 5 expression through the up-regulation of microRNA-124 in 3T3-L1 cells. *Exp. Cell Res.* **2013**, *319*, 2125–2134. [CrossRef] [PubMed]
43. Cao, Z.; Umek, R.M.; McKnight, S.L. Regulated expression of three C/EBP isoforms during adipose conversion of 3T3-L1 cells. *Genes Dev.* **1991**, *5*, 1538–1552. [CrossRef] [PubMed]
44. Lu, N.Z.; Cidlowski, J.A. Translational Regulatory Mechanisms Generate N-terminal Glucocorticoid Receptor Isoforms with Unique Transcriptional Target Genes. *Mol. Cell* **2005**, *18*, 331–342. [CrossRef] [PubMed]
45. Infante, M.; Armani, A.; Mammi, C.; Fabbri, A.; Caprio, M. Impact of Adrenal Steroids on Regulation of Adipose Tissue. *Compr. Physiol.* **2011**, *7*, 1425–1447. [CrossRef]

MDPI AG
Grosspeteranlage 5
4052 Basel
Switzerland
Tel.: +41 61 683 77 34

Genes Editorial Office
E-mail: genes@mdpi.com
www.mdpi.com/journal/genes